AF353104

Remote Sensing in Agriculture: State-of-the-Art

Remote Sensing in Agriculture: State-of-the-Art

Editors

Enrico Borgogno-Mondino
Eufemia Tarantino
Alessandra Capolupo

MDPI • Basel • Beijing • Wuhan • Barcelona • Belgrade • Manchester • Tokyo • Cluj • Tianjin

Editors
Enrico Borgogno-Mondino
Department of Agricultural, Forest and Food
Sciences (DISAFA), University of Torino
Italy

Eufemia Tarantino
Department of Civil, Environmental, Land,
Construction and Chemistry (DICATECh),
Politecnico di Bari
Italy

Alessandra Capolupo
Department of Civil, Environmental, Land,
Construction and Chemistry (DICATECh),
Politecnico di Bari
Italy

Editorial Office
MDPI
St. Alban-Anlage 66
4052 Basel, Switzerland

This is a reprint of articles from the Special Issue published online in the open access journal *Remote Sensing* (ISSN 2072-4292) (available at: https://www.mdpi.com/journal/remotesensing/special_issues/agriculture_RS).

For citation purposes, cite each article independently as indicated on the article page online and as indicated below:

LastName, A.A.; LastName, B.B.; LastName, C.C. Article Title. *Journal Name* **Year**, *Volume Number*, Page Range.

ISBN 978-3-0365-5483-9 (Hbk)
ISBN 978-3-0365-5484-6 (PDF)

Contents

About the Editors

Enrico Borgogno-Mondino

Enrico Borgogno-Mondino (Associate Professor), Since 2015 Associate Professor in Geomatics at the Department of Agricultural, Forest and Food Sciences (DISAFA) - University of Torino. 1996 Master Degree in Environmental Engineering by Politecnico di Torino (Italy); in 2004 Ph.D. in Geodesy and Geomatics by Politecnico di Milano (Italy). Since 2002 lecturer in BSc, MS, post-graduate masters and PhD courses. Main research topics are related to agro-forestry application of Geomatics, included optical and SAR remote sensing, digital photogrammetry, LiDAR, GIS and survey. Author of more than 150 papers in National and International Scientific Proceedings, Journals and Books. Editorial board member in MDPI Remote Sensing, MDPI Agronomy, MDPI Geomatics.

Guest editor of Special Issues: MDPI Land, MDPI Remote Sensing, Frontiers in Forests and Global Change. Member of the Board of the Italian Association of Remote Sensing since 2011. 2017–2021 Vice-President of the Italian Confederation of Scientific Associations for Territorial and Environmental Information (ASITA). Since 2022 President of the ASITA Scientific Council and Vice-president of the ASITA Board. Scientific Responsible of various research projects since 2007 and tutor of more than 40 BS, MS and Ph.D. theses since 2010.

Eufemia Tarantino

Eufemia Tarantino (Full Professor) is Full Professor of Geomatics at the Polytechnic University of Bari and teaches Cartography, GIS and Remote Sensing disciplines at the BSc and MSc degree courses in Environmental Engineering and the MSc degree course in Building Systems, as well as Advances in Geomatic Engineering at the PhD Course.

Currently, she plays some academic roles (i.e., coordinator of the MSc Course of Environmental and Territorial Engineering, Member of the Executive Council, Responsible of student career fair) at the Department of Civil, Environmental, Land, Building Engineering and Chemistry (DICATECh).

Her Research Interests are related to:

Analysis of metric characteristics and methodologies for the extraction of 2D/3D geometric primitives from satellite/aerial/UAV data with a high spatial resolution to update Technical/Thematic Cartography.

Methodologies for multi-temporal analysis (Change detection) and classification of optical satellite sensor data aimed at environmental and cultural heritage monitoring and documentation.

Development and implementation of GIS published in the Web for interactive geospatial analysis.

The research topics were carried out from 2001 to date as part of National and European Research Projects developed as Supervisor and Member, testified by more than 130 scientific contributions (journals articles, conference proceedings and books chapters). A total of 57 scientific publications (22 Journal Articles, 29 Conference Proceedings, 5 Editorials and 1 Books chapters) stand out in international SCOPUS and Thomson Reuters ISI Web of Knowledge databases with a relative quality index. The internationalization of her scientific activity includes the collaboration as a reviewer of up to 18 international indexed journals JCR (SCI), as well as member of the editorial board of "Remote Sensing" journal and "ISPRS International Journal of Geo-Information" as well as international scientific committees of conferences.

Alessandra Capolupo

Alessandra Capolupo (Assistant Professor) is Assistant Professor of Geomatics (ICAR/06) at the Politecnico di Bari and she teaches "Cartography and GIS" course at the BSc degree course in Environmental Engineering. Her research activity started in 2013 as a Ph.D. student and it lasts until today.

Her research interests concern geomatics. Indeed, she has focused on new technologies, such as cartography, photogrammetry, remote sensing, that have been applied to different fields, like environment monitoring and management. She developed innovative algorithms and methods exploiting the potentials of geospatial data acquired by several satellite platforms, which were later on integrated and investigated using the information provided by other sensors, such as, for instance, UAV, airplane, GNSS. Such algorithms, as demonstrated through scientific papers published in ISI-SCOPUS journals, were applied in European and National projects such as CLARITY (http://clarity-h2020.eu/), LIFE ECOREMED (http://www.ecoremed.it/) (awarded the title "Best LIFE project"), PON BiOPOLIS (PON03PE_00107_01), Campania Trasparente (http://www. campaniatrasparente.it/). Moreover, she won some academic awards (Best paper award, Young scientist award, Top reviewers, STAR program).

Preface to "Remote Sensing in Agriculture: State-of-the-Art"

Abstract: The Special Issue on "Remote Sensing in Agriculture: State-of-the-Art" gives an exhaustive overview of the ongoing remote sensing technology transfer into the agricultural sector. It consists of 10 high-quality papers focusing on a wide range of remote sensing models and techniques to forecast crop production and yield, to map agricultural landscape and to evaluate plant and soil biophysical features. Satellite, RPAS, and SAR data were involved. This preface describes shortly each contribution published in such Special Issue.

Keywords: free satellite data; RPAS/UAV; copernicus; prescription maps; crop management; crop monitoring; crop water requirements; services in agriculture; crop productivity; data process standardization

1. Overview and Aim

Agriculture has been crucial in human life and, over the years, to meet the emerging challenges, production activities and rural landscapes have been gradually moulded and adapted to society demands [1]. For instance, diversifying crop production and optimizing its yield should satisfy the increasing food need due to the continuous global population growing and its changed eating habits [2]. Similarly, minimizing the environmental cost of new agricultural productive activities plays a key role in improving human well-being and preserving biodiversity and ecosystem health [3]. Indeed, the use of inorganic pesticides and chemical fertilizers over recent decades has been recognized as one of the main factors responsible for pollutants diffusion in agricultural soils [4,5], water [6,7], and air [8]. Moreover, the common agricultural policy (CAP) of the European Union, supplying grants for farming activities, requires a continued control of farmers' declarations, to ensure the legitimacy of grants. Satellite missions, such as Sentinel-1 and Sentinel-2, are going to support this control action, thus heavily entering into the new agriculture trend [9,10]. Moreover, insurance companies operating in the agricultural field are looking at the new satellite missions as potential tools for supporting their activities [11,12].

The above-mentioned issues are only a few examples of the problems caused by the agricultural sector which are currently being faced. Thus, a big effort is still required to detect best practices to operate through. Nevertheless, it is evident that a large amount of data concerning both temporal and spatial variations in crop conditions, rural landscapes and, more generally, in the overall agricultural systems, should be collected and integrated. Remote sensing relevance in supporting agricultural applications has been recognized since the 1970s [13]. Nevertheless, because of the reduced spatial and spectral resolutions of the first satellite sensors and their unsuitable revisiting times, remote sensing-based applications were limited for a long time. The new satellite missions, as the Sentinels of the European Space Agency (ESA), and the introduction of new tools, as Remotely Piloted Aircraft Systems (RPAS), has turned on the footlights on such techniques again, offering new opportunities to explore [14–17].

This Special Issue (SI) on "Remote Sensing in Agriculture: State-of-the-Art" moves within this framework and was aimed at gathering contributions useful to delineate the ongoing trends of remote sensing technology transfer to the agricultural sector. Reviewing of conventional methods, proposing of novel data-collecting tools and handling techniques were expected to populate this volume, included focuses about eventual limitations and challenges. Finally, 10 high-quality contributions were selected to be included in this SI, making it possible to significantly account for most of the above-mentioned expectations. Some highlights of the presented contributions are given in Section 2.

2. Highlights of Research Articles

The SI collected documents satisfyingly deal with the most of issues that the ongoing technology transfer process of remote sensing to agriculture is proposing. Contributions [12,18–20] present some applications based on Synthetic Aperture Radar (SAR) data. Specifically, Ajadi et al. [18] introduce a new method, based on Hidden Markov Random Field (HMRF), to identify crop lodging and to map its extension. Iowa and Illinois were the pilot sites selected to test the new approach. Research results will impact on future use of SAR-based information for operational crop lodging assessment. De Petris et al. [9] propose to map apple orchards damaged by a stormy event by adopting H-α-A polarimetric decomposition technique. Thus, a probability map of potentially storm-damaged orchards was produced. This result may support local funding restoration policies. Hoskera et al., [19] used Sentinel 1 data to estimate soil moisture by adopting both localized and generalized linear models. Particularly, the authors derive 39 localized linear models and 9 generalized linear models. Such models were validated using in situ data and all of them showed promising results. Lastly, Sun et al. [20] propose a novel approach to merge time series Sentinel-1 (S1) and Sentinel-2 (S2) data to map different crop kinds over oasis agricultural areas. A statistically homogeneous pixel (SHP) distributed scatterer interferometry (DSI) algorithm was applied to handle Sentinel 1 data while the random forest technique has been applied to exploit optical properties. The resultant map of five major crop types were generated by integrating the outcomes produced by both methods.

A single contribution proposes a review about the adoption of thermal data acquired by RPAS in supporting precision agriculture. After reporting their main applications and exploring their potentialities, this it offers a potential outlook of development [21].

Contributions [22–24], instead, deal with RPAS-based applications. Each paper refers to an acquisition experience operated with a different sensor, namely LiDAR, hyperspectral and multispectral sensors. LiDAR data were used to estimate fresh biomass and crop height for three different crops (potato, sugar beet, and winter wheat) [21]. Fresh biomass and crop height were assessed using 3DPI algorithm and the mean height of a variable number of points selected for each m2, respectively. The approach showed promising outcomes, albeit the authors outlined that results are strongly dependent on flight conditions. Ref. [23] presents a work where a hyperspectral imaging sensor was mounted on a ground-based vehicle and a RPAS to explore their potentialities in detecting and quantifying yellow rust in wheat at ground canopy and plot scale, respectively. It is the first time that such an experiment was conducted. The authors pinpointed limitations and challenges of such an approach. In [24], the authors adopted two different RPAS equipped with two multispectral mini sensors to analyse vegetated areas. Additionally, the authors explored the opportunity to integrate such sensors to detect vegetation changes.

Lastly, the studies proposed by [25,26] involve the application of satellite data to the agricultural sector. The former [25] derives nine vegetation indices from Moderate Resolution Imaging Spectroradiometer (MODIS) data to predict crop yield in Mongolia. The authors concluded that the Normalized Difference Water Index (NDWI) and Visible and Shortwave Infrared Drought Index (VSDI) are the optimal indicators to meet research purposes. Additionally, the end of June and the beginning of July have been recognised as the best timings to forecast the production yield. Conversely, [26] assess and compare the performance of MODIS, Landsat, and blended images in evaluating crop yield over a period of about 6 years (2009–2015). This contribution is aimed at detecting the best data to accurately monitor biophysical processes and yields by using freely available data.

Conflicts of Interest: The authors declare no conflict of interest.

References

1. Gevaert, C.M.; Suomalainen, J.; Tang, J.; Kooistra, L. Generation of Spectral–Temporal Response Surfaces by Combining Multispectral Satellite and Hyperspectral UAV Imagery for Precision Agriculture Applications. *IEEE J. Sel. Top. Appl. Earth Obs. Remote Sens.* **2015**, *8*, 3140–3146, doi:10.1109/JSTARS.2015.2406339.

2. Gebbers, R.; Adamchuk, V.I. Precision Agriculture and Food Security. *Science* **2010**, *327*, 828–831, doi:10.1126/science.1183899.

3. Capolupo, A.; Pindozzi, S.; Okello, C.; Boccia, L. Indirect field technology for detecting areas object of illegal spills harmful to human health: Application of drones, photogrammetry and hydrological models. *Geospat. Health* **2014**, *8*, S699–S707, doi:10.4081/gh.2014.298

4. Capolupo, A.; Nasta, P.; Palladino, M.; Cervelli, E.; Boccia, L.; Romano, N. Assessing the Ability of Hybrid Poplar for In-Situsitu Phytoextraction of Cadmium by Using UAV-Photogrammetry and 3D Flow Simulator. *Int. J. Remote Sens.* **2018**, *39*, 5175–5194, doi:10.1080/01431161.2017.1422876.

5. Tarantino E, Figorito B. Mapping rural areas with widespread plastic covered vineyards using true color aerial data. *Remote Sens.* **2012**, *4*, 1913–1928, doi:10.3390/rs4071913.

6. Close, M.E.; Humphries, B.; Northcott, G. Outcomes of the First Combined National Survey of Pesticides and Emerging Organic Contaminants (EOCs) in Groundwater in New Zealand 2018. *Sci. Total. Environ.* **2021**, *754*, 142005, doi:10.1016/j.scitotenv.2020.142005.

7. Siad, S.M.; Gioia, A.; Hoogenboom, G.; Iacobellis, V.; Novelli, A.; Tarantino, E.; Zdruli, P. Durum wheat cover analysis in the scope of policy and market price changes: A case study in Southern Italy. *Agriculture* **2017**, *7*, 12, doi:10.3390/agriculture7020012.

8. Weldeslassie, T.; Naz, H.; Singh, B.; Oves, M. Chemical Contaminants for Soil, Air and Aquatic Ecosystem. In *Modern Age Environmental Problems and Their Remediation*; Oves, M., Zain Khan, M., M.I. Ismail, I., Eds.; Springer International Publishing: Cham, Switzerland, 2018; pp. 1–22, ISBN 978-3-319-64501-8.

9. Sarvia, F.; Xausa, E.; De Petris, S.; Cantamessa, G.; Borgogno-Mondino, E. A Possible Role of Copernicus Sentinel-2 Data to Support Common Agricultural Policy Controls in Agriculture. *Agronomy* **2021**, *11*, 110, doi:10.3390/agronomy11010110.

10. Sarvia, F.; De Petris, S.; Ghilardi, F.; Xausa, E.; Cantamessa, G.; Borgogno-Mondino, E. The Importance of Agronomic Knowledge for Crop Detection by Sentinel-2 in the CAP Controls Framework: A Possible Rule-Based Classification Approach. *Agronomy* **2022**, *12*, 1228, doi:10.3390/agronomy12051228.

11. Borgogno-Mondino, E.; Sarvia, F.; Gomarasca, M.A. Supporting Insurance Strategies in Agriculture by Remote Sensing: A Possible Approach at Regional Level. In *Proceedings of the Computational Science and Its Applications—ICCSA 2019*; Misra, S., Gervasi, O., Murgante, B., Stankova, E., Korkhov, V., Torre, C., Rocha, A.M.A.C., Taniar, D., Apduhan, B.O., Tarantino, E., Eds.; Springer International Publishing: Cham, Switzerland, 2019; pp. 186–199.

12. De Petris, S.; Sarvia, F.; Gullino, M.; Tarantino, E.; Borgogno-Mondino, E. Sentinel-1 Polarimetry to Map Apple Orchard Damage after a Storm. *Remote Sens.* **2021**, *13*, 1030, doi:10.3390/rs13051030.

13. Aguilar, M.Á.; Jiménez-Lao, R.; Nemmaoui, A.; Aguilar, F.J.; Koc-San, D.; Tarantino, E.; Chourak, M. Evaluation of the Consistency of Simultaneously Acquired Sentinel-2 and Landsat 8 Imagery on Plastic Covered Greenhouses. *Remote Sens.* **2020**, *12*, 2015, doi:10.3390/rs12122015.

14. Borgogno Mondino, E.; Gajetti, M. Preliminary Considerations about Costs and Potential Market of Remote Sensing from UAV in the Italian Viticulture Context. *Eur. J. Remote Sens.* **2017**, *50*, 310–319.

15. Mandal, D.; Kumar, V.; Lopez-Sanchez, J.M.; Bhattacharya, A.; McNairn, H.; Rao, Y.S. Crop Biophysical Parameter Retrieval from Sentinel-1 SAR Data with a Multi-Target Inversion of Water Cloud Model. *Int. J. Remote Sens.* **2020**, *41*, 5503–5524, doi:10.1080/01431161.2020.1734261.

16. De Petris, S.; Sarvia, F.; Borgogno-Mondino, E. RPAS-Based Photogrammetry to Support Tree Stability Assessment: Longing for Precision Arboriculture. *Urban For. Urban Green.* **2020**, *55*, 126862, doi:10.1016/j.ufug.2020.126862.

17. Borgogno Mondino, E. Remote Sensing from RPAS in Agriculture: An Overview of Expectations and Unanswered Questions. *Mech. Mach. Sci.* **2018**, *49*, 483–492, doi:10.1007/978-3-319-61276-8_51.

18. Ajadi, O.A.; Liao, H.; Jaacks, J.; Delos Santos, A.; Kumpatla, S.P.; Patel, R.; Swatantran, A. Landscape-Scale Crop Lodging Assessment across Iowa and Illinois Using Synthetic Aperture Radar (SAR) Images. *Remote Sens.* **2020**, *12*, 3885, doi:10.3390/rs12233885.

19. Hoskera, A.K.; Nico, G.; Irshad Ahmed, M.; Whitbread, A. Accuracies of Soil Moisture Estimations Using a Semi-Empirical Model over Bare Soil Agricultural Croplands from Sentinel-1 SAR Data. *Remote Sens.* **2020**, *12*, 1664, doi:10.3390/rs12101664.

20. Sun, L.; Chen, J.; Guo, S.; Deng, X.; Han, Y. Integration of Time Series Sentinel-1 and Sentinel-2 Imagery for Crop Type Mapping over Oasis Agricultural Areas. *Remote Sens.* **2020**, *12*, 158, doi:10.3390/rs12010158.

21. Messina, G.; Modica, G. Applications of UAV Thermal Imagery in Precision Agriculture: State of the Art and Future Research Outlook. *Remote Sens.* **2020**, *12*, 1491, doi:10.3390/rs12091491.

22. ten Harkel, J.; Bartholomeus, H.; Kooistra, L. Biomass and Crop Height Estimation of Different Crops Using UAV-Based Lidar. *Remote Sens.* **2020**, *12*, 17, doi:10.3390/rs12010017.

23. Bohnenkamp, D.; Behmann, J.; Mahlein, A.-K. In-Field Detection of Yellow Rust in Wheat on the Ground Canopy and UAV Scale. *Remote Sens.* **2019**, *11*, 2495, doi:10.3390/rs11212495.

24. Lu, H.; Fan, T.; Ghimire, P.; Deng, L. Experimental Evaluation and Consistency Comparison of UAV Multispectral Minisensors. *Remote Sens.* **2020**, *12*, 2542, doi:10.3390/rs12162542.

25. Tuvdendorj, B.; Wu, B.; Zeng, H.; Batdelger, G.; Nanzad, L. Determination of Appropriate Remote Sensing Indices for Spring Wheat Yield Estimation in Mongolia. *Remote Sens.* **2019**, *11*, 2568, doi:10.3390/rs11212568.

26. Chen, Y.; McVicar, T.R.; Donohue, R.J.; Garg, N.; Waldner, F.; Ota, N.; Li, L.; Lawes, R. To Blend or Not to Blend? A Framework for Nationwide Landsat–MODIS Data Selection for Crop Yield Prediction. *Remote Sens.* **2020**, *12*, 1653, doi:10.3390/rs12101653.

Enrico Borgogno-Mondino, Eufemia Tarantino, and Alessandra Capolupo
Editors

 *remote sensing*

Article

Sentinel-1 Polarimetry to Map Apple Orchard Damage after a Storm

Samuele De Petris [1], Filippo Sarvia [1], Michele Gullino [2], Eufemia Tarantino [3] and Enrico Borgogno-Mondino [1,*

[1] Department of Agriculture, Forest and Food Sciences, University of Torino, L.go Braccini 2, 10095 Grugliasco, Italy; samuele.depetris@unito.it (S.D.P.); filippo.sarvia@unito.it (F.S.)
[2] Az. Agr. Fessia Franca, v. Lagnasco 6, 12030 Manta, Italy; michele.gullino@edu.unito.it
[3] DICATECh, Politecnico di Bari, Via Orabona 4, 70125 Bari, Italy; eufemia.tarantino@poliba.it
* Correspondence: enrico.borgogno@unito.it

Abstract: Climate change increases extreme whether events such as floods, hailstorms, or storms, which can affect agriculture, causing damages and economic loss within the agro-food sector. Optical remote sensing data have been successfully used in damage detections. Cloud conditions limit their potential, especially while monitoring floods or storms that are usually related to cloudy situations. Conversely, data from the Polarimetric Synthetic Aperture Radar (PolSAR) are operational in all-weather conditions and are sensitive to the geometrical properties of crops. Apple orchards play a key role in the Italian agriculture sector, presenting a cultivation system that is very sensitive to high-wind events. In this work, the H-α-A polarimetric decomposition technique was adopted to map damaged apple orchards with reference to a stormy event that had occurred in the study area (NW Italy) on 12 August 2020. The results showed that damaged orchards have higher H (entropy) and α (alpha angle) values compared with undamaged ones taken as reference (Mann–Whitney one-tailed test U = 14,514, $p < 0.001$; U = 16604, $p < 0.001$ for H and α, respectively). By contrast, A (anisotropy) values were significantly lower for damaged orchards (Mann–Whitney one-tailed test U = 8616, $p < 0.001$). Based on this evidence, the authors generated a map of potentially storm-damaged orchards, assigning a probability value to each of them. This map is intended to support local funding restoration policies by insurance companies and local administrations.

Keywords: Sentinel-1; apple orchard damage; polarimetric decomposition; entropy; anisotropy; alpha angle; storm damage mapping; economic loss; insurance support

Citation: De Petris, S.; Sarvia, F.; Gullino, M.; Tarantino, E.; Borgogno-Mondino, E. Sentinel-1 Polarimetry to Map Apple Orchard Damage after a Storm. *Remote Sens.* **2021**, *13*, 1030. https://doi.org/10.3390/rs13051030

Academic Editor: Mario Cunha

Received: 9 February 2021
Accepted: 5 March 2021
Published: 9 March 2021

Publisher's Note: MDPI stays neutral with regard to jurisdictional claims in published maps and institutional affiliations.

1. Introduction

Climate change and related natural disasters affect several sectors [1]. Agriculture is one of the most vulnerable [2,3]. Between 2005 and 2015, the impact of natural disasters on the agricultural sector was estimated to be 96 billion dollars in damaged, or completely lost, crops [4]. Climate change-related effects (e.g., temperature and precipitation increasing in terms of level, time, and variability) are expected to reduce the yield and quality of many crops, especially cereals and fodder cereals [5].

Storms and hail also can cause serious damage to crops [6]. Hurricanes can cause much damage, with grass lodging, uprooting of orchards, and falling trees [7,8]. These critical events, potentially highly impacting farmers' income, must be carefully accounted for in the context of risk management in agriculture.

Fruits and vegetables represent (year 2018) about 14% of the total value of European (EU) agricultural production [9,10]. These crops are very important for many EU member states, in particular for Mediterranean countries such as Spain, Italy, and France. Italy is one of the main European leaders in the apple sector [11]. Consequently, the yield loss risks concerning the fruit and vegetable sector must be minimized. Major threats concern diseases, insects, and natural disasters such as hail, drought, frost, and storms.

Apple cultivation is very intensive today, with a plant density around 2000 plants per hectare [12]. Such density allows a very high yearly production (about 45 tons per hectare) [13], which is obtained by a row-based cultivation strategy where young plants begin to be productive after the third year. The adoption of low-vigor rootstocks enables an increase in planting density and rapid fruiting. Unfortunately, this kind of cultivation determines a very underdeveloped root system, not enough to guarantee plant stability under unfavorable conditions. The situation is more critical during extreme weather events, especially when there are many weighty fruits, i.e., before harvesting [14,15]. Steel cables anchored to concrete or wooden poles are used to improve row stability.

Within this context, when a stormy event occurs, it is important to assess the spatial level and extent of damage to start remedial actions and minimize crop loss. Farmers are interested in damage estimation especially when a refund is due by insurance companies [16,17]. In this case, damage is assessed through on-the-spot checks by an expert surveyor from the insurance company, who determines the extent, type, and quality of damage. Such an approach depends on a high level of subjectivity related to the expert's skill and experience. Moreover, these operations require a lot of time and are expensive, especially where large areas have been affected by the event.

In this operative context concerning crop damage analysis, a more objective monitoring could play a key role, providing more robust forecasts about potential yield or yield losses. Many agricultural stakeholders, such as farmers, consortia, agronomists, insurance companies, and local administrations, require a continuous monitoring of crops over large spatial extents.

A method based on free Earth Observation (EO) data can certainly represent an effective support [18] and the consequent technological transfer desirable [19–24].

In particular, optical remote sensing data have been successfully used in several operational frameworks, as proved by many works [25–32]; unfortunately, cloud conditions limit the nominal temporal resolution of this type of data, especially while monitoring natural disasters (e.g., floods or storms) that ordinarily occur when clouds are present. Data from synthetic aperture radar (SAR) systems can operate during all-weather conditions, and, while exploring agronomical issues, they can be used to analyze the moisture and geometrical conditions of crops [33–35]. In particular, dual-polarimetric SAR acquisitions from Copernicus Sentinel-1 mission (S1) provide unique opportunities to disseminate operational monitoring for several application communities [36,37]. Dual-pol acquisition mode has a larger swath and a lower data volume compared with full-pol acquisitions, thus improving data collection and processing for operational activities [38,39]. Polarimetric data can provide information about polarization amplitude and phase, allowing scattering mechanism definition (i.e., single-bounce, double-bounce, or volume scattering) induced by target properties. SAR polarimetry (PolSAR) is a technique that analyzes SAR polarization with respect to the vector of polarized electromagnetic waves. When a signal passes through a medium, the refraction index changes, or when it strikes an object, it is reflected; the so-called backscattering matrix [40] contains information about the reflectivity, shape, and orientation of the reflecting target. An important improvement in the extraction of physical information from the ordinary coherent backscattering matrix was achieved by Cloude and Pottier [41,42], who proposed the composition of system vectors. Most studies have assessed the sensitivity of polarimetric indicators derived from the C-band space-borne SAR to derive crop parameters [43]. The PolSAR technique was successfully applied to monitor crop growth and give estimates of yield. For example, Betberder [44] analyzed temporal trends of polarimetric indicators, proving their high potential to detect crop growth changes. Valcarce [45] used polarimetric data time series for land-cover classification, adopting a decision tree classification algorithm performing high crop class detection accuracies. Mercier and Qi [46,47] used PolSAR to support/integrate vegetation phenology monitoring based on optical data.

Only few works referring to PolSAR application in crop damage analysis are present in the literature [48,49], denoting a lack of scientific production about this issue. Nevertheless, hailstorms and storms are known to change vegetation structure, resulting in lodging or tree uprooting/breaking. Therefore, this peculiar effect changes polarimetric response and could be used to detect and characterize tree structure [50]. In general, it can be said that decomposition techniques offer a new insight into PolSAR data for describing vegetation structural proprieties [51].

The polarimetric decomposition technique decomposes the signal into its individual scattering components, permitting identification of the dominant scattering type [42,52]; this information is related to the target structural properties [18,53,54]. Various decomposition techniques have been proposed, and Lee and Cloude provided a comprehensive review about this topic [42,55]. Model-based [56] and eigenvector-based [41] algorithms have been preferred by many researchers [51]. According to Ji and his collaborators [57], the Cloude–Pottier H-α-A decomposition seems to be the most promising approach. It is based on second-order statistics extracted by a set of neighbor pixels that are used to calculate the local entropy H and the α angle (related to average scattering mechanisms). These are used to define a Cartesian space, H-α, that is linearly divided into nine zones describing the main scattering mechanisms. Recently, eigenvector decomposition has been widely applied in several applications [55,58–61]. The method was originally developed for quad-polarization data. Nevertheless, it was also adapted to work with dual-pol data [57,62,63], and consequently, it can be successfully used to retrieve polarimetric information also from S1 data that are unable to collect quad-pol data.

In this work, the applicability of the H-α-A polarimetric decomposition technique to the detection and mapping of damages from storms affecting fruit orchards was tested. In particular, the proposed case study refers to the stormy event that occurred in Northwest Italy on 12 August 2020. Consequently, a map of potentially damaged orchards was generated with the aim of supporting insurance companies and local administrations to address their funding restoration policies.

2. Materials and Methods

2.1. Study Area

On 12 August 2020, an exceptional storm affected the Northwest of Italy. In particular, the storm uprooted many apple orchards in the province of Cuneo (Piemonte region, NW Italy). Moreover, it occurred in a critical period of the year, when the main fruits (apples, pears, and peaches) were still to be harvested (Figure 1). Because in this period the farmers are focused on harvesting, no early recovery efforts were performed in the damaged fields. Therefore, the majority of the uprooted trees were not removed until October.

Figure 1. An apple orchard (cultivar "Gala") with hail nets uprooted by the storm on 12 August 2020. At the bottom, many mature apples can be noted, suggesting the economic loss caused by the storm.

The study area includes four municipalities: Saluzzo, Verzuolo, Manta, and Lagnasco (Figure 2). The area of interest (AOI) is sized about 132.23 km^2. It plays a crucial economic role in Piemonte fruit production. In fact, this zone is suitable for this cultivation: the loose soil without water stagnation, sunny and dry atmosphere, and strong temperature difference between day and night allow the correct ripening and coloring of fruits. Apples represent the primary crop in Manta. Since August is a droughty period in the AOI, no significative previous precipitations had occurred before the event; 1.2 mm had cumulated in the previous week, as reported by the regional environmental agency (www.arpa. piemonte.it). Therefore, the authors supposed that moisture-related conditions cannot significantly affect the SAR signal.

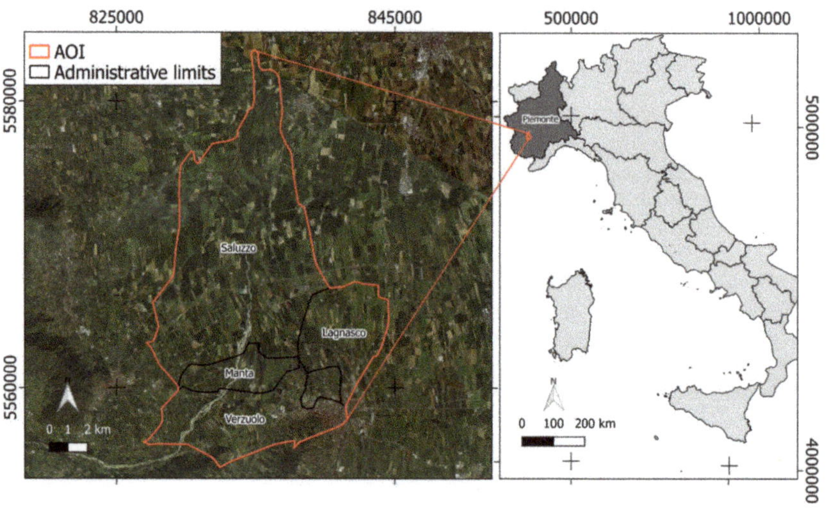

Figure 2. Italian regions (light gray) and the Piemonte region (dark gray). (Red) The AOI includes the Saluzzo, Manta, Lagnasco, and Verzuolo municipalities (reference frame: WGS84 UTM32N).

2.2. Data and Data Collection

2.2.1. Sentinel-1 Data

Sentinel-1 is currently one of the largest space-borne missions providing free and openly accessible SAR data. The S1 mission relies on a constellation of two satellites (Sentinel-1A and Sentinel-1B) operating in the C-band (5.54 cm wavelength). The main acquisition mode over land is the Interferometric Wide (IW) swath, recording approximately 250 km in length at 5×20 m spatial resolution in a single look. Ordinarily, S1 records data in a dual pole mode (VV and VH), where electromagnetic waves are polarized vertically (V) for transmission and horizontally/vertically for reception. The data are recorded as complex values (I/Q components) and in SAR geometry (range and azimuth). A descending single-look complex (SLC) IW image (relative orbit no. 139), acquired after the storm (14 August 2020), was obtained from the Copernicus Open Access Hub (https://scihub.copernicus.eu/dhus/#/home, accessed: 20 December 2020).

2.2.2. Cadastral Data

A cadastral map coupled with farmers' applications for EU Common Agricultural Policy (CAP) incentives was used in this work to classify the orchards in the AOI. The correspondent map (hereafter called orchard map (OM)) was consequently generated. The damaged orchards were analyzed at cadastral parcel level. The cadastral map was obtained for free from the regional geoportal in vector format georeferenced in the WGS84 UTM zone 32N reference frame and updated in 2018 (nominal scale was 1:2000). Databases containing farmers' applications for EU CAP incentives of 2019 were used to map orchard types in the AOI (2020 data are not yet available). Every year, farmers support their activities with CAP incentives. These data were obtained for free from the regional public information system for agriculture. CAP applications contain the cadastral parcel code and the declaration of the most relevant crops as communicated by farmers. In this way, it is possible to couple the cadastral map with crop type information at parcel level by an ordinary join operation available in the Geographical Information System (GIS) software. In this work, 2040 (about 1136 ha) apple orchards were selected from the joined data to test the procedure.

2.2.3. Ground Dataset

A ground survey was conducted to gather the field data needed to calibrate and validate the PolSAR-based mapping procedure. In total, 72 apple orchards were surveyed (about 3.5% of the apple orchards in the AOI) during a ground campaign aimed at labeling damaged (22) and undamaged (50) fields. Specifically, the surveyed fields have an average size of about 0.92 ha, fitting well with the S1 geometrical resolution. In fact, about 40 S1 pixels can characterize each field. In particular, a visual assessment aimed at recognizing the following conditions was performed: if the majority of the trees were uprooted, the field was labeled as damaged; otherwise, it was labeled undamaged, and the related cadastral parcel was selected from the OM layer.

The dataset was split in a training (60%) and a test set (40%) by random selection from the surveyed parcels. In total, 13 damaged fields (hereafter called DTFs) and 28 undamaged ones (hereafter called UTFs) were assigned to the training set. Conversely, 10 damaged fields (hereafter called DVFs) and 21 undamaged ones (hereafter called UVFs) were assigned to the test set. The training and test set parcels are shown in Figure 3.

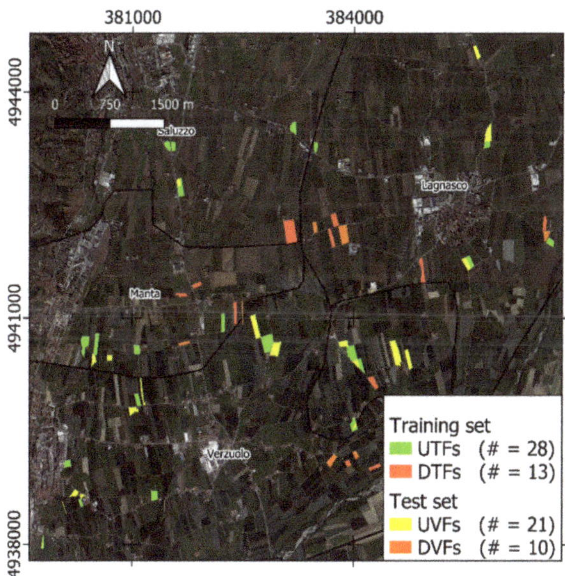

Figure 3. Parcels belonging to the training and test sets. Colors (see legend) define the state of the surveyed parcel (damaged/undamaged). Reference frame is WGS84 UTM 32N.

This dataset was provided by local farmers. The authors found that the supplied sample includes 72 fields corresponding to about 3.5% of the apple orchards in the AOI. The authors had just the opportunity of comparing the sample size with the expected total number of apple orchards in the AOI (about 2050). The authors are aware that this sample size does not perfectly fit statical requirements. Nevertheless, it well represents ordinary availability of ground data from farmers when working with actual data not directly managed by scientists. This situation well represents a common operational condition when working with technology transfer issues, especially in the agronomic sector. In fact, the most of data from farmers, generally, rely on their autonomous collections and decision of making them public. Moreover, the private property of parcels is an objective limiting factor for all the analyses, since free access is not guaranteed. With these premises, we proceed to process the data.

A preliminary economic assessment was also performed since the storm occurred close to the apple harvesting period, determining a significant problem for local apple yield in 2020. This was obtained considering, for damaged parcels, a potential yield equal to the average one in the Piemonte region (31 t·ha^{-1}) and a reference unitary price of 380 €·t^{-1}. These values were obtained from the Italian Statistics Institute (ISTAT) [64].

2.3. Data Processing

2.3.1. Polarimetric Decomposition

The available S1 IW SLC image was processed to compute the polarimetric decomposition parameters. The adopted workflow is shown in Figure 4 and proposed by [65]. The target polarimetric analysis is ordinarily performed starting from the coherency matrix [66,67] or from the 2×2 covariance matrix ($\mathbf{C_2}$). Preprocessing steps were managed using the ESA SNAP v. 7.0.0 software [68].

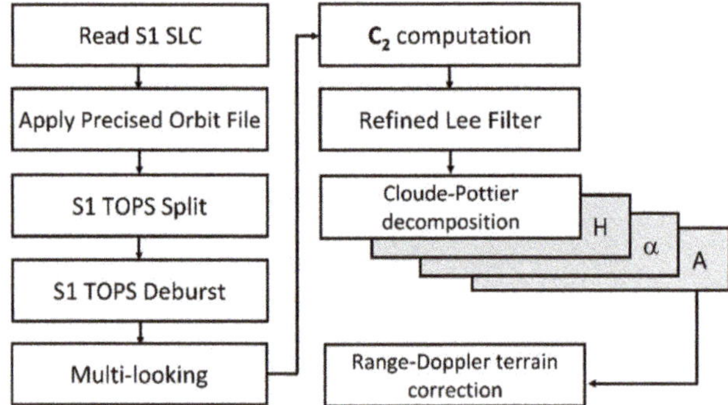

Figure 4. The adopted workflow. All steps were managed in SNAP ESA v. 7.0.

First, the precise orbit state vector data were downloaded from the ESA archive (https://qc.sentinel1.eo.esa.int/, accessed: 20 December 2020) and applied to refine the satellite position. Precise orbit files are delivered within 20 days after data acquisition and provide accurate satellite position and velocity information. Using the TOPS split module, 1 sub-swath and 2 bursts were selected based on AOI coverage. A radiometric calibration was applied and the result saved in a complex-valued format needed to compute $\mathbf{C_2}$. TOPS deburst was applied by merging different bursts into a single SLC image. A spatial subset was then generated covering the AOI. The subset was multi-looked by 4×1 (range and azimuth direction, respectively) to generate squared pixels. The resulting multi-looked image, with a geometrical resolution equal to 15 m, was used to generate the local $\mathbf{C_2}$ at pixel level. With respect to quad polarization, dual-polarimetric SAR sensors generate a matrix showing the half of the totally occurring scattering components involved in fully polarimetric imagery [69]. In particular, the covariance matrix for dual polarization (e.g., Sentinel-1) is often calculated with reference to a second-order scattering information [18] generated from the spatial averaging of the scattering vector k = [S$_{VV}$, S$_{VH}$]^T as expressed in Equation (1):

$$C_2 = \begin{bmatrix} C_{11} & C_{12} \\ C_{21} & C_{22} \end{bmatrix} = \begin{bmatrix} \langle |S_{VV}| \rangle & \langle S_{VH}S_{VV}^* \rangle \\ \langle S_{VH}S_{VV}^* \rangle & \langle |S_{VH}| \rangle \end{bmatrix} \tag{1}$$

where $*$ denotes the complex conjugate and $\langle\,\rangle$ the local mean value in a 5×5 moving window. Each $\mathbf{C_2}$ element (C_{11}, C_{22}, $\Re(C_{12})$, and $\Im(C_{12})$) is stored individually and successively refined by Lee filtering (5×5 kernel size) to minimize speckle-related noise. H-α-A polarimetric decomposition was obtained by eigenvector computation as proposed by different authors [57,62,63]. The modified formula for dual-pol data, as proposed by [66], is reported in Equations (2) and (3).

$$\langle C_2 \rangle = [U] \begin{bmatrix} \lambda_1 & 0 \\ 0 & \lambda_2 \end{bmatrix} [U]^{*T} = \lambda_1 u_1 u_1^{*T} + \lambda_2 u_2 u_2^{*T} \tag{2}$$

$$[U] = \begin{bmatrix} U_{11} & U_{12} \\ U_{21} & U_{22} \end{bmatrix} = [u_1 \; u_2] = \begin{bmatrix} \cos\alpha & -\sin\alpha\,e^{-j\delta} \\ \sin\alpha\,e^{j\delta} & \cos\alpha \end{bmatrix} \tag{3}$$

where $\lambda_1 \geq \lambda_2 \geq 0$ are the local eigenvalues, $[U]$ is the orthogonal unitary matrix, $*$ and T represents the complex conjugate and transpose matrices, respectively. The angles α and δ define the orientation and size of the polarization ellipse of the recorded signal [62]. The eigenvector dual-pol decomposition results in three roll-invariant parameters: polarimetric scattering entropy (H), mean scattering angle (α), and scattering anisotropy (A).

H was calculated from Equation (4):

$$H = -\sum_{i=1}^{2}(-P_i \log_2 P_i) \tag{4}$$

where

$$P_i = \frac{\lambda_i}{\lambda_1 + \lambda_2}$$

H defines scatter randomness; it can vary between 0 and 1 and is related to the number of dominant scattering mechanisms, being proportional to the degree of depolarization [70]. H = 0 means that the coherency matrix shows only one eigenvalue and, therefore, the relative orientation of the correspondent pixel elements is quite simplified (e.g., single-bounce reflection).

Anisotropy A (Equation (5)) provides additional information about H in terms of the difference between scattering mechanisms.

$$A = \frac{\lambda_1 - \lambda_2}{\lambda_1 + \lambda_2} \tag{5}$$

The anisotropy quantifies the relative strength between first and second dominant scattering mechanisms. It is strictly related to the degree of signal polarization [18,71,72]. According to Mandal [18], the state of polarization of an electromagnetic (EM) wave is characterized in terms of the degree of polarization ($0 \leq A \leq 1$). The latter is defined as the ratio between the average intensity of the polarized portion of the signal and its total intensity [73]. A = 1 and A = 0 for a completely polarized and completely unpolarized wave, respectively. The unpolarized part of the received wave, $(1 - A)$, is assumed to represent the volume scattering component from the distributed targets [74].

Average scattering mechanisms (i.e., surface, double-bounce, and volume scattering) can be identified with respect to the α parameter, which is computed according to Equation (6):

$$\alpha = \sum_{i=1}^{2} P_i \cos^{-1}\left(\frac{|\lambda_1 + \lambda_2|}{\sqrt{2}\sqrt{|\lambda_1|^2 + |\lambda_2|^2}}\right) \tag{6}$$

The α angles close to $0°$ denote a diffuse surface scattering, α close to $45°$ means dipole scattering (caused by volumes), and α close to $90°$ means double-bounce scattering mechanisms.

With these premises, the raster layer mapping local H, α, and A values was computed from the pre-processed SLC image. It was projected onto the WGS84 UTM 32N reference frame, applying the range–Doppler terrain correction. The adopted digital terrain model (DTM) needed for this step was the one freely obtainable from the Piemonte region geoportal [75]. It is supplied with a 5 m grid size and a height accuracy of ± 0.30 m and was generated in 2011. The nearest-neighbor resampling method was adopted during the range–Doppler terrain correction.

2.3.2. Testing H-α-A Values after the Storm

To assess how the storm changed the orchards' polarimetric behavior, a preliminary analysis was performed with reference to the training set. In particular, DTF and UTF pixels distributions were compared using the Mann–Whitney (MW) nonparametric test (one-tailed) [76]. The MW null hypothesis is that DTFs and UTFs have an identical distribution. The one-sided alternative "greater" was set, assuming that the DTF cumulated frequency distribution was expected to have shifted to the right of the UTF one (i.e., DTFs were greater than UTFs) [77].

The authors preliminary explored the polarimetric indices' behavior using reference ground data. In particular, the frequency distributions were perceptively assessed using boxplots (see Section 3.1). The median value of distribution highlights a shift between damaged and undamaged fields. Therefore, to test these perceptive differences, the authors performed one tail test since the direction of changes is a priori known.

Three MW tests were performed to test if the DTF distributions of the H-α-A pixels within the parcels were statistically different from the UTF ones. All statistical analyses were performed using R software v. 3.6.3 [78]; conversely, spatial analysis was done using SAGA GIS 7.0 [79].

2.3.3. Detection of Damaged Orchards

The main goal of this work was to test the capability of the PolSAR technique to recognize damaged orchards. For this task, UTFs were assumed as representatives of the state of undamaged orchards. Samples were sized about 23 ha and represented about 2% of OM. In spite of this small sample size, the UTFs preliminarily resulted in a good dataset, whose reliability was confirmed by ground surveys. With these premises, the H-α-A distributions within UTFs were used to represent the reference distributions of the undamaged orchards. All H-α-A distributions from the AOI mapped parcels were tested against undamaged ones by the MW test, checking the following conditions: (i) parcel H distribution was greater than that of the UTFs; (ii) parcel α distribution was greater than that of the UTFs; (iii) parcel A distribution was lower than that of the UTFs. The resulting MW U-statistic and related p-value were then mapped for each orchard parcel. Moreover, the compound probability (CP) [80] was also calculated according to Equation (7) using R software v. 3.6.3. CP represents the probability that the previously mentioned three conditions were simultaneously satisfied.

$$CP = (1 - p_H)(1 - p_\alpha)(1 - p_A) \tag{7}$$

where p_H is the p-value resulting from the MW test under condition (i), p_α is the p-value resulting from the MW test under condition (ii), and p_A is the p-value resulting from the MW test under condition (iii). The resulting CP was then mapped for all OM parcels, representing its compound probability to have been damaged by the storm. A threshold value of CP able to separate damaged fields from undamaged ones has to be necessarily selected by final users, e.g., the insurance company or local public administration, according to their specific policies and strategies. Nevertheless, a possible solution is proposed here, relying on the standard error of the mean (SEM) of the CP distributions of the DTFs and UTFs. The estimated threshold value was used to generate the map of damaged orchards (DM): parcels showing a CP value lower than the threshold was classified as "undamaged," otherwise as "damaged." The DMs were then tested against the previously mentioned test set and the correspondent confusion matrix calculated to assess the accuracy of detection.

3. Results

3.1. H-α-A Analysis

The statistical distributions of H-α-A were computed with reference to DTFs and UTFs (Figure 5).

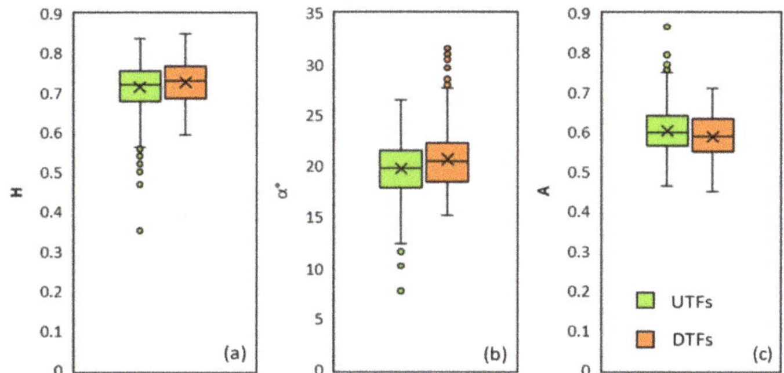

Figure 5. Boxplots of H-α-A distributions for UTFs and DTFs. The boxplot values are from bottom to top, respectively, 5th, 25th, 50th—cross is mean value—75th, and 95th percentiles. (**a**) Entropy pixel distribution; (**b**) alpha angle pixel distribution; (**c**) anisotropy pixel distribution.

The MW test results (Table 1) show that the H and α distributions of DTFs presented values significantly greater than UTFs; conversely, the A distribution of the DTFs was lower than that of the UTFs.

Table 1. MW test results obtained by comparing the H-α-A pixel distributions of DTFs and UTFs.

	U	*p*-Value
H	14,514	0.000159
α	16,604	3.83×10^{-10}
A	8616	0.000161

3.2. Damaged Orchards' Mapping

Based on the assumption that a storm can change the polarimetric behavior of orchards according to previously mentioned dynamics, a map of CP representing the parcel probability of being recognized as damaged was generated using the UTF dataset as reference (Figure 6).

With reference to CP, a threshold value was estimated to separate damaged fields from undamaged ones based on the SEM of CP statistic distributions of the DTFs and UTFs (Figure 7b). The DTFs showed a CP mean and SEM value of 0.715 and 0.125, respectively; consequently, one can assume that the CP mean value of all damaged orchards reasonably falls in the range 0.715 ± 0.125, about 0.6 being the lower boundary. A threshold equal to 0.6 was therefore selected to generate the DM binary classification (Figure 7a).

A total of 217 ha (430 orchards) of potentially damaged apple orchards were detected in the AOI. According to the OM layer, 19% of the apple orchards were damaged after the event.

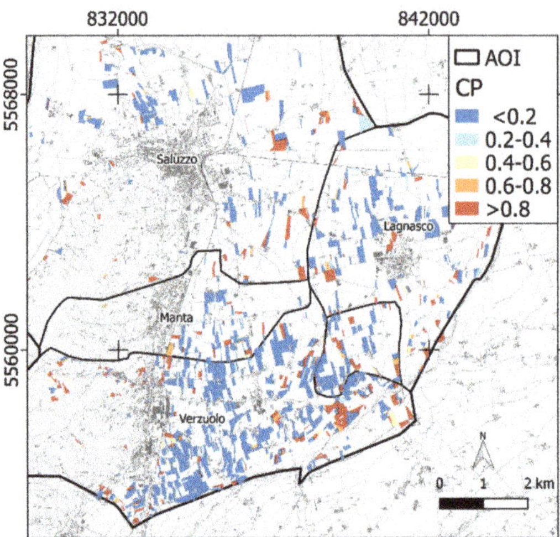

Figure 6. A CP map of apple orchards in the AOI (Reference frame: WGS84 UTM32N).

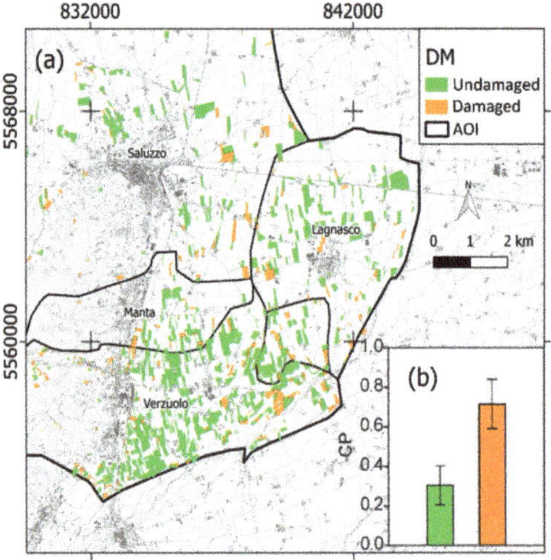

Figure 7. (**a**) DM binary classification of the OM in the AOI (reference frame is WGS84 UTM 32N); (**b**) bar chart representing mean and 1 SEM of the CP for DTFs and UTFs.

The DM was validated with respect to the test set, and the correspondent confusion matrix computed (Table 2). Classification accuracy is defined here as the one for binary classification of imbalanced data [81–83] since, in the test set, the number of undamaged fields was significantly greater than that of damaged fields. The resulting precision and specificity were pretty high (0.80 and 0.71, respectively), while balanced accuracy was found to be 0.75. Overall accuracy was 0.74, while F1 score (harmonic mean of the precision and recall) and G-mean (geometric mean of sensitivity and precision) were both about 0.67.

Table 2. Metrics derived from the confusion matrix of the DM with respect to the test set. True positives (TPs): number of damaged elements predicted as damaged; false positives (FPs): number of undamaged elements predicted as damaged; false negatives (FNs): number of damaged elements predicted as undamaged; true negatives (TNs): number of undamaged elements predicted as undamaged.

		Classification	
		Damaged	Undamaged
Reference	Damaged	8	2
	Undamaged	6	15
		Accuracies	
Measure		Value	Formula
Sensitivity		0.80	TPR = TP/ (TP + FN)
Specificity		0.71	SPC = TN/ (FP + TN)
Precision		0.57	PPV = TP/ (TP + FP)
Negative Predictive Value		0.88	NPV = TN/ (TN + FN)
False Positive Rate		0.28	FPR = FP/ (FP + TN)
False Discovery Rate		0.42	FDR = FP/ (FP + TP)
False Negative Rate		0.20	FNR = FN/ (FN + TP)
Overall Accuracy		0.74	OA = (TP + TN) / (TP + TN + FP + FN)
F1 Score		0.66	F1 = 2TP/ (2TP + FP + FN)
Balanced Accuracy		0.75	BA = TPR + TNR/2
G-Mean		0.67	G-mean= sqrt (TPR *PPV)

Furthermore, it is worth stressing that the storm occurred close to the harvesting period, determining a significant problem for local apple yield in 2020. With reference to the AOI, a preliminary estimate of economic loss was computed to be about €2,500,000. Reported estimates could certainly vary according to the apple orchards' age, apple variety, plant density, agronomic management, and local soil properties. Nevertheless, these estimates constituted a preliminary assessment of storm damage that occurred on 12 August 2020. Future validation is expected to test these economic deductions.

4. Discussions

Concerning the damaged orchards' H-α-A distributions (Figure 5 and Table 1), higher values of H and α in the damaged parcels could be attributed to the changes in vegetation structure (Figure 8). In fact, the inter-row spaces of the damaged orchards, after the storm, were completely covered with the crowns of the broken or uprooted plants, which determined a different scattering geometry. Pre-event plant row geometry was characterized by a regular pattern, which drastically changed to a more disordered one, where the fallen crown elements increased the H values. Since the pre-event scattering mechanism was determined by regularly aligned and spaced plants (rows) alternating with bare soil/grass (inter-rows), it determined intermediate α values. After the storm, it can be assumed that the scattering mechanism was strongly influenced by crown volume, inducing an increase in the α values. Conversely, A appeared to reduce after the event. This could be possibly related to a reduction in the eigenvalue difference $\lambda_1 - \lambda_2$ related to the slightly different scattering mechanism after the storm. The volumetric mechanism appeared to be the prevailing one in the damaged parcels, as proved by the H increase. Since the canopy causes a strong depolarization of the SAR signal, the degree of depolarization (i.e., 1-A) tends to increase with crown closure [18]. Given these interpretation keys, the results obtained seem to support the idea that, after a relevant event able to significantly change vegetation structure, the orchards' polarimetric behavior significantly changes. Based on the collected reference data, damaged orchards tend to show (i) higher values of H and α due to the increased contribution of the volume scattering mechanism, and (ii) lower A values, possibly due to the inter-row closure generated by broken/fallen trees, which increase signal depolarization.

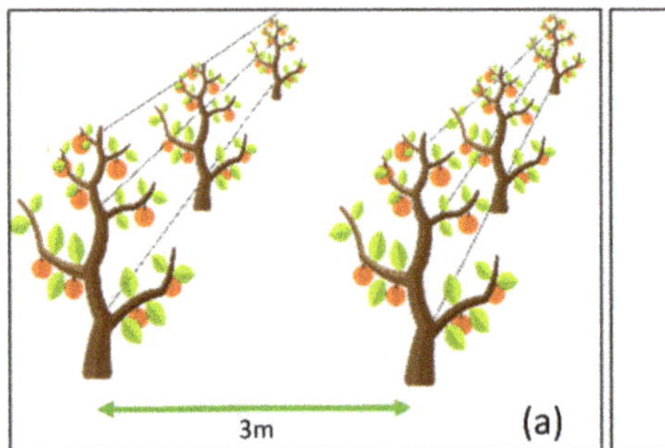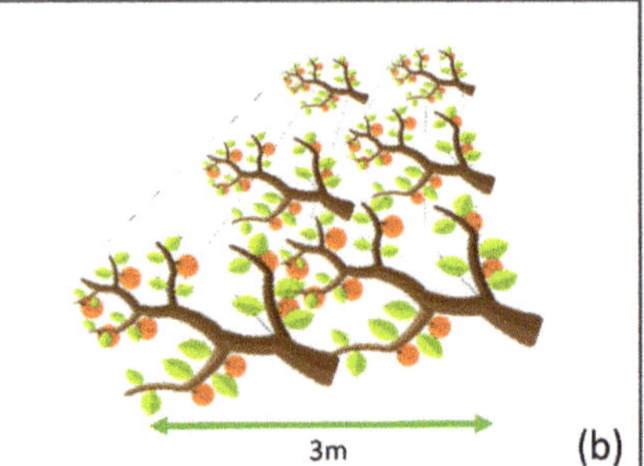

Figure 8. A sketch representing orchard condition before (**a**) and after (**b**) the storm. In (**a**) the pattern row/inter-row is well defined; (**b**) after the storm, apple tree uprooting occurred, altering the row/inter-row pattern, and crowns covering the ground increased volumetric scattering.

Concerning the mapping of damaged orchards, the results reported in Table 2 suggest that polarimetric decomposition of S1 data is an effective approach to map orchards affected by a storm, especially during cloudy weather situations. Nevertheless, it is worth stressing that some limitations still persist while working with dual-pol decomposition. In comparison to quad polarization, dual-pol SAR sensors collect half of the scattering matrix components involved in fully polarimetric imagery. Therefore, dual-pol derived products may vary from the classical Wishart distribution. In fact, [57] highlighted that entropy/alpha decomposition using one co-polarization and one cross-polarization does not adequately extract scattering mechanisms in the H-α plane. Nevertheless, Cloude [62] proved how these differences result similarly to the conventional quad-pol one while working with vegetation. In spite of these differences, many operative frameworks were proposed proving how information lost during the dual-pol acquisition can be compensated for enhancing image swath and satellite revisit frequency. Moreover, often quad-pol SAR data are not available free of charge and not readily available for operative purposes. S1 is currently one of the largest space-borne missions providing free and open-access SAR data having high temporal resolution, fitting well with vegetation dynamics monitoring requirement.

Future developments are expected to test if pre- and post-H-α-A differences can be used to semi-automatically detect significance changes. It is worth highlighting that the majority of apple orchards in the study area are covered by plastic nets to protect the trees against hail. Probably, plastic nets can influence the complex permittivity of the analyzed volume and therefore affect the polarimetric response of the observed uprooted trees. Since in the study area, a few fields do not have hail nets, the authors did not survey such orchards, and therefore no assessment looking for the effects of nets on polarimetric response was performed. A specific research should be addressed to assess how plastic hail nets can affect backscattered signal.

5. Conclusions

In this work, a preliminary assessment about the polarimetric behavior of orchards after a storm was performed. The analysis was aimed at proposing a first methodological approach to detect orchard damage by a storm based on the PolSAR decomposition technique using S1 data. The joint adoption of free accessible S1 data, institutional free auxiliary data (a cadastral map and farmers' CAP application database), and open software (SNAP) constituted a peculiar trait of the proposed approach. It moves in the direction of technological transfer, aiming at making SAR data/techniques an operational tool for agronomic applications, with special concern about weather-related damages to crops, which could be of interest to insurance companies or public administrations. The results proved that storm damages significantly increase the H and α parameters. By contrast, the

A parameter tends to be lower in the damaged orchards. This phenomenon is possibly related to the changes affecting vegetation structure in the damaged fields, where the crowns and branches of fallen/broken plants fill the inter-row space, changing the regular pattern ordinarily characterizing apple orchards. Based on this evidence, the authors proposed a methodology to map possibly damaged orchards that relies on the knowledge about the behavior of witness (and neighboring) undamaged orchards. The method permitted the mapping of the probability that an orchard is damaged or not, constituting a new free tool able to improve orchard monitoring after a calamitous event by regional agencies and insurance companies. It is worth reminding that only apple orchards were considered for this case study. Future developments are expected to test the effectiveness of this method in other orchard types, as pear or peach, which are very diffuse in the AOI.

Author Contributions: Conceptualization, S.D.P.; data curation, S.D.P.; formal analysis, S.D.P. and F.S.; investigation, E.B.-M.; methodology, S.D.P. and E.B.-M.; software, S.D.P.; validation, M.G.; visualization, F.S.; writing—original draft, S.D.P., F.S. and M.G.; writing—review and editing, E.T. and E.B.-M. All authors have read and agreed to the published version of the manuscript.

Funding: This research received no external funding.

Institutional Review Board Statement: Not applicable.

Informed Consent Statement: Not applicable.

Data Availability Statement: Data sharing not applicable.

Acknowledgments: We would like to thank Az. Agr. Fessia Franca for having provided ground control data and fundamental operational information useful to reach the results presented in this work.

Conflicts of Interest: The authors declare no conflict of interest.

References

1. Hsiang, S.; Kopp, R.; Jina, A.; Rising, J.; Delgado, M.; Mohan, S.; Rasmussen, D.J.; Muir-Wood, R.; Wilson, P.; Oppenheimer, M. Estimating Economic Damage from Climate Change in the United States. *Science* **2017**, *356*, 1362–1369. [CrossRef]
2. Nelson, G.C.; Valin, H.; Sands, R.D.; Havlík, P.; Ahammad, H.; Deryng, D.; Elliott, J.; Fujimori, S.; Hasegawa, T.; Heyhoe, E. Climate Change Effects on Agriculture: Economic Responses to Biophysical Shocks. *Proc. Natl. Acad. Sci. USA* **2014**, *111*, 3274–3279. [CrossRef]
3. Moore, F.C.; Baldos, U.; Hertel, T.; Diaz, D. New Science of Climate Change Impacts on Agriculture Implies Higher Social Cost of Carbon. *Nat. Commun.* **2017**, *8*, 1–9. [CrossRef]
4. FAO The Impact of Natural Hazards and Disasters on Agriculture and Food Security and Nutrition–A Call for Action to Build Resilient Livelihoods. Available online: http://www.fao.org/emergencies/resources/documents/resources-detail/zh/c/280784/ (accessed on 20 December 2020).
5. Welbergen, J.A.; Klose, S.M.; Markus, N.; Eby, P. Climate Change and the Effects of Temperature Extremes on Australian Flying-Foxes. *Proc. R. Soc. B. Biol. Sci.* **2008**, *275*, 419–425. [CrossRef]
6. Wang, E.; Bertis, B.L.; Jimmy, R.W.; Yu, Y. Simulation of Hail Effects on Crop Yield Losses for Corn-Belt States in USA. *Trans. Chin. Soc. Agric. Eng.* **2012**, *28*, 177–185.
7. Reighard, G.L.; Parker, M.L.; Krewer, G.W.; Beckman, T.G.; Wood, B.W.; Smith, J.E.; Whiddon, J. Impact of Hurricanes on Peach and Pecan Orchards in the Southeastern United States. *HortScience* **2001**, *36*, 250–252. [CrossRef]
8. Crane, J.; Balerdi, C.; Campbell, R.; Campbell, C.; Goldweber, S. Managing Fruit Orchards to Minimize Hurricane Damage. *HortTechnology* **1994**, *4*, 21–27. [CrossRef]
9. Notarnicola, B.; Tassielli, G.; Renzulli, P.A.; Castellani, V.; Sala, S. Environmental Impacts of Food Consumption in Europe. *J. Clean. Prod.* **2017**, *140*, 753–765. [CrossRef]
10. Bos-Brouwers, H.E.J.; Graf, V.; Aramyan, L.; Oberc, B. Food Redistribution in the EU–Mapping and Analysis of Existing Regulatory and Policy Measures Impacting Food Redistribution from EU Member States. Available online: https://research.wur.nl/en/publications/food-redistribution-in-the-eu-mapping-and-analysis-of-existing-re (accessed on 20 December 2020).
11. FAO FAOSTAT. Available online: http://www.fao.org/faostat/en/#home (accessed on 13 November 2020).
12. Eccher, T.; Granelli, G. Fruit Quality and Yield of Different Apple Cultivars as Affected by Tree Density. *Acta Hortic.* **2006**, 535–540. [CrossRef]
13. Childers, N.F.; Morris, J.R.; Sibbett, G.S. *Modern Fruit Science: Orchard and Small Fruit Culture*; Horticultural Publications: Gainesville, GA, USA, 1995.

14. Lordan, J.; Gomez, M.; Francescatto, P.; Robinson, T.L. Long-Term Effects of Tree Density and Tree Shape on Apple Orchard Performance, a 20 Year Study–Part 2, Economic Analysis. *Sci. Hortic.* **2019**, *244*, 435–444. [CrossRef]

15. Lauri, P. Apple Tree Architecture and Cultivation-a Tree in a System. In *Proceedings of the I International Apple Symposium 1261*; ISHS Acta Horticulturae: Yangling, China, 2016; pp. 173–184.

16. Prabhakar, M.; Gopinath, K.A.; Reddy, A.G.K.; Thirupathi, M.; Rao, C.S. Mapping Hailstorm Damaged Crop Area Using Multispectral Satellite Data. *Egypt. J. Remote Sens. Space Sci.* **2019**, *22*, 73–79. [CrossRef]

17. Botzen, W.J.W.; Bouwer, L.M.; Van den Bergh, J. Climate Change and Hailstorm Damage: Empirical Evidence and Implications for Agriculture and Insurance. *Resour. Energy Econ.* **2010**, *32*, 341–362. [CrossRef]

18. Mandal, D.; Kumar, V.; Ratha, D.; Dey, S.; Bhattacharya, A.; Lopez-Sanchez, J.M.; McNairn, H.; Rao, Y.S. Dual Polarimetric Radar Vegetation Index for Crop Growth Monitoring Using Sentinel-1 SAR Data. *Remote Sens. Environ.* **2020**, *247*, 111954. [CrossRef]

19. De Petris, S.; Berretti, R.; Guiot, E.; Giannetti, F.; Motta, R.; Borgogno-Mondino, E. Detection And Characterization of Forest Harvesting In Piedmont Through Sentinel-2 Imagery: A Methodological Proposal. *Ann. Silvic. Res.* **2020**, *45*, 92–98. [CrossRef]

20. Sarvia, F.; De Petris, S.; Borgogno-Mondino, E. Remotely sensed data to support insurance strategies in agriculture. In *Remote Sensing for Agriculture, Ecosystems, and Hydrology XXI*; SPIE-Intl. Soc. Opt. Eng.: Strasbourg, France, 2019.

21. Sarvia, F.; De Petris, S.; Borgogno-Mondino, E. Multi-Scale Remote Sensing to Support Insurance Policies in Agriculture: From Mid-Term to Instantaneous Deductions. *GIScience Remote Sens.* **2020**, *57*, 770–784. [CrossRef]

22. Borgogno-Mondino, E.; Sarvia, F.; Gomarasca, M.A. Supporting Insurance Strategies in Agriculture by Remote Sensing: A Possible Approach at Regional Level. In *Proceedings of the Lecture Notes in Computer Science*; Springer Science and Business Media LLC: Berlin/Heidelberg, Germany, 2019; pp. 186–199.

23. Sarvia, F.; De Petris, S.; Borgogno-Mondino, E. A Methodological Proposal to Support Estimation of Damages from Hailstorms Based on Copernicus Sentinel 2 Data Times Series. In *Proceedings of the International Conference on Computational Science and Its Applications*; Springer: Berlin/Heidelberg, Germany, 2020; pp. 737–751.

24. Sarvia, F.; Xausa, E.; Petris, S.D.; Cantamessa, G.; Borgogno-Mondino, E. A Possible Role of Copernicus Sentinel-2 Data to Support Common Agricultural Policy Controls in Agriculture. *Agronomy* **2021**, *11*, 110. [CrossRef]

25. Boryan, C.; Yang, Z.; Mueller, R.; Craig, M. Monitoring US Agriculture: The US Department of Agriculture, National Agricultural Statistics Service, Cropland Data Layer Program. *Geocarto Int.* **2011**, *26*, 341–358. [CrossRef]

26. López-Lozano, R.; Duveiller, G.; Seguini, L.; Meroni, M.; García-Condado, S.; Hooker, J.; Leo, O.; Baruth, B. Towards Regional Grain Yield Forecasting with 1 Km-Resolution EO Biophysical Products: Strengths and Limitations at Pan-European Level. *Agric. For. Meteorol.* **2015**, *206*, 12–32. [CrossRef]

27. Chipanshi, A.; Zhang, Y.; Kouadio, L.; Newlands, N.; Davidson, A.; Hill, H.; Warren, R.; Qian, B.; Daneshfar, B.; Bedard, F. Evaluation of the Integrated Canadian Crop Yield Forecaster (ICCYF) Model for in-Season Prediction of Crop Yield across the Canadian Agricultural Landscape. *Agric. For. Meteorol.* **2015**, *206*, 137–150. [CrossRef]

28. de Wit, A.; Duveiller, G.; Defourny, P. Estimating Regional Winter Wheat Yield with WOFOST through the Assimilation of Green Area Index Retrieved from MODIS Observations. *Agric. For. Meteorol.* **2012**, *164*, 39–52. [CrossRef]

29. Doraiswamy, P.C.; Sinclair, T.R.; Hollinger, S.; Akhmedov, B.; Stern, A.; Prueger, J. Application of MODIS Derived Parameters for Regional Crop Yield Assessment. *Remote Sens. Environ.* **2005**, *97*, 192–202. [CrossRef]

30. Chen, Y.; Zhang, Z.; Tao, F. Improving Regional Winter Wheat Yield Estimation through Assimilation of Phenology and Leaf Area Index from Remote Sensing Data. *Eur. J. Agron.* **2018**, *101*, 163–173. [CrossRef]

31. Patel, N.R.; Bhattacharjee, B.; Mohammed, A.J.; Tanupriya, B.; Saha, S.K. Remote Sensing of Regional Yield Assessment of Wheat in Haryana, India. *Int. J. Remote Sens.* **2006**, *27*, 4071–4090. [CrossRef]

32. Orusa, T.; Orusa, R.; Viani, A.; Carella, E.; Borgogno Mondino, E. Geomatics and EO Data to Support Wildlife Diseases Assessment at Landscape Level: A Pilot Experience to Map Infectious Keratoconjunctivitis in Chamois and Phenological Trends in Aosta Valley (NW Italy). *Remote Sens.* **2020**, *12*, 3542. [CrossRef]

33. Ulaby, F.T.; Moore, R.K.; Fung, A.K. *Microwave Remote Sensing: Active and Passive. Volume 1-Microwave Remote Sensing Fundamentals and Radiometry*; Artech House: Norwood, MA, USA, 1981.

34. Ulaby, F. Radar Response to Vegetation. *IEEE Trans. Antennas Propag.* **1975**, *23*, 36–45. [CrossRef]

35. McNairn, H.; Shang, J. A review of multitemporal synthetic aperture radar (SAR) for crop monitoring. In *Multitemporal Remote Sensing*; Springer: Berlin, Germany, 2016; pp. 317–340.

36. Karagiannopoulou, A.; Tsiakos, C.; Tsimiklis, G.; Tsertou, A.; Amditis, A.; Milcinski, G.; Vesel, N.; Protic, D.; Kilibarda, M.; Tsakiridis, N. An integrated service-based solution addressing the modernised common agriculture policy regulations and environmental perspectives. In *Proceedings Volume 11528, Remote Sensing for Agriculture, Ecosystems, and Hydrology XXII*; International Society for Optics and Photonics: Bellingham, WA, USA, 2020.

37. Kanjir, U.; DJurić, N.; Veljanovski, T. Sentinel-2 Based Temporal Detection of Agricultural Land Use Anomalies in Support of Common Agricultural Policy Monitoring. *ISPRS Int. J. Geo Inf.* **2018**, *7*, 405. [CrossRef]

38. Lee, J.-S.; Grunes, M.R.; Pottier, E. Quantitative Comparison of Classification Capability: Fully Polarimetric versus Dual and Single-Polarization SAR. *IEEE Trans. Geosci. Remote Sens.* **2001**, *39*, 2343–2351.

39. Ainsworth, T.L.; Kelly, J.P.; Lee, J.-S. Classification Comparisons between Dual-Pol, Compact Polarimetric and Quad-Pol SAR Imagery. *ISPRS J. Photogramm. Remote Sens.* **2009**, *64*, 464–471. [CrossRef]

40. Boerner, W.-M.; Mott, H.; Luneburg, E. Polarimetry in Remote Sensing: Basic and Applied Concepts. In Proceedings of the IGARSS'97, 1997 IEEE International Geoscience and Remote Sensing Symposium Proceedings, Remote Sensing-A Scientific Vision for Sustainable Development Singapore, 3–8 August 1997; Volume 3, pp. 1401–1403.
41. Cloude, S.R.; Pottier, E. An Entropy Based Classification Scheme for Land Applications of Polarimetric SAR. *IEEE Trans. Geosci. Remote Sens.* **1997**, *35*, 68–78. [CrossRef]
42. Cloude, S.R.; Pottier, E. A Review of Target Decomposition Theorems in Radar Polarimetry. *IEEE Trans. Geosci. Remote Sens.* **1996**, *34*, 498–518. [CrossRef]
43. Nasirzadehdizaji, R.; Balik Sanli, F.; Abdikan, S.; Cakir, Z.; Sekertekin, A.; Ustuner, M. Sensitivity Analysis of Multi-Temporal Sentinel-1 SAR Parameters to Crop Height and Canopy Coverage. *Appl. Sci.* **2019**, *9*, 655. [CrossRef]
44. Betbeder, J.; Fieuzal, R.; Philippets, Y.; Ferro-Famil, L.; Baup, F. Contribution of Multitemporal Polarimetric Synthetic Aperture Radar Data for Monitoring Winter Wheat and Rapeseed Crops. *J. Appl. Remote Sens.* **2016**, *10*, 026020. [CrossRef]
45. Valcarce-Diñeiro, R.; Arias-Pérez, B.; Lopez-Sanchez, J.M.; Sánchez, N. Multi-Temporal Dual-and Quad-Polarimetric Synthetic Aperture Radar Data for Crop-Type Mapping. *Remote Sens.* **2019**, *11*, 1518. [CrossRef]
46. Mercier, A.; Betbeder, J.; Baudry, J.; Le Roux, V.; Spicher, F.; Lacoux, J.; Roger, D.; Hubert-Moy, L. Evaluation of Sentinel-1 & 2 Time Series for Predicting Wheat and Rapeseed Phenological Stages. *ISPRS J. Photogramm. Remote Sens.* **2020**, *163*, 231–256.
47. Qi, Z.; Yeh, A.G.-O.; Li, X. A Crop Phenology Knowledge-Based Approach for Monthly Monitoring of Construction Land Expansion Using Polarimetric Synthetic Aperture Radar Imagery. *ISPRS J. Photogramm. Remote Sens.* **2017**, *133*, 1–17. [CrossRef]
48. Zhao, L.; Yang, J.; Li, P.; Shi, L.; Zhang, L. Characterizing Lodging Damage in Wheat and Canola Using Radarsat-2 Polarimetric SAR Data. *Remote Sens. Lett.* **2017**, *8*, 667–675. [CrossRef]
49. Yang, H.; Chen, E.; Li, Z.; Zhao, C.; Yang, G.; Pignatti, S.; Casa, R.; Zhao, L. Wheat Lodging Monitoring Using Polarimetric Index from RADARSAT-2 Data. *Int. J. Appl. Earth Obs. Geoinf.* **2015**, *34*, 157–166. [CrossRef]
50. Dickinson, C.; Siqueira, P.; Clewley, D.; Lucas, R. Classification of Forest Composition Using Polarimetric Decomposition in Multiple Landscapes. *Remote Sens. Environ.* **2013**, *131*, 206–214. [CrossRef]
51. Trisasongko, B.H. The Use of Polarimetric SAR Data for Forest Disturbance Monitoring. *Sens. Imaging* **2010**, *11*, 1–13. [CrossRef]
52. Le Toan, T.; Beaudoin, A.; Riom, J.; Guyon, D. Relating Forest Biomass to SAR Data. *IEEE Trans. Geosci. Remote Sens.* **1992**, *30*, 403–411. [CrossRef]
53. Ruiz, J.S.; Ordonez, Y.F.; McNairn, H. Corn Monitoring and Crop Yield Using Optical and Microwave Remote Sensing. *Geosci. Remote Sens.* **2008**, *10*, 405–420.
54. Mandal, D.; Kumar, V.; Lopez-Sanchez, J.M.; Bhattacharya, A.; McNairn, H.; Rao, Y.S. Crop Biophysical Parameter Retrieval from Sentinel-1 SAR Data with a Multi-Target Inversion of Water Cloud Model. *Int. J. Remote Sens.* **2020**, *41*, 5503–5524. [CrossRef]
55. Lee, J.-S.; Grunes, M.R.; Ainsworth, T.L.; Du, L.-J.; Schuler, D.L.; Cloude, S.R. Unsupervised Classification Using Polarimetric Decomposition and the Complex Wishart Classifier. *IEEE Trans. Geosci. Remote Sens.* **1999**, *37*, 2249–2258.
56. Freeman, A.; Durden, S.L. A Three-Component Scattering Model for Polarimetric SAR Data. *IEEE Trans. Geosci. Remote Sens.* **1998**, *36*, 963–973. [CrossRef]
57. Ji, K.; Wu, Y. Scattering Mechanism Extraction by a Modified Cloude-Pottier Decomposition for Dual Polarization SAR. *Remote Sens.* **2015**, *7*, 7447. [CrossRef]
58. McNairn, H.; Shang, J.; Jiao, X.; Champagne, C. The Contribution of ALOS PALSAR Multipolarization and Polarimetric Data to Crop Classification. *IEEE Trans. Geosci. Remote Sens.* **2009**, *47*, 3981–3992. [CrossRef]
59. Lopez-Sanchez, J.M.; Cloude, S.R.; Ballester-Berman, J.D. Rice Phenology Monitoring by Means of SAR Polarimetry at X-Band. *IEEE Trans. Geosci. Remote Sens.* **2011**, *50*, 2695–2709. [CrossRef]
60. Ramsey III, E.; Rangoonwala, A.; Suzuoki, Y.; Jones, C.E. Oil Detection in a Coastal Marsh with Polarimetric Synthetic Aperture Radar (SAR). *Remote Sens.* **2011**, *3*, 2630. [CrossRef]
61. Yonezawa, C.; Watanabe, M.; Saito, G. Polarimetric Decomposition Analysis of ALOS PALSAR Observation Data before and after a Landslide Event. *Remote Sens.* **2012**, *4*, 2314. [CrossRef]
62. Cloude, S.R. The Dual Polarisation Entropy/Alpha Decomposition. In Proceedings of the 3rd International Workshop on Science and Applications of SAR Polarimetry and Polarimetric Interferometry, Noordwijk, The Netherlands, 2007; pp. 22–26.
63. Ghods, S.; Shojaedini, S.V.; Maghsoudi, Y. A Modified H- α Classification Method for Dcp Compact Polarimetric Mode by Reconstructing Quad h and α Parameters from Dual Ones. *IEEE J. Sel. Top. Appl. Earth Obs. Remote Sens.* **2016**, *9*, 2233–2241. [CrossRef]
64. ISTAT ISTAT Data. Available online: http://dati.istat.it/ (accessed on 13 November 2020).
65. Mandal, D.; Vaka, D.S.; Bhogapurapu, N.R.; Vanama, V.S.K.; Kumar, V.; Rao, Y.S.; Bhattacharya, A. Sentinel-1 SLC Preprocessing Workflow for Polarimetric Applications: A Generic Practice for Generating Dual-Pol Covariance Matrix Elements in SNAP S-1 Toolbox. 2019. Available online: https://www.preprints.org/manuscript/201911.0393/v1 (accessed on 9 February 2021).
66. Lee, J.-S.; Pottier, E. *Polarimetric Radar Imaging: From Basics to Applications*; CRC Press: Boca Raton, FL, USA, 2017.
67. Shan, Z.; Wang, C.; Zhang, H.; Chen, J. H-Alpha Decomposition and Alternative Parameters for Dual Polarization SAR Data. In Proceedings of the PIERS Proceedings, SuZhou, China, 12–16 September 2011.
68. Veci, L.; Prats-Iraola, P.; Scheiber, R.; Collard, F.; Fomferra, N.; Engdahl, M. The Sentinel-1 Toolbox. In Proceedings of the IEEE International Geoscience and Remote Sensing Symposium (IGARSS), Quebec City, QC, Canada, 13–18 July 2014; pp. 1–3.

69. Ainsworth, T.L.; Kelly, J.; Lee, J.-S. Polarimetric Analysis of Dual Polarimetric SAR Imagery. In Proceedings of the 7th European Conference on Synthetic Aperture Radar; VDE, Friedrichshafen, Germany, 2–5 June 2008; 2008; pp. 1–4.
70. Crabbe, R.A.; Lamb, D.W.; Edwards, C.; Andersson, K.; Schneider, D. A Preliminary Investigation of the Potential of Sentinel-1 Radar to Estimate Pasture Biomass in a Grazed Pasture Landscape. *Remote Sens.* **2019**, *11*, 872. [CrossRef]
71. Chang, J.; Shoshany, M. Radar Polarization and Ecological Pattern Properties across Mediterranean-to-Arid Transition Zone. *Remote Sens. Environ.* **2017**, *200*, 368–377. [CrossRef]
72. Chang, J.G.; Shoshany, M.; Oh, Y. Polarimetric Radar Vegetation Index for Biomass Estimation in Desert Fringe Ecosystems. *IEEE Trans. Geosci. Remote Sens.* **2018**, *56*, 7102–7108. [CrossRef]
73. Cloude, S. *Polarisation: Applications in Remote Sensing*; Oxford University Press: Oxford, UK, 2009.
74. Raney, R.K.; Cahill, J.T.; Patterson, G.W.; Bussey, D.B.J. The M-Chi Decomposition of Hybrid Dual-Polarimetric Radar Data with Application to Lunar Craters. *J. Geophys. Res. Planets* **2012**, *117*. [CrossRef]
75. Borgogno Mondino, E.; Fissore, V.; Lessio, A.; Motta, R. Are the New Gridded DSM/DTMs of the Piemonte Region (Italy) Proper for Forestry? A Fast and Simple Approach for a Posteriori Metric Assessment. *iFor. Biogeosci. For.* **2016**, *9*, 901–909. [CrossRef]
76. Nachar, N. The Mann-Whitney U: A Test for Assessing Whether Two Independent Samples Come from the Same Distribution. *Tutor. Quant. Methods Psychol.* **2008**, *4*, 13–20. [CrossRef]
77. Hollander, M.; Wolfe, D.A.; Chicken, E. *Nonparametric Statistical Methods*; John Wiley & Sons: Hoboken, NJ, USA, 2013; Volume 751.
78. R Development Core Team, R. *R: A Language and Environment for Statistical Computing*; R Foundation for Statistical Computing: Vienna, Austria, 2013.
79. Conrad, O.; Bechtel, B.; Bock, M.; Dietrich, H.; Fischer, E.; Gerlitz, L.; Wehberg, J.; Wichmann, V.; Böhner, J. System for Automated Geoscientific Analyses (SAGA) v. 2.1. 4. *Geosci. Model Dev. Discuss.* **2015**, *8*, 1991–2007. [CrossRef]
80. Wallis, W.A. Compounding Probabilities from Independent Significance Tests. *Econom. J. Econom. Soc.* **1942**, *10*, 229–248. [CrossRef]
81. Zliobaite, I. On the Relation between Accuracy and Fairness in Binary Classification. *arXiv* **2015**, arXiv:1505.05723.
82. Akosa, J. Predictive Accuracy: A Misleading Performance Measure for Highly Imbalanced Data. Available online: https://www.semanticscholar.org/paper/Predictive-Accuracy-%3A-A-Misleading-Performance-for-Akosa/8eff162ba887b6ed3091d5b6aa1a89e23342cb5c (accessed on 9 February 2021).
83. Guo, X.; Yin, Y.; Dong, C.; Yang, G.; Zhou, G. On the Class Imbalance Problem. In *Proceedings of the 2008 Fourth International Conference on Natural Computation*; IEEE: Jinan, China, 2008; Volume 4, pp. 192–201.

Technical Note

Landscape-Scale Crop Lodging Assessment across Iowa and Illinois Using Synthetic Aperture Radar (SAR) Images

Olaniyi A. Ajadi [1,*], **Heming Liao** [1], **Jason Jaacks** [1], **Alfredo Delos Santos** [1], **Siva P. Kumpatla** [1], **Rinkal Patel** [2] **and Anu Swatantran** [1]

[1]   Research & Development Corteva Agriscience™, 7000 NW 62nd Avenue, Johnston, IA 50131, USA; heming.liao@corteva.com (H.L.); jason.jaacks@corteva.com (J.J.); alfredo.delossantos@corteva.com (A.D.S.); siva.kumpatla@corteva.com (S.P.K.); anuradha.swatantran@corteva.com (A.S.)

[2]   Granular 8700 Crescent Chase, Johnston, IA 50131, USA; rinkalpatel@granular.ag

*   Correspondence: olaniyi.ajadi@corteva.com; Tel.: +1-(515)-535-2983

Received: 30 October 2020; Accepted: 25 November 2020; Published: 27 November 2020

Abstract: Crop lodging, the tilting of stems from their natural upright position, usually occurs after a heavy storm event. Since lodging of a crop seriously affects its yield, rapid assessment of crop lodging is valuable for farmers, policymakers, agronomists, insurance companies, and relief workers. Synthetic Aperture Radar (SAR) sensors have been recognized as valuable data sources for mapping lodging extent because of their good penetrating power and high-resolution remote sensing ability. Compared to other sources, SAR's weather and illumination independence and large area coverage at fine spatial resolution (3 m to 20 m) support frequent and detailed observations. Because of these advantages, SAR has the potential in supporting near real-time monitoring of lodging in fields when combined with automated image processing. In this study, a method based on change detection using modified Hidden Markov Random Field (HMRF) and Sentinel-1A data were utilized to identify lodging and map its extent. Results obtained have shown that when lodging occurs, the VH polarization's backscatter (σVH) increases between the pre-lodging event image and the post-lodging event image. The increase in σVH is due to the increase in volume scattering and vegetation-soil double bounce scattering resulting from the structural changes in the crop canopy. Using Sentinel-1A images and applying our proposed approach across several fields in Iowa and Illinois, we mapped the extent of the 2020 Derecho (wind storm) lodging disaster. In addition, we separated lodged regions into severely and moderately lodged areas. We estimated that approximately 2.56 million acres of corn and 1.27 million acres of soybean were lodged. Further analysis also showed the separation between un-lodged (healthy) fields and lodged fields. The observations in this study can guide future use of SAR-based information for operational crop lodging assessment.

Keywords: Synthetic Aperture Radar; SAR; lodging; Hidden Markov Random Field; HMRF; CDL; corn; soybean; crop Monitoring; crop management

---

1. Introduction

With the increase in global population and the increase in food demand, the monitoring of agricultural activities has been of utmost importance. The increasing frequency and intensity of extreme weather events have also made monitoring of agricultural fields very critical. Lodging, the tilting of plant stems from their natural upright position, is a major yield limiting factor to crops such as corn, wheat, and barley [1–3]. Corn is vulnerable to lodging during its growing stages, particularly between early to late vegetative period [4]. Lodging in corn could be as a result of insufficient root growth due to soil compaction, or due to increased occurrence of heavy rain and derechos (wind storms). According to the National Oceanic and Atmospheric Administration (NOAA), derechos are straight-line windstorms that are associated with a fast-moving group of severe thunderstorms. The winds are destructive and can be as strong as those found in tornadoes and hurricanes. Derecho lodging results in serious damage to crop growth and development as it impedes the circulation of water and nutrients in the plant which in turn suppresses photosynthesis, leading to deterioration of grain quality and total yield loss [5]. For a more comprehensive overview on the mechanics of lodging, factors affecting lodging, and crop yield response to lodging, the reader is referred to Chauhan, et al. [6]. Lodging also reduces grower's profitability and, for this reason, it is important to detect lodging quickly, map its extent, and quantitatively measure its severity. Accurate and timely mapping of lodged fields can help guide farmers during harvest operations, help crop insurance companies during crop loss assessment, and can improve crop yield forecasts [7].

Mapping of lodged fields is typically based on visual inspection (field-based approach). However, this approach is extremely laborious, time consuming, and is infeasible for large areas. In recent years, remote sensing has been used for mapping lodging since it offers a more scalable approach and is also cost effective. Multi-temporal images acquired by optical [8,9] and radar [10,11] sensors have routinely been applied for lodging identification. Since these two sensor types have their unique sensitivity and imaging characteristics, their performance in the mapping of lodging varies.

Recently, synthetic aperture radar (SAR) data have gained considerable interest in lodging applications because SAR is an active sensor, operating without regard to weather, smoke, cloud cover, or daylight [12]. SAR sensors offer a clear advantage because of their unique scattering sensitivity to crop structure and large area coverage.

In Chauhan, et al. [8], the authors used time series of SAR backscatter, SAR coherence, and spectral reflectance derived from Sentinel-1 and Sentinel-2 data to detect lodging incidence and understand the effect of lodging in wheat. The most reliable discriminators for differentiating lodged wheat from healthy wheat were Sentinel-2 red edge band (740 nm), Sentinel-2 near infrared band (865 nm), and Sentinel-1 VH backscatter. Shu, et al. [13] have used the dual-polarization of Sentinel-1A data to develop a change detection method using plant height before and after lodging in maize to calculate the lodging angle and monitor the lodging degree. The results showed that VV backscatter was sensitive to lodged maize while the ratio of VH to VV backscatter was sensitive to non-lodged maize. In a similar study, Chauhan, et al. [14], explored the advantage of multi-sensor SAR data (Sentinel-1 and RADARSAT-2) to develop a quantitative approach to detect crop lodging stages (moderate, severe, and very severe) based on the crop angle of inclination. Quantitative relationships using support vector regression (SVR) models were established between remote sensing derived metrics from Sentinel-1 and RADARSAT-2 timeseries and field measured crop angle of inclination values [14].

While several researchers have predominantly used the single and dual-polarization of Sentinel-1 to address crop lodging, others have focused on using the multi-configuration (multi-polarization and multi-incidence angle) data from RADARSAT-2. For example, Yang, et al. [15] used a time series of Radarsat-2 images and target decomposition techniques to derive a set of polarimetric features and backscattering intensity features to compare typical lodged fields and normal fields. In their study, they found that polarimetric ratios (especially those based on odd/double scattering) were sensitive in distinguishing lodged and normal fields. In a different study, Chen, et al. [16] used polarimetric features from Polarimetric SAR (PolSAR) to identify sugarcane lodging. The authors found that several polarimetric features, such as horizontal transmit and vertical receive (HV) intensity, double-bounce scattering, and volume scattering derived from RADARSAT-2 data were helpful in sugarcane lodging identification.

Despite these and other studies carried out throughout the last decade, the integration of SAR remote sensing into routine mapping of lodging and severity assessment remains difficult for the following reasons: (a) The acquisition of SAR dataset to coincide with the specific date of lodging is not always feasible; (b) The heterogeneous distribution of lodging makes it difficult to be detected with SAR and also, in particular, in order to map lodging precisely, the size of the lodged field must be larger than the spatial resolution of the SAR sensor; (c) The mapping of lodging over large spatial extent and determination of lodging rate; (d) The acquisition of ground truth data (known lodged fields) to evaluate lodging severity can itself be an overwhelming task as it is extremely labor intensive and time consuming; (e) The accuracy assessment of the lodging severity also has some shortcomings since there are no standard scales to quantify lodging into categories such as severe, moderate, or mild.

To address and overcome some of these challenges, we used the 2020 Derecho event in Midwest U.S. as a case-study to evaluate SAR for large-scale lodging detection and mapping. The main objectives and novelty of this study are to:

(1) Understand the changes in backscatter over large-scale lodged fields and how to use the backscatter to classify lodging into severe or moderate categories.
(2) Generate large-scale spatial extent maps of lodging using a change detection approach modified from our previous study [17] and determine the lodging rate (lodged crop per unit area) using the USDA's Crop Data Layer (CDL) map.
(3) Qualitatively explore the relationship between high wind speed and lodged fields.
(4) Explore the capability of Sentinel-1A over an optical dataset like the one from Landsat-8.
(5) Explore if lodged and un-lodged (healthy) fields differ between the pre-lodging event image and post-lodging event image.

In this paper, the proposed change detection method for lodging utilizes the concept of ratio image generation. The generation of ratio images suppresses background information while enhancing change information [17]. The generated ratio images were filtered using a non-local means filter [18,19] and classified into different classes using the Hidden Markov Random Field (HMRF). The HMRF considers the contextual information of neighboring pixels (i.e., a neighboring pixel is expected to have similar intensities and similar class labels) during classification. Published methods differ in their approach to extract change detection map. In the work done by Kasetkasem and Varshney [20], the authors used a MRF to model noiseless images for an optimal change image using the maximum a posteriori probability computation and the simulated annealing (SA) algorithm. In Zhao, et al. [21], the authors combined Voronoi Tessellation (VT) and HMRF based Fuzzy C-Means (FCM) algorithm (VTHMRF-FCM) for texture image segmentation. Similarly, in Yang and Yi [22], a novel method based on applying HMRF and generative adversarial network (GAN) on high-resolution SAR images was used for ship detection. To our knowledge and according to Chauhan, et al. [6], the use of satellite-based change detection for crop lodging is sparse. To date, there is no method available for corn and soybean lodging over large spatial areas.

2. Study Area and Data

2.1. Study Area

This study was carried out in Iowa and Illinois, spanning across 52 counties (Figure 1), covering the region affected by the 2020 Derecho lodging disaster which occurred on 10 August 2020. Iowa and Illinois are located in the Western part of the Corn Belt region of the United States and are part of the top agriculture production states. Iowa has a total agricultural land of 30.6 million acres, accounting for 92% of the state's total land area while Illinois has a total agricultural land of 27 million acres, accounting for 75% of the state's total land area. The two-dominant crops in both states are corn and soybean. Other crops like wheat and winter wheat are also planted but they account for a very small portion of all the cropland. In this study, the estimation of lodging rate (lodged crop per unit area) was only focused on corn and soybean. In Iowa and Illinois, the climate is cold and temperate. In Iowa, the wettest month with the most precipitation is June with an average precipitation of 107 mm and an average temperature of 82 °F, while the driest month is January with an average precipitation of 23 mm of rainfall and an average temperature of 31 °F. In Illinois, the wettest month with the most precipitation is August in the Northern portion of Illinois with an average precipitation of 89 mm and an average temperature of 82 °F, while the driest month is February with an average precipitation of 49 mm of rainfall and an average temperature of 35 °F.

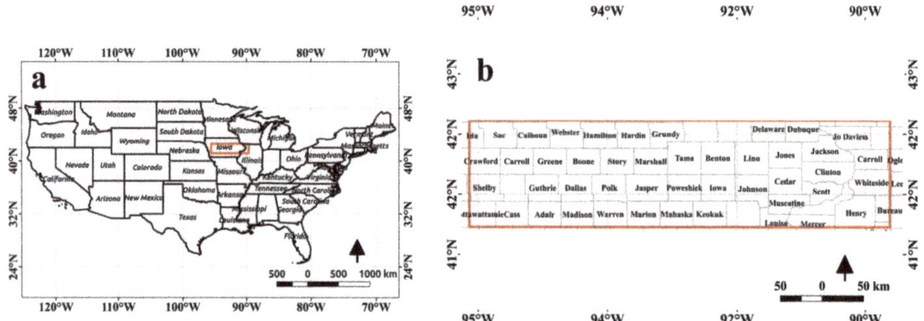

Figure 1. Study area covering (**a**) the region affected by the 2020 Derecho lodging disaster in the states of Iowa and Illinois as shown by the red outline and (**b**) the 52 counties in the affected region.

2.2. Field Data

In order to generate lodged crop per unit area, we used the 2018 CDL map because the 2020 CDL map is not yet available. Another reason for using the 2018 CDL map is that most farmers in Iowa and Illinois perform crop rotation. Fields where farmers consistently performed crop rotations for the last eight years were kept while fields where crop rotations were not consistent for the last eight years were removed. By doing this, we believe that corn and soybeans areas in 2018 should be similar to those of corn and soybean areas in 2020. The CDL data have a spatial resolution of 30 m, includes 132 detailed class labels, and was created from Landsat dataset using a decision tree algorithm trained on field samples [23]. The overall accuracy of the CDL dataset is about 95% for the United States Corn Belt. Apart from the CDL, we collected some ground truth data by field survey. Lodged fields were selected as samples for field observation. Mobile phone GPS was used for positioning the sampling points.

2.3. Remote Sensing Data

During 10 August 2020 and 11 August 2020, a derecho swept through Midwest that caused severe and widespread windstorms with some areas experiencing low-class tornadoes and heavy rain. As part of this, a windstorm was observed in Iowa and Illinois on 10 August 2020. The wind speed in both states ranged between 60 and 100 mph (Figure 2). Wind speed data for this study were acquired from the National Weather Service (NWS). NWS used several sources like Iowa DOT, personal weather station, Automated Surface Observing Systems (ASOS), and Automated Weather Observing Systems (AWOS) to estimate the wind speed data. Since corn and soybean are already in their reproductive stages at the time of the windstorm, high wind speed led to severe lodging.

In order to cover the agricultural areas damaged by the windstorm, we acquired six Sentinel-1A images in Interferometric Wideswath (IW) instrument mode. Our Sentinel-1A images were acquired between 29 July 2020, and 22 August 2020. Each Sentinel-1A image collection had a resolution of 10 m, a dual polarization (VH and VV; V = Vertical, H = Horizontal), and consisted of Level-1 Ground Range Detected (GRD) scenes. As shown in Table 1 and Figure 3, three images were acquired before the lodging event (pre-lodging event) and three images were acquired after the lodging event (post-lodging event).

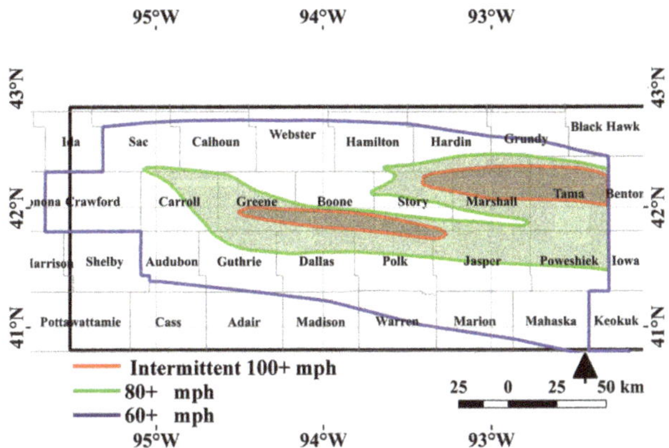

Figure 2. Wind speed data acquired from the National Weather Service (NWS).

Table 1. Details of Sentinel-1A images acquired over Iowa and Illinois.

Date 2015	Flight Direction	Local Standard Time	Incidence Angle	Polarizations	Lodging Event
29 July 2020	ascending	T00:05:15.604Z	30°–46°	VH & VV	Pre-
3 August 2020	ascending	T00:13:42.330Z	30°–46°	VH & VV	Pre-
4 August 2020	ascending	T23:57:18.422Z	30°–46°	VH & VV	Pre-
15 August 2020	ascending	T00:13:42.984Z	30°–46°	VH & VV	Post-
16 August 2020	ascending	T23:57:19.120Z	30°–46°	VH & VV	Post-
22 August 2020	ascending	T00:05:16.823Z	30°–46°	VH & VV	Post-

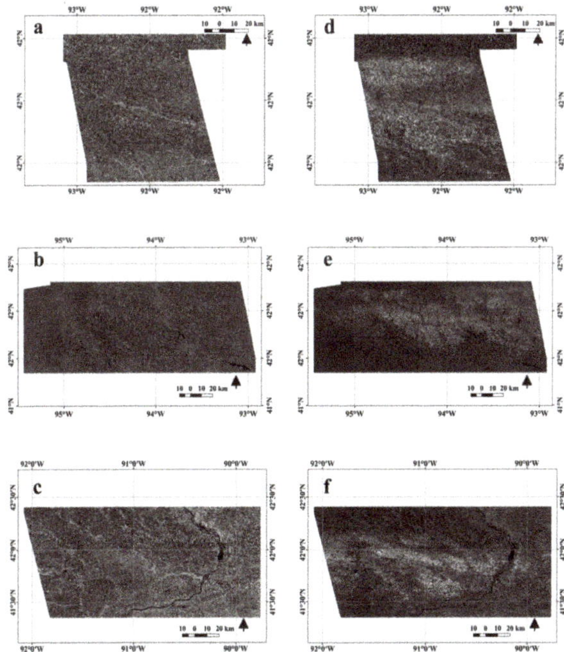

Figure 3. Pre-lodging images acquired on (**a**) 29 July 2020; (**b**) 3 August 2020; and (**c**) 4 August 2020. Post-lodging images acquired on (**d**) 22 August 2020; (**e**) 15 August 2020; and (**f**) 16 August 2020.

3. Methods

A graphical overview of the framework used for this study is illustrated in Figure 4. Major steps of the framework include preprocessing, ratio image formation, and change detection classification approach. All the steps in this paper can be reproduced quickly and all analyses for the study were carried out using python programming language.

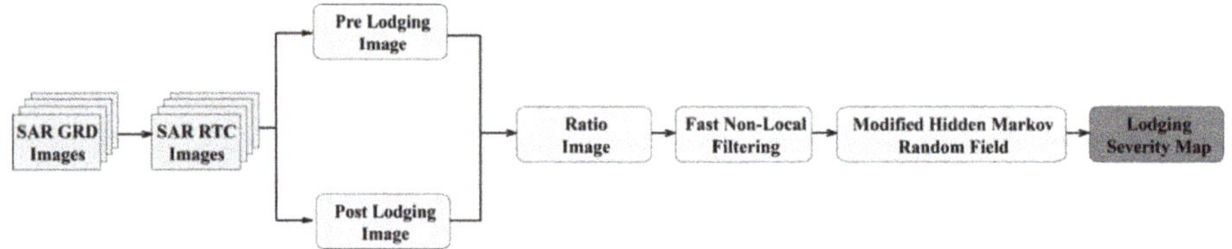

Figure 4. Approach for mapping the 2020 Derecho lodging disaster.

In this paper, our proposed approach requires some parameters to be set beforehand. The parameters that need to be set include (i) the neighborhood size of the non-local means filtering step; (ii) the kernel size of the majority filter; (iii) the structuring element of the morphological filter; and, finally, (iv) the maximum number of allowed change classes. Please note that while we identified optimal settings for these parameters, we found that the performance of our algorithm does not critically depend on the exact choice for these variables. This is true for the following reasons: (i) as non-local means filtering is performed very early in the workflow, the impact of changes in the neighborhood size is mitigated by subsequent processing steps such as the application of mathematical morphology. Hence, we found that varying the neighborhood size from its optimal value changed system performance only slowly; (ii) the increase or decrease in the kernel size of the majority filter slowly decreases our change detection performance, yet this reduction of performance does not become significant unless the kernel size is increased tremendously; (iii) from an analysis of a broad range of data from different change detection projects we found that (1) a 5 × 5 pixel-sized structuring element of the morphological filter led to the most consistent results; and that (2) change detection performance changed slowly with deviation from the 5 × 5 pixel setting. Hence, while 5 × 5 pixel was found to be optimal, the exact choice of the window size is not critical for change detection success; finally, (iv) the maximum number of allowable change classes is a very uncritical variable as it merely sets an upper bound for a subsequent algorithm that automatically determines the number of distinguishable classes in a data set. By presetting this variable to 3 classes we ensure that changes as a result of lodging are captured.

3.1. Image Preprocessing

Sentinel-1A preprocessing was carried out using SeNtinel Application Platform (SNAP) software version 6.0 (https://step.esa.int/main/toolboxes/snap/), an open source common architecture provided by European Space Agency (ESA). The preprocessing step includes orbit file correction, GRD border noise removal, thermal noise removal, calibration, filtering using refined Lee filter, radiometric terrain correction (RTC), and geometric terrain correction (GTC). For more details on the preprocessing step and the importance of RTC, the reader is referred to Ajadi, Meyer and Webley [17].

3.2. Logarithmic Scaling and Ratio Image Formation

While both VH polarization and VV polarization were sensitive to crop lodging assessment, we only used the VH polarization of Sentinel-1A dataset in this study because it depicts crop phenology very well and, moreover, it is very sensitive to crop canopy structure. In order to increase the detectability of lodging and to suppress background information from SAR data, ratio images $(X_{Ris} = [X_{Ri1}, X_{Ri2}, X_{Ri3}])$ were formed (Figures 5 and 6) using the pre-lodging event image and of the post-lodging event image of similar geometry, respectively (Table 1 and Figure 3). Note that, due to the performed radiometric correction steps, images are not required to come from identical geometries. Also, because the pre-lodging and post-lodging event images are logarithmically scaled, the creation of ratio images is performed as a subtraction operation. Afterwards, a fast-non-local means filtering procedure was applied to all ratio images in order to filter out the speckle noise while preserving the details. The fast-non-local means uses redundant information to reduce noise and restore the original noise-free image by performing a weighted average of pixel values, considering the spatial and intensity similarities between pixels [17].

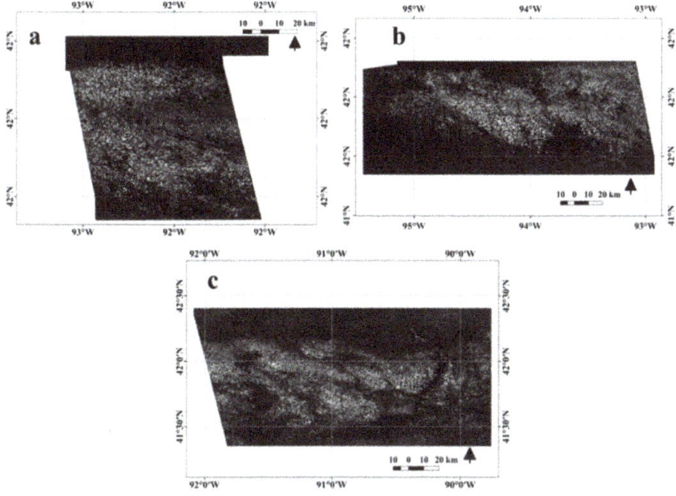

Figure 5. Ratio images generated between (**a**) 29 July 2020 and 22 August 2020; (**b**) 3 August 2020 and 15 August 2020; (**c**) 4 August 2020 and 16 August 2020.

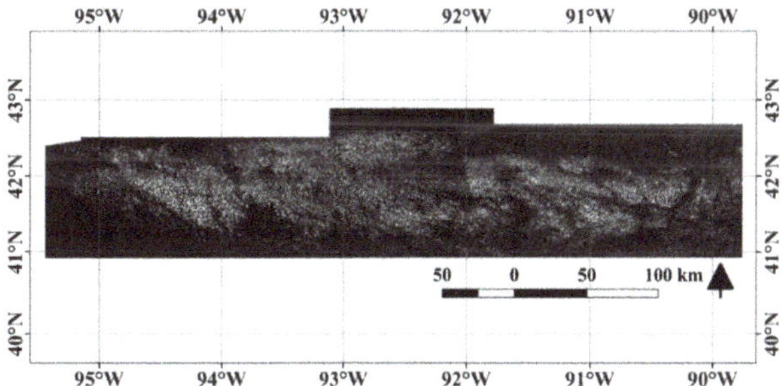

Figure 6. Spatial mosaic of all ratio maps in Figure 5a–c.

3.3. Change Detection Classification Approach

In this study, we employed the Hidden Markov Random Field (HMRF) approach to perform our lodging-based change detection classification. The HMRF approach fully utilizes and enhances our previous change detection classification approach described in [17] by improving its robustness and sensitivity to false alarms. The method in [17] employed the Finite Gaussian Mixture (FGM) model for image classification. For example, if a ratio image X_{Ri} contains N dimensional vector of pixels with $I = \{1, 2, \ldots\ldots\ldots, N\}$ being the set of pixel indices, then for each pixel i in X_{Ri} a class label x_i is inferred using the conditional probability as shown in [17]. Each pixel in the FGM is independent from their neighboring pixels, meaning they do not consider the relationship of pixels within its neighboring system. To improve the FGM, we modified the HMRF approach proposed by Zhang, et al. [24] and employed it. By assuming a Gaussian distribution, the HMRF model is given by

$$p(X_{Ri}|x_{Ni};\theta) = \sum_{l\in L} f(X_{Ri};\theta_l)\, q(l|x_{Ni}) \tag{1}$$

where $q(l|x_{Ni})$ is a conditional probability mass function (pmf) for a class label given that x_{Ni} are N neighbors for a pixel x_i. The segmentation process of HMRF requires an initial estimate for the class labels x_0 and initial parameters $(\theta_0 = mean(\mu_0),\ variance\ (\sigma_0^2))$. In this research, we used K-means clustering approach to provide these initial labels and initial parameters. The update for parameter θ_0 was estimated iteratively using our Expectation-maximization (EM) algorithm [17]. The initial labels were updated iteratively using the maximum a posteriori (MAP) algorithm. The updated label is now used to solve for $\hat{x}$ that minimizes the total posterior energy

$$\hat{x} = \underset{x\in\chi}{argmin}\{U(X_{Ri}|x) + U(x)\} \tag{2}$$

where χ is the set of all possible configurations of labels, $\hat{x}$ is the estimated class label, $U(X_{LR}|x)$ is the likelihood energy, and $U(x)$ is the prior energy function which is defined by clique potentials. A clique is a subset of nodes in which every node is connected to every other node. The clique potential was defined on pairs of neighboring pixels. We assumed that one pixel has at most eight neighboring pixels. The MAP algorithm iteration stops when Equation (2) converges or when the maximum iteration we set is reached. In this study, we classified each ratio image in X_{Ris} into three classes namely no change, moderate change, and severe change. Finally, the classified ratio image was filtered using a majority filter with mathematical morphology to remove small isolated misclassified pixels. Majority filter with mathematical morphology works by first replacing the pixels in a neighborhood using majority of their adjacent neighboring pixels and then applying opening by reconstruction followed with closing by reconstruction. The order of first doing opening by reconstruction followed by closing by reconstruction was designed to reduce noise while preserving the geometric details in the image. In this study, we set the kernel size of our majority filter to 3 and the size of our morphological kernel to 5. In the final classification map, we removed the no change class, and we multiplied each changed class (moderate change and severe change) by each crop (corn and soybean) mask derived from the CDL to estimate lodged crops per unit area.

4. Results

4.1. Sentinel-1A Backscatter Analysis for Lodging Detection

The first instances of lodging were observed in the fields on 11 August 2020 due to the windstorm on 10 August 2020. Because of the temporal frequency of Sentinel-1, lodging assessment was done on acquisitions after the lodging event. As seen in the ratio images (Figures 5 and 6), the VH polarization's backscatter (σVH) increased between the pre-lodging image and the post-lodging image. We predict the increase in value of σVH is based on two conditions:

(1) The first condition is due to volume scattering from the crops. In a healthy crop, the orientation of the crop canopy is erect. When crop lodging occurs, the orientation of the canopy elements changes, the ears and the stems bend downwards and incline against each other. Since σVH is sensitive to structural changes, when lodging occurs, the volume scattering increases.

(2) The second condition is due to the high sensitivity between the interaction of vegetation and soil (double bounce effect). When a field is lodged, the vegetation-soil double scattering increases and the σVH increases as well.

The approach for the mapping of lodging as detailed in Section 3 captures the increase in backscatter as a result of the lodging event. When the lodging is severe, condition 1 and condition

2 hold and changes in σVH are very high but when lodging is moderate, condition 1 holds and changes in σVH are moderate. The reader should note that both severe and moderate lodging are very detrimental and they will lead to yield loss. In this study, ratio images in X_{Ris} were classified into three classes. The first class, which depicts no change, comprises mostly of urban areas, healthy crops, and standing water. The second class, which depicts moderate change, contains moderately lodged fields while the third class, which depicts severe change, contains severely lodged fields (Figures 7 and 8). After classification of the lodging areas, we then quantified the amount of corn and soybean that were moderately and severely lodged (Table 2) using the CDL map.

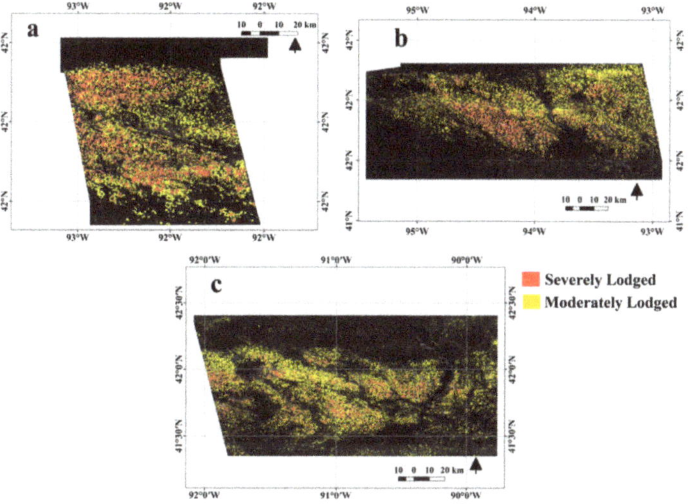

Figure 7. Lodging maps generated between (**a**) 29 July 2020 and 22 August 2020; (**b**) 3 August 2020 and 15 August 2020; (**c**) 4 August 2020 and 16 August 2020.

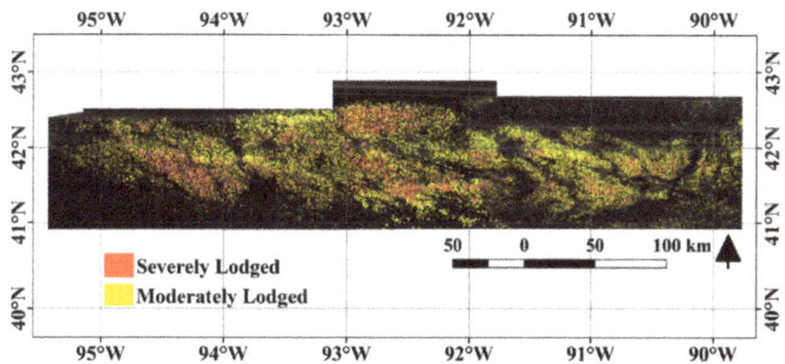

Figure 8. Spatial mosaic of all lodging maps in Figure 7a–c.

Table 2. Amount of lodged corn and soybean per unit area.

	Corn		Soybean		
	Severe (Acres)	Moderate (Acres)	Severe (Acres)	Moderate (Acres)	Total (Acres)
Figure 7a	287,238	398,821	79,246	307,482	1,072,787
Figure 7b	466,129	637,524	121,415	318,600	1,543,668
Figure 7c	326,819	442,113	120,602	318,308	1,207,842
Total (Figure 8)	1,080,186	1,478,458	321,263	944,390	
	2,558,644		1,265,653		3,824,297

4.2. Qualitative Relationship between High Wind Speed and Lodged Fields

To qualitatively check if there is a relationship between high wind speed and lodged fields, we overlaid Figures 5b and 7b with the wind speed map (Figure 2). These overlay is shown in Figure 9a,b respectively. The result showed that our estimated lodged fields followed areas with increased wind speed (Figure 9a,b). In particular, areas with wind speed of 80+ mph captured most of our estimated severely and moderately lodged fields while areas with wind speed of 60+ mph captured few lodged fields.

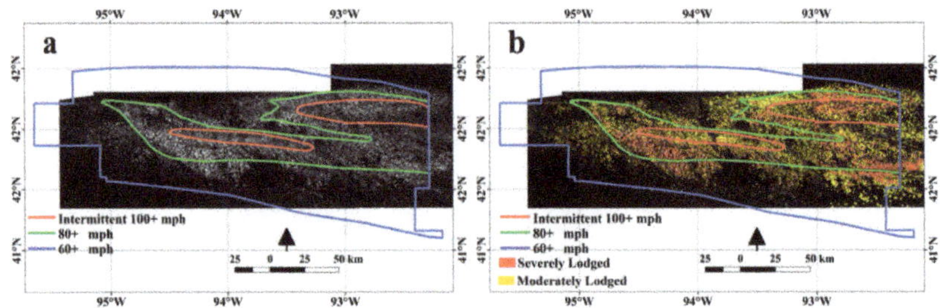

Figure 9. Example of (**a**) ratio map (Figure 5b) and (**b**) severity map (Figure 7b) overlaid with wind speed map in Figure 2.

4.3. Reliability of the Approach Employed in Mapping Lodging

To qualitatively access the performance of the Derecho lodged area mapped, we acquired eight ground truth data from field observation along the interstate I-80W. Due to limited amount of ground truth data collected, we were unable to perform extensive quantitative analysis of our method. While overlaying the eight ground truth points acquired with our lodged map, we observed that each point overlaid perfectly on an area that was lodged (Figure 10).

Furthermore, we compared an example of the lodged change detection map generated from Sentinel-1A ratio image and optical Landsat-8 ratio image by using our classification approach. The Landsat-8 ratio image was generated from the green band of pre-lodging event image acquired on 10 July 2020, and the post-lodging event image acquired on 11 August 2020 (Figure 11).

We chose the green band over every other band of Landsat-8 because the green band has a larger dynamic range which allowed for better separation of lodged and un-lodged areas. As can be seen from Figure 12, the lodging map generated from Landsat-8 ratio image and the Sentinel-1A ratio image were able to clearly detect lodging. While our approach was able to classify the Sentinel-1A ratio image into severely lodged and moderately lodged fields, we were unable to do so in the Landsat-8 ratio image. This observation showed that Sentinel-1A radar sensor has the advantage in reflecting the structural changes of lodging fields, while the Landsat-8 optical sensor only has the advantage in detecting the biochemical changes of lodging fields. All in all, for this study, we believe that the obvious feature for lodging is the structural change and the distribution pattern of lodged fields (severe and moderate) from Sentinel-1A image were similar to the distribution pattern of the lodged fields in the Landsat-8 image.

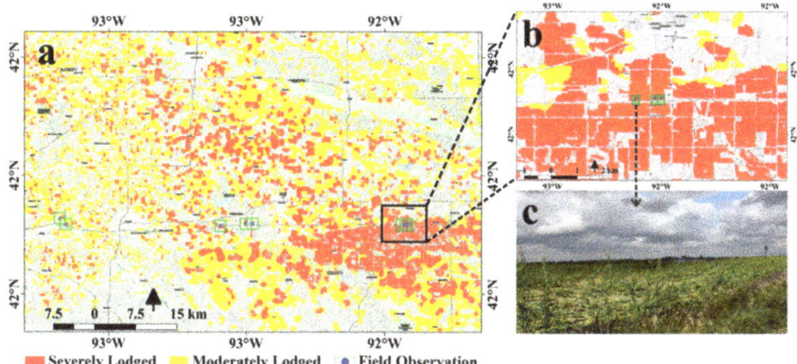

Figure 10. Location of field observations for (**a**) all the eight points and (**b**) a zoomed location for three points. (**c**) Photo for one of the field observations in "b".

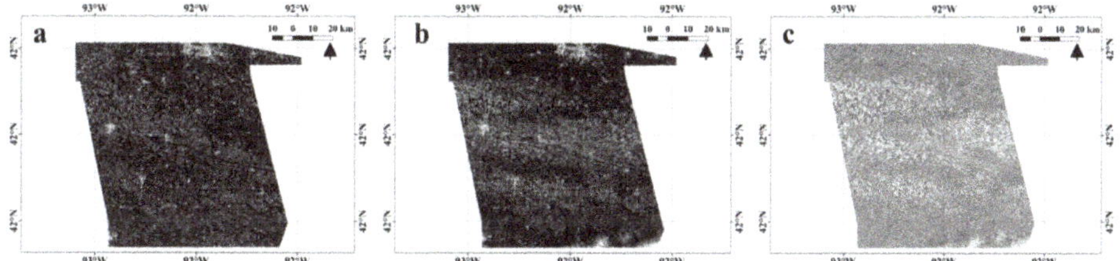

Figure 11. (**a**) Landsat-8 pre-lodging event image acquired on 10 July 2020. (**b**)Landsat-8 post-lodging event image acquired on 11 August 2020. (**c**) Ratio image generated between 10 July 2020 and 11 August 2020.

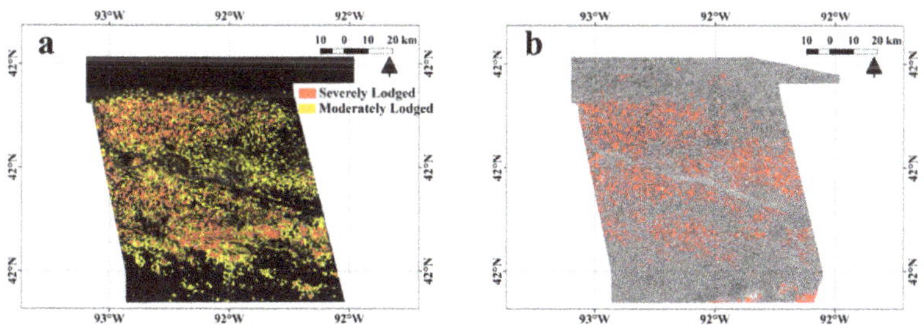

Figure 12. Severity map generated from (**a**) Sentinel-1A and (**b**) Landsat-8.

5. Discussion

5.1. Spatial-Extent of Lodging and Lodging Rate

Based on the reports by USDA Risk Management Agency (RMA), 57 counties in Iowa and Illinois were in the path of the storm which led to the Derecho lodging disaster. In the report by RMA, there are roughly 14 million acres of insured crops within those 57 counties. Also, based on the Storm Prediction Center preliminary storm reports and assessment from MODIS satellite imagery, the Iowa Department of Agriculture and Land Stewardship estimated that about 3.57 million acres of corn and 2.5 million acres of soybeans were likely to have been impacted by the severe wind on 10 August 2020. Based on our findings using Sentinel-1A data across 52 counties, we estimated that a total of approximately 2.56 million acres of corn and approximately 1.27 million acres of soybean were impacted during the Derecho lodging disaster (Figure 8 and Table 2). Furthermore, we observed that out of the 52 counties, only 32 counties have more than 10 thousand acres per county impacted by the storm. Within these 32 impacted counties, we leveraged our proprietary yield prediction capability to quantify the yield potential prior to the storm, and we were able to quantify the total bushels lost following the windstorm. We observed the following in our findings:

(a)   Corn acres impacted: ~2.36 million acres impacted out of ~4.90 million acres planted (48% acre impact).

(b)   Corn bushels impacted: ~442.37 million bushels impacted out of ~918.36 million bushels expected prior to the storm (48% reduction).

(c)   Soybean acres impacted: ~1.27 million acres impacted out of ~4.69 million acres planted (28% acre impact).

(d)   Soybean bushels impacted: ~80.90 million bushels impacted out of ~279.00 million bushels expected prior to the storm (29% yield reduction)

As shown above and in Table 2, fields with moderate lodging were more frequent than fields with severe lodging and corn fields lodged more than soybean fields. We believe the reason for this difference is due to relatively more resistance of soybean to wind than corn.

5.2. Temporal Behavior of Un-Lodged (Healthy) and Lodged Fields throughout the Observation Period

The main difference in a field before and after lodging is the reduction of plant height. Therefore, the backscattering coefficients for lodged and un-lodged field will be different because for a lodged field, either one of the two conditions stated in Section 4.1 will hold. In this study, we compared nine randomly selected un-lodged fields to nine randomly selected lodged fields (Figure 13) using both the VH mean polarization backscatter and VV mean polarization backscatter. While observing Figure 13, on the one hand, un-lodged fields showed no change at all in both the VH and VV mean polarization backscatter. On the other hand, we saw an increase of approximately 6 dB in the VH mean polarization backscatter and 5 dB in the VV mean polarization backscatter for all the fields between the pre-lodging event date and the post-lodging event date respectively. Even though there is a slightly higher difference in the VH mean polarization backscatter, the sensitivity of VH polarization to lodging is similar to the sensitivity of VV polarization to lodging. In this study, the increase in both the VH and VV mean polarization backscatters follows the reported observations in [8]. In Chauhan, et al. [8], the authors observed a clear linear trend of increasing VH polarization backscatter and VV polarization backscatter with the increase in the lodging severity. It is worth mentioning that corn and soybean in all the fields are in the grain-filling period or plateau stage. At this stage, the height and greenness of the crop should be fairly constant, and this is why we see no change in the un-lodged fields. To further show how lodged fields and un-lodged fields differ on a large scale, we generated the VH and VV mean polarization backscatter over Dallas county in Iowa (Figure 14). By qualitatively observing the mean backscatter between the pre-lodging event date and the post-lodging event date in Figure 14, we saw an increase of approximately 3 dB in the VH mean polarization backscatter and 1 dB in the VV mean polarization backscatter. Based on the difference observed between the lodged fields and un-lodged fields in Dallas county, it is quite evident that our approach is sensitive enough to detect lodging.

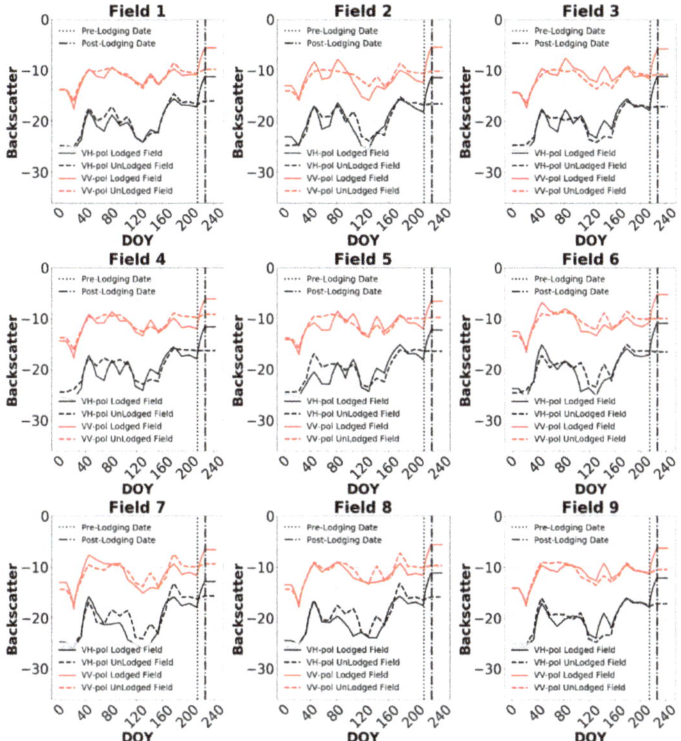

Figure 13. Time series of lodged and un-lodged fields.

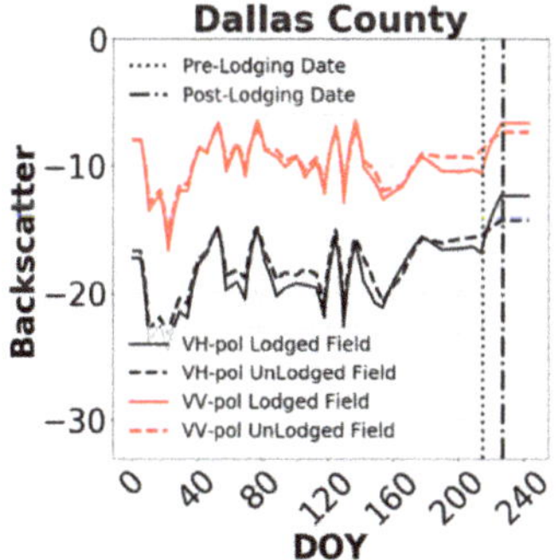

Figure 14. Time series of lodged and un-lodged fields in Dallas county, Iowa.

6. Conclusions

The potential and feasibility of using SAR data for monitoring the Derecho lodging disaster was demonstrated in this study. Though crop lodging assessment using SAR data have been shown in a few studies earlier, studies that have assessed crop lodging using satellite-based data at large spatial scale are still sparse and knowledge relating to the changes in SAR signatures was lacking in literature. With the advent of dense SAR time series from Sentinel-1 and proposed missions like NASA ISRO Synthetic Aperture Radar (NISAR), there is an increased interest in exploring SAR for lodging identification. The main conclusions of this study are summarized below.

(1)   The modified change detection approach used was shown to be capable of providing near real-time monitoring of the Derecho lodging disaster by generating detailed parameters, such as backscatter changes, lodging extent, and lodging rate in corn and soybean. The generated lodging extent maps from SAR showed both severely and moderately damaged fields. The use of CDL also allowed the estimation of lodged crop per unit area, showing relatively more lodging in corn fields than soybean fields. We believe the sensitivity of corn to lodging was caused by its unique structural characteristics (long vertical orientation of its stalk). We estimated that a total of approximately 2.56 million acres of corn and approximately 1.27 million acres of soybean were impacted during the Derecho lodging disaster.

(2)   The modified change detection approach used was reliable and the reliability can be seen by the similar distribution patterns of lodged fields in the Sentinel-1A imagery and Landsat-8 imagery. Furthermore, the generated lodged field maps show correlation with areas of extreme wind speed.

(3)   The backscatter difference between the timeseries of lodged and un-lodged (healthy) fields differ. Our analyses from nine fields showed almost no change between the pre-lodging event image and post-lodging event image of an un-lodged field while we noted an approximately 6 dB increase in the VH mean polarization backscatter and 5 dB increase in the VV mean polarization backscatter for all the fields between the pre-lodging event date and the post-lodging event date. When we aggregated all the un-lodged fields and the lodged fields across Dallas county, we saw an increase of approximately 3 dB in the VH mean polarization backscatter and 1 dB in the VV mean polarization backscatter between the pre-lodging event date and the post-lodging event date. Taken together, these results suggest that differences in VH polarization and VV polarization can serve as useful lodging indicators at parcel- and landscape-level and enable rapid mapping of widespread lodging events using SAR data.

Taken together, the results and the insights from this study demonstrate the practicability of using high resolution SAR remote sensing data for large scale identification of crop lodging. The approach employed is simple, effective, and can help decision-makers obtain critical information on lodging identification to support precision management in order to provide effective emergency relief. Future studies will evaluate if there is any improvement in the lodging detection performance when SAR and multispectral datasets are fused. We will also undertake quantitative measures to further validate our results. Using the validated crop lodging area in this study as a reference, we will investigate the use of deep learning models to generate a standard near-real time framework for crop lodging assessment. We will also explore the use of Gamma distribution for fitting the EM algorithm in the HMRF rather than Gaussian distribution. With adequate ground truth information and further validation, large-scale crop lodging assessment can improve crop yield forecasting and crop insurance, ultimately benefitting food production and food security initiatives.

Author Contributions: Conceptualization, O.A.A.; methodology, O.A.A.; software, O.A.A.; validation, O.A.A. and H.L.; formal analysis, O.A.A.; investigation, O.A.A. and H.L.; data curation, O.A.A. and A.D.S.; writing—original draft preparation, O.A.A.; writing—review and editing, O.A.A., H.L., J.J., R.P., A.S., and S.P.K.; visualization, O.A.A.; supervision, A.S.; project administration, R.P. and A.S.; funding acquisition, A.S. All authors have read and agreed to the published version of the manuscript.

Funding: This work was supported by Corteva Agriscience™.

Conflicts of Interest: The authors declare no conflict of interest.

References

1.   Pinthus, M.J. Lodging in wheat, barley, and oats: The phenomenon, its causes, and preventive measures. In *Advances in Agronomy*; Elsevier: Amsterdam, The Netherlands, 1974; Volume 25, pp. 209–263.

2.   Wu, W.; Ma, B.L. A new method for assessing plant lodging and the impact of management options on lodging in canola crop production. *Sci. Rep.* **2016**, *6*, 31890. [CrossRef] [PubMed]

3.   Xue, J.; Gou, L.; Zhao, Y.; Yao, M.; Yao, H.; Tian, J.; Zhang, W. Effects of light intensity within the canopy on maize lodging. *Field Crop. Res.* **2016**, *188*, 133–141. [CrossRef]

4. Nielsen, B.; Colville, D. Stalk Lodging in corn: Guidelines for Preventive Management. Available online: https://www.extension.purdue.edu/extmedia/ay/ay-262.html (accessed on 30 October 2020).

5. Kong, E.; Liu, D.; Guo, X.; Yang, W.; Sun, J.; Li, X.; Zhan, K.; Cui, D.; Lin, J.; Zhang, A. Anatomical and chemical characteristics associated with lodging resistance in wheat. *Crop J.* **2013**, *1*, 43–49. [CrossRef]

6. Chauhan, S.; Darvishzadeh, R.; Boschetti, M.; Pepe, M.; Nelson, A. Remote sensing-based crop lodging assessment: Current status and perspectives. *Isprs J. Photogramm. Remote Sens.* **2019**, *151*, 124–140. [CrossRef]

7. Shah, S.; Chang, X.; Martin, P. Effect of dose and timing of-application of different plant growth regulators on lodging and grain yield of a scottish landrace of barley (bere) in orkney, scotland. *Int. J. Environ. Agric. Biotechnol.* **2017**, *2*, 238871. [CrossRef]

8. Chauhan, S.; Darvishzadeh, R.; Lu, Y.; Boschetti, M.; Nelson, A. Understanding wheat lodging using multi-temporal sentinel-1 and sentinel-2 data. *Remote Sens. Environ.* **2020**, *243*, 111804. [CrossRef]

9. Chauhan, S.; Darvishzadeh, R.; Lu, Y.; Stroppiana, D.; Boschetti, M.; Pepe, M.; Nelson, A. Wheat lodging assessment using multispectral uav data. *Int. Arch. Photogramm. Remote Sens. Spat. Inf. Sci.* **2019**, 235–240. [CrossRef]

10. Han, D.; Yang, H.; Yang, G.; Qiu, C. 2017 SAR in Big Data Era: Models, Methods and Applications (BIGSARDATA). In *Monitoring Model of Corn Lodging Based on Sentinel-1 Radar Image*; IEEE: New York, NY, USA, 2017; pp. 1–5.

11. Zhao, L.; Yang, J.; Li, P.; Shi, L.; Zhang, L. Characterizing lodging damage in wheat and canola using radarsat-2 polarimetric sar data. *Remote Sens. Lett.* **2017**, *8*, 667–675. [CrossRef]

12. Meyer, F.; McAlpin, D.; Gong, W.; Ajadi, O.; Arko, S.; Webley, P.; Dehn, J. Integrating sar and derived products into operational volcano monitoring and decision support systems. *Isprs J. Photogramm. Remote Sens.* **2015**, *100*, 106–117. [CrossRef]

13. Shu, M.; Zhou, L.; Gu, X.; Ma, Y.; Sun, Q.; Yang, G.; Zhou, C. Monitoring of maize lodging using multi-temporal sentinel-1 sar data. *Adv. Space Res.* **2020**, *65*, 470–480. [CrossRef]

14. Chauhan, S.; Darvishzadeh, R.; Boschetti, M.; Nelson, A. Estimation of crop angle of inclination for lodged wheat using multi-sensor sar data. *Remote Sens. Environ.* **2020**, *236*, 111488. [CrossRef]

15. Yang, H.; Chen, E.; Li, Z.; Zhao, C.; Yang, G.; Pignatti, S.; Casa, R.; Zhao, L. Wheat lodging monitoring using polarimetric index from radarsat-2 data. *Int. J. Appl. Earth Obs. Geoinf.* **2015**, *34*, 157–166. [CrossRef]

16. Chen, J.; Li, H.; Han, Y. Fifth International Conference on Agro-Geoinformatics (Agro-Geoinformatics). In *Potential of Radarsat-2 Data on Identifying Sugarcane Lodging Caused by Typhoon*; IEEE: New York, NY, USA, 2016; pp. 1–6.

17. Ajadi, O.A.; Meyer, F.J.; Webley, P.W. Change detection in synthetic aperture radar images using a multiscale-driven approach. *Remote Sens.* **2016**, *8*, 482. [CrossRef]

18. Baier, G.; Rossi, C.; Lachaise, M.; Zhu, X.X.; Bamler, R. A nonlocal insar filter for high-resolution dem generation from tandem-x interferograms. *IEEE Trans. Geosci. Remote Sens.* **2018**, *56*, 6469–6483. [CrossRef]

19. Deledalle, C.-A.; Denis, L.; Tupin, F.; Reigber, A.; Jäger, M. Nl-sar: A unified nonlocal framework for resolution-preserving (pol)(in) sar denoising. *IEEE Trans. Geosci. Remote Sens.* **2014**, *53*, 2021–2038. [CrossRef]

20. Kasetkasem, T.; Varshney, P.K. An image change detection algorithm based on markov random field models. *IEEE Trans. Geosci. Remote Sens.* **2002**, *40*, 1815–1823. [CrossRef]

21. Zhao, Q.-H.; Li, X.-L.; Li, Y.; Zhao, X.-M. A fuzzy clustering image segmentation algorithm based on hidden markov random field models and voronoi tessellation. *Pattern Recognit. Lett.* **2017**, *85*, 49–55. [CrossRef]

22. Yang, M.; Yi, C. A novel method of ship detection in high-resolution sar images based on gan and hmrf models. *Prog. Electromagn. Res.* **2019**, *83*, 77–82. [CrossRef]

23. Boryan, C.; Yang, Z.; Mueller, R.; Craig, M. Monitoring us agriculture: The us department of agriculture, national agricultural statistics service, cropland data layer program. *Geocarto Int.* **2011**, *26*, 341–358. [CrossRef]

24. Zhang, Y.; Brady, M.; Smith, S. Segmentation of brain mr images through a hidden markov random field model and the expectation-maximization algorithm. *IEEE Trans. Med Imaging* **2001**, *20*, 45–57. [CrossRef] [PubMed]

Publisher's Note: MDPI stays neutral with regard to jurisdictional claims in published maps and institutional affiliations.

Article

Accuracies of Soil Moisture Estimations Using a Semi-Empirical Model over Bare Soil Agricultural Croplands from Sentinel-1 SAR Data

Anil Kumar Hoskera [1,*], Giovanni Nico [2,3], Mohammed Irshad Ahmed [1] and Anthony Whitbread [1]

[1] International Crops Research Institute for the Semi-Arid Tropics, Patancheru 502324, India; irshad@cgiar.org (M.I.A.); a.whitbread@cgiar.org (A.W.)

[2] Consiglio Nazionale delle Ricerche, Istituto per le Applicazioni del Calcolo, Via Amendola 122 O, 70126 Bari, Italy; g.nico@iac.cnr.it

[3] Institute of Earth Sciences, Saint Petersburg State University, 199034 Saint Petersburg, Russia

* Correspondence: a.hoskera@cgiar.org

Received: 24 March 2020; Accepted: 21 May 2020; Published: 22 May 2020

Abstract: This study describes a semi-empirical model developed to estimate volumetric soil moisture (ϑ_v) in bare soils during the dry season (March–May) using C-band (5.42 GHz) synthetic aperture radar (SAR) imagery acquired from the Sentinel-1 European satellite platform at a 20 m spatial resolution. The semi-empirical model was developed using backscatter coefficient (σ dB) and in situ soil moisture collected from Siruguppa *taluk* (sub-district) in the Karnataka state of India. The backscatter coefficients σ_{VV}^0 and σ_{VH}^0 were extracted from SAR images at 62 geo-referenced locations where ground sampling and volumetric soil moisture were measured at a 10 cm (0–10 cm) depth using a soil core sampler and a standard gravimetric method during the dry months (March–May) of 2017 and 2018. A linear equation was proposed by combining σ_{VV}^0 and σ_{VH}^0 to estimate soil moisture. Both localized and generalized linear models were derived. Thirty-nine localized linear models were obtained using the 13 Sentinel-1 images used in this study, considering each polarimetric channel Co-Polarization (VV) and Cross-Polarization (VH) separately, and also their linear combination of VV + VH. Furthermore, nine generalized linear models were derived using all the Sentinel-1 images acquired in 2017 and 2018; three generalized models were derived by combining the two years (2017 and 2018) for each polarimetric channel; and three more models were derived for the linear combination of σ_{VV}^0 and σ_{VH}^0. The above set of equations were validated and the Root Mean Square Error (RMSE) was 0.030 and 0.030 for 2017 and 2018, respectively, and 0.02 for the combined years of 2017 and 2018. Both localized and generalized models were compared with in situ data. Both kind of models revealed that the linear combination of $\sigma_{VV}^0 + \sigma_{VH}^0$ showed a significantly higher R^2 than the individual polarimetric channels.

Keywords: volumetric soil moisture; synthetic aperture radar (SAR); Sentinel-1; soil moisture semi-empirical model; soil moisture Karnataka India

---

1. Introduction

Soil moisture estimation across space and time has become possible with the advent of microwave remote sensing [1]. The amount of moisture in the soil is a function of physical, chemical, and management practices. Soil moisture is highly dynamic across space and correlated in time. The radar backscattering coefficient is a function of soil characteristics such as dielectric constant, texture, and surface roughness, and depends on the wavelength, polarization, and angle of incidence of the radar [1]. Shorter wavelength C-band radar backscatter has shown sensitivity to surface soil moisture at a depth of about 5 cm [2–4]. The launch of the Sentinel-1 mission of the European Space

Agency has made a huge amount of C-band data acquired since 2014 from all over the Earth's surface accessible. This opened up new perspectives on studying soil moisture in semi-arid regions, as was undertaken in Karnataka, India, in this work. Large scale soil moisture monitoring will provide greater insights into energy fluxes, which can result in improved meteorological and climatic projections [5] that will provide critical inputs for agriculture.

There have been studies based on physical, empirical, and semi-empirical models that estimate soil moisture over bare soils through radar remote sensing [6–8]. Physical approaches require many input parameters such as surface roughness and slope, which are not available under practical conditions [8]. Empirical models are only data driven, whereas semi-empirical models, while being data driven, also support theoretical considerations. In soil studies, they are site-specific and generally valid for specific soil characteristics [3]. Previous semi-empirical studies have considered single polarization to build a relationship between soil moisture and a backscatter model at 10 cm depth [9] and estimated ϑ_v with a root mean square error (RMSE) of 3–6% [10–12] using C-band data. There have also been studies that have used the SAR interferometry technique and Sentinel-1 data to estimate soil moisture and compare them with in situ measurements [13]. Even though SAR interferometry is less frequently used in the remote sensing community to estimate soil moisture, its advantage lies in its ability to disentangle moisture and terrain roughness contributions. Most SAR-based soil moisture estimation studies have covered small areas limited to a few hundred square kilometers [11–17]. Estimating soil moisture over a wider area and at a higher resolution using SAR imagery will provide information on managing water resources and irrigation scheduling that can benefit a large number of farmers [14].

The aim of this study was to estimate soil moisture in bare rice agricultural soils. While SAR images have been used to estimate rice phenology using X-band TerraSAR-X images [15], there have been limited studies to estimate the soil moisture in bare rice agricultural soils using Sentinel-1 C-band images. Bare soils in Siruguppa are rice growing areas that lie bare after the rice crop has been harvested in March, with rice stubble and weeds that have dried up during summer (March–June). By the time the monsoon rains start, it is extremely critical to estimate the amount of soil moisture in the top 10 cm, which will help farmers decide when to start preparing the land and start sowing the next crop. Surface roughness, soil status, soil moisture, and crop residue distribution affect radar backscatter [16]. It is well established that σ^0_{VV} is more sensitive to variation in soils and σ^0_{VH} is more suited to the identification of dry crop residue [17]. Utilizing both together can improve the accuracy of soil moisture estimates [18]. Nevertheless, soil moisture studies using σ^0_{VV} and σ^0_{VH} together, especially using Sentinel 1 SAR data, are limited. The need for such studies over significantly large agricultural fields is very important to study agriculture, water, and food security. The major goal of this study was to estimate soil moisture over bare soils using both σ^0_{VV} and σ^0_{VH} polarization and compare it with in situ measurements at a 10 cm (0–10 cm) depth. At the time of measurement, soil moisture to 10 cm is at the steady state and consistent across that top surface layer and therefore the C-band can be assumed to detect the top 10 cm layer. However, it is known that C-band SAR signals cannot penetrate to a 10 cm depth.

The contribution of standing stubble to total backscattering coefficient is comparable with that of the soil surface when the stubble has more than 75% water content. Backscatter coefficient decreases with a decrease in water content in the stubble. However, when the water content in the stubble is less than 40%, the contribution to the total backscattering coefficient is negligible [19]. We investigated both localized and generalized linear models to try to disentangle the stubble and soil moisture contributions. The linear coefficients of localized models were derived using in situ data acquired on a specific Sentinel-1 day. In contrast, generalized models were built using all in situ measurements acquired in the study period, thus adding the temporal dimension to the analysis of Sentinel-1 data. The question we wanted to answer is: can semi-empirical models estimate soil moisture, getting rid of the stubble contribution to the backscattering coefficient? We tried to answer this question by studying the effects of each variable, time, and polarization, separately. A localized model does not take into account the temporal evolution of backscattering, while a generalized model includes the time variable when estimating the linear coefficients. Furthermore, for each model, it is possible to keep the polarimetric channels separated or merge them. In this work, we used a large dataset of in situ measurements of soil-moisture acquired across a 2-year period to answer the above question. The issues of the stability of results and of collinearity of data are crucial and will be used to assess the results of this experiment.

The rest of this paper is organized as follows: Section 2 is devoted to materials and methods, Section 3 to the results, and Section 4 presents the discussion. Finally, a few conclusions are drawn in Section 5.

2. Materials and Methods

2.1. Study Area

The study was conducted in Siruguppa *taluk* (sub-district) in the Bellary district of Karnataka state, India (Figure 1). Siruguppa is located between 15.35°N to 15.83°N latitudes and 76.69°E to 76.71°E longitudes covering an area of 1048 sq. km. Its climate is moderate and dry most of the year. It experiences high temperatures ranging from 23.2 °C to 42.4 °C from March to May and an annual rainfall of 645 mm. Irrigation from canal discharges cater to 60% of the cropped area, and the rest is either rainfed or irrigated through groundwater. Most of the crops are grown in predominantly black-clay, red-loamy, and red-sandy soils.

The River Tungabhadra runs diagonally across Siruguppa from the northwest, providing water for irrigation. The major crops grown are paddy, sorghum, pearl millet, sunflower, groundnut, cotton and sugarcane. The last decade saw a fall in kharif (rainy season) crop production due to deficit rainfall during the monsoon in some places in the *taluk*, leading to a shift from paddy and millets to cash crops such as cotton and sugarcane. The Deccan Plateau region is frequently prone to drought, making information on soil moisture critical for allocating water resources and scheduling irrigation. The date of sowing is a critical decision farmers make after the initial rainfall has occurred. This is done based on traditional knowledge and the physical assessment of soil moisture by hand or using a push probe. A scientific estimation of soil moisture can help farmers to decide the sowing date. This study was conducted on "bare agriculture fields" of Siruguppa to estimate soil moisture using radar remote sensing.

2.2. In Situ Data

2.2.1. Soil Sampling and Ground Data Collection

The soils of Siruguppa are classified into Vertisols (covering 720.9 km^2), Aridisols (146.8 km^2), Inceptisols (65.1 km^2), Alfisols (34.1 km^2), and other land cover such as rock outcrops (21.5 km^2). The locations for soil sample collection were based on random sampling, taking into account the fractions of different soil types. This mitigates the effects of variation from sampling error and increases the precision of the measured variable [20]. Soil samples were collected using a 10 cm standard metallic cylinder for a soil type to account for vertical and horizontal homogeneity [21], and weighed on site using a Mettler Toledo electronic balance. A handheld GPS (Garmin etrex) was used to georeference the locations immediately with an average accuracy of 2.5 meters as we collected it after a good almanac was received. Sixty-two locations were sampled spread across the four soil types. Forty-eight locations were sampled in Vertisols, eight in Inceptisols, four in Aridisols, and two in Alfisols. This was repeated for two years (2017 and 2018) over 13 dates of satellite overpasses, bringing the total data points to 806 (Figure 1).

Bulk density (BD) samples were collected simultaneously using standard cylindrical cores on site to estimate volumetric soil moisture (ϑ_v). The sampling was carried out from March to May in bare agricultural soils with crop residue from paddy and weeds.

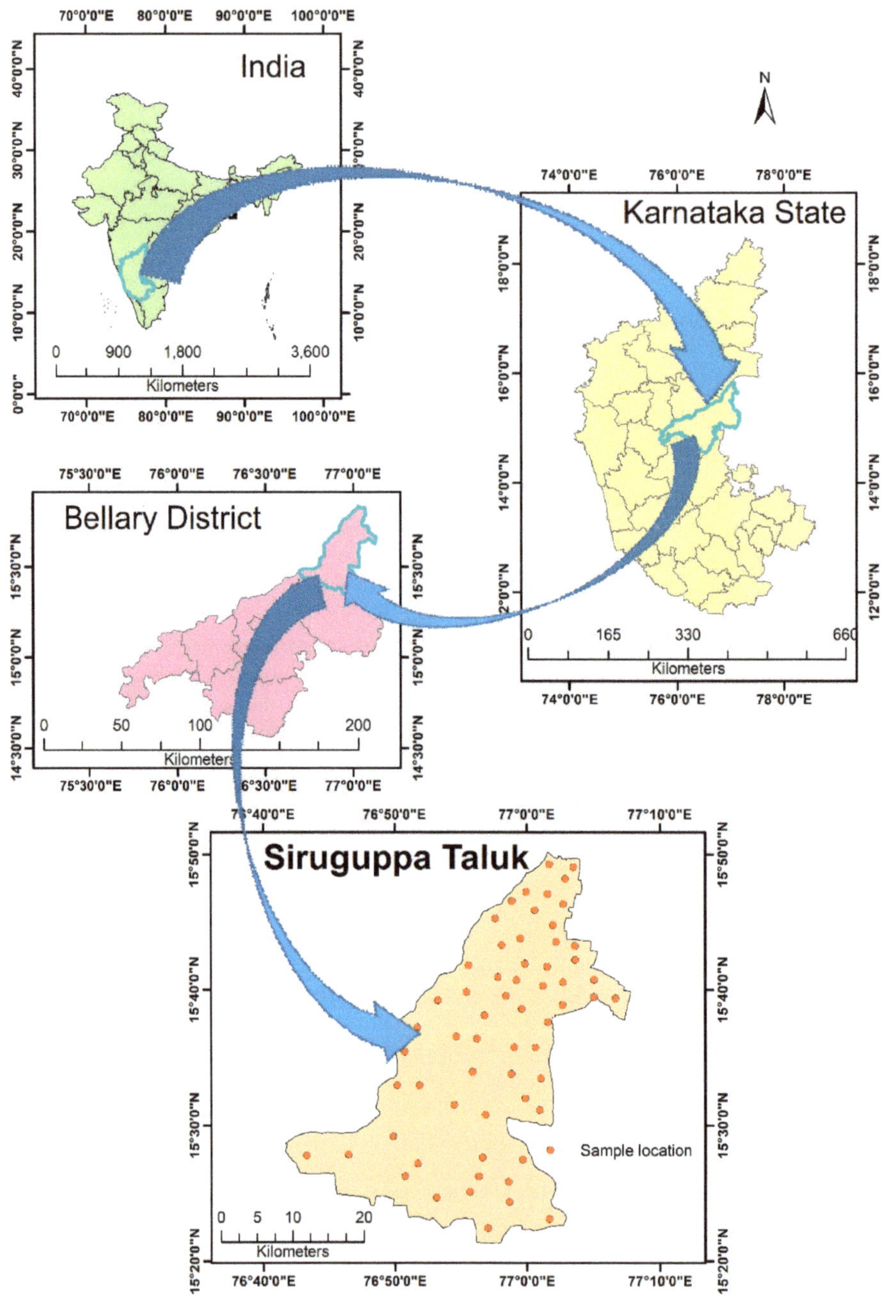

Figure 1. Map of the study area with sampling locations.

2.2.2. Laboratory Analysis

Volumetric soil moisture was measured in two steps. First, the gravimetric method was used to estimate soil moisture from field samples over bare agricultural land [22]. Global Positioning System

(GPS) coordinates were taken at each sample location to allow the approximate identification of the soil sample location with the image pixel. The soil collected from the ground after measuring the wet weight (ϑ_w) was filled in airtight polythene bags and numbered with their corresponding GPS ID. The polythene bags were brought to the soil laboratory to measure their dry weight (ϑ_d) using a standard drying process. Each sample was transferred to a microwave bowl and placed in the oven at 105 °C for 24 h, and the weight measured as dry weight. The following formula was used to estimate gravimetric soil moisture:

$$\vartheta_0 = \frac{\vartheta_w - \vartheta_d}{\vartheta_d} \left[\frac{g}{g} \right] \tag{1}$$

The second step involved collecting the soil cores to estimate bulk density (BD). The drying process was repeated for each sample and the following formula was used to estimate BD:

$$\vartheta_0 = \frac{\vartheta_d}{V} \left[\frac{g}{cm^3} \right] \tag{2}$$

where V is the volume of the core.

Volumetric soil moisture was expressed as:

$$\vartheta_v = \frac{\vartheta_0 \cdot BD}{\rho_{H_2O}} \left[\frac{cm^3}{cm^3} \right] \tag{3}$$

where ρ_{H_2O} is the water density.

2.3. Data Collection and Pre-Processing

Thirteen Sentinel-1 images were used, six acquired between 4 March 2017 and 27 May 2017 and seven between 11 March 2018 and 22 May 2018 (see Table 1). The incidence angle varied from 30° to 35° covering the study area in Co-Polarization (VV) and Cross-Polarization (VH) polarization. The frequency of the acquisition of imagery over India is very low, and a cycle of low and high number of acquisitions in alternating months was seen from the data portal (Table 1). Pre-processing of SAR imagery was carried out using SNAP software developed by the European Space Agency (ESA). Radiometric calibration, thermal noise removal, and terrain correction (using the Range Doppler terrain correction operator) algorithms were applied to obtain the backscattering coefficient (σ dB) [23]. A Lee speckle filter was applied to reduce speckle noise. Linear σ^0_{VV} and σ^0_{VH} were converted to dB values.

Sentinel-2 Level-1C S2 imagery with less than 10% cloud cover was downloaded for the years 2016 to 2018. These were converted to Level 2A to obtain bottom of atmosphere reflectance using SNAP software provided by ESA under a GNU General Public License V3. Visible and Near Infrared Radiation (NIR) bands B4 and B8 were used to generate normalized difference vegetation index (NDVI) to delineate the agricultural area.

Table 1. Acquisition dates of Sentinel-1 images: Interferometric Wide (IW) swath mode, relative orbit 63, descending. Ground Range Detected (GRD) product type (VV and VH polarization). Images were downloaded from the European Space Agency (ESA) portal https://scihub.copernicus.eu/.

Year	Acquisition Date						
2017	4 March	28 March	21 April	3 May	15 May	27 May	
2018	11 March	23 March	4 April	16 April	28 April	10 May	22 May

2.4. Methodology

The study began with pre-processing of Sentinel-1 C-band data (described in Section 2.3) to obtain σ^0 from both polarizations after applying appropriate corrections and speckle reduction. The in situ data collected during the field missions were used to extract σ^0_{VV} and σ^0_{VH} values in dB from the respective images of different dates (Table 1). The in situ data and σ° data were compiled to analyze and build a semi-empirical model. Agricultural land was derived using band B4 and B8 of a time series of Sentinel-2 images used to calculate the NDVI for the date for which an image was available in the season during each year. Random forest (RF) classification was applied to the set of nine NDVI images covering the study area and training dataset. This is useful to mask out non-agricultural areas when visualizing soil moisture estimates. An evaluation of the semi-empirical model was conducted to assess the accuracy of soil moisture (Figure 2).

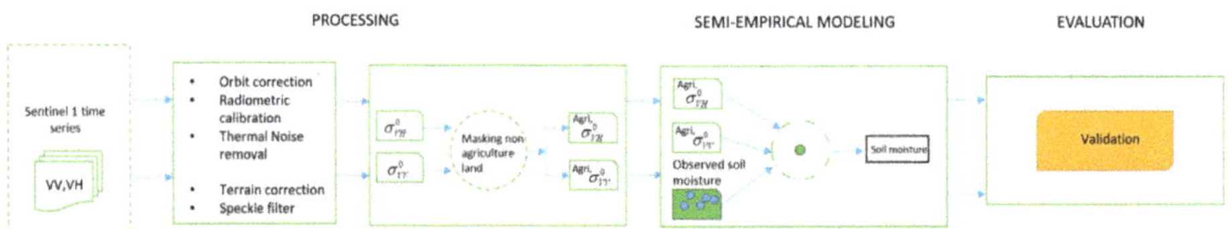

Figure 2. The process of estimating soil moisture using Sentinel-1 Co-Polarization (VV) and Cross-Polarization (VH) imagery.

2.4.1. Semi-Empirical Model

A semi-empirical model was proposed to estimate soil moisture over bare soils in agricultural areas from the backscatter coefficient based on a linear relationship. The linear equation captures the backscatter from bare soil, which constitutes soil moisture and surface roughness (as crop residue) and includes both VV and VH backscattering coefficients as:

$$f(\vartheta_v) = A\,\sigma\,VV + B\,\sigma(VH) + T \tag{4}$$

where ϑ_v is the volumetric soil moisture; A, B, and T are empirical constants; and σ^0_{VV} and σ^0_{VH} are the VV and VH backscattering coefficients, respectively.

On bare soil, σ^0_{VV} and σ^0_{VH} are mainly influenced by soil moisture. Since the major crop in the study area is rice, there is a crop residue as rice stubble on the ground. The rice stubble at 75% water content also contributes to the σ^0_{VV}, but decreases as the water content decreases and is negligible in both polarizations [19,24]. A linear combination including both polarizations was found to better estimate soil moisture from bare soil.

2.4.2. Delineation of Agricultural Fields

The estimation of soil moisture is more meaningful when linked to the purpose for which it is used. The ideal domain for use of such information is agricultural lands. Ideally, NDVI [25] is used to understand changes in crop phenology as the growing season progresses. Since the target class was only agricultural land, time series NDVI during the cropping season was best suited for the delineation using Sentinel-2 imagery. A set of nine NDVI images during the three crop seasons was used to estimate land cover using the RF algorithm [26]. The training dataset included land use in the soil sample locations (62). Additionally, 200 training samples were used: 100 from agricultural land and 100 from non-agricultural land. This product was used as a base for mapping soil moisture in agricultural lands.

2.4.3. Evaluation of Semi-Empirical Model

Basic information like maximum, minimum and mean in situ soil moisture were generated (Table 2). Linear regression was used to understand the relationship between Sentinel-1 backscattering coefficients and in situ soil moisture data. The P value, which indicates the significance of the accuracy assessment was significant ($\leq$0.05) and not significant ($\geq$0.05). The RMSE of the modeled soil moisture was estimated using the equation:

$$RMSE = \sqrt{\frac{1}{N} \sum_{i=1}^{N} (Y_{mes} - Y_{est})^2}$$ (5)

To understand the contribution of each polarization and sum of both polarizations to the accuracy of the model, residual standard error (RSE) of the estimated soil moisture was calculated using equation:

$$\text{Residual standard error (RSE)} = \sqrt{\frac{\sum (y_0 - y_e)^2}{n-2}}$$ (6)

where y_0 is the observed soil moisture; y_e is the predicted soil moisture; and n is the degree of freedom.

3. Results

A well distributed sampling scheme and data collected over two years yielded a well calibrated model to estimate soil moisture in the bare agricultural soils during the dry season (March–May). Linear and multi-linear regression was used to find the relationship between observed soil moisture and backscatter coefficients by deriving the model constants for each date and a combination of dates.

3.1. Field Measurements and Laboratory Analysis

Soil moisture was estimated using the gravimetric method for all 62 samples spread over Siruguppa *taluk* (Figure 1) for each date of satellite pass. Mean volumetric soil moisture (ϑ_v) in the samples ranged from 0.22 m^3/m^3 to 0.28 m^3/m^3 from 4 March 2017 to 27 May 2017. Minimum ϑ_v varied from 0.12 m^3/m^3 to 0.17 m^3/m^3 from March 2017 to May 2017 and the maximum ϑ_v varied from 0.30 m^3/m^3 to 0.34 m^3/m^3, respectively (Table 2). Figure 3 illustrates the range of values that each point in the population takes above and below the mean for six dates of satellite passes during 2017. It is worth noting that Figure 3 displays the soil moisture values measured the day of the satellite passes and for this reason, the ranges of the variation of soil moisture appeared as different from those reported in Table 2. Similarly, measurements were made during 2018 at the same locations. The minimum ϑ_v varied from 0.11 m^3/m^3 to 0.15 m^3/m^3 and the maximum varied from 0.32 to 0.34 m^3/m^3 from March 2018 to May 2018, respectively. The mean ϑ_v was measured between 0.23 m^3/m^3 and 0.26 m^3/m^3 (Table 2). Figure 4 shows the range of values during 2018 for the seven dates of satellite passes during 2018.

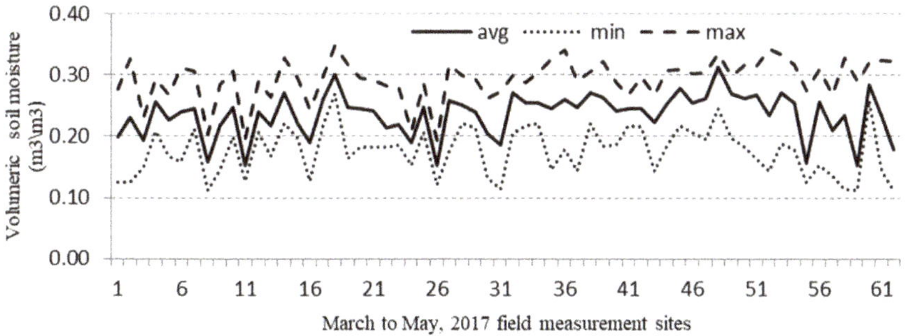

Figure 3. Observed soil moisture of each point during six passes of the satellite estimated using the gravimetric method during 2017.

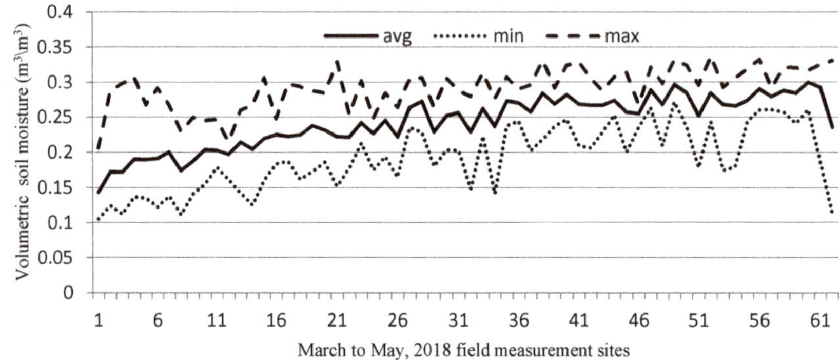

Figure 4. Observed soil moisture of each point during seven passes of the satellite estimated using the gravimetric method during 2018.

Table 2. Observed soil moisture of all 62 samples combined during each pass of the satellite during 2017 and 2018 collected at Siruguppa *taluk*.

Field Measurement	Soil Moisture (m³/m³)		
	Min	Max	Mean
4 March 2017	0.14	0.31	0.22
28 March 2017	0.12	0.30	0.23
21 April 2017	0.14	0.30	0.22
3 May 2017	0.12	0.30	0.23
15 May 2017	0.17	0.34	0.28
27 May 2017	0.15	0.30	0.23
11 March 2018	0.15	0.33	0.24
23 March 2018	0.15	0.34	0.26
4 April 2018	0.13	0.33	0.24
16 April 2018	0.13	0.34	0.25
28 April 2018	0.14	0.33	0.25
10 May 2018	0.11	0.32	0.24
22 May 2018	0.11	0.32	0.23

3.2. Localized and Generalized Relationships

The concepts of localized and generalized relationships were used in the in situ measurements of soil moisture and SAR estimates. A relationship was localized if it was obtained using single date data points in the study area, collected both in 2017 and 2018. A generalized relationship was obtained when all the dates data points were considered in the study area (Figure 5).

The relationship for localized models showed R^2 ranging from 0.62 to 0.75 between σ_{VV}^0 and ϑ_v, revealing a significantly strong relationship in 2017 (Table 3). As far as σ_{VH}^0 is concerned, it was found to have a lower R^2, ranging from 0.43 to 0.70. During 2018, R^2 values ranged from 0.56 to 0.69 for σ_{VV}^0 and from 0.31 to 0.62 for σ_{VH}^0. The linear combination of σ_{VV}^0 and σ_{VH}^0 showed higher R^2 values, ranging from 0.71 to 0.88 during 2017, and from 0.60 to 0.86 during 2018 (see Table 3).

Generalized relationships attempted to study the impact of seasonal effects observed in the study area due to different agroecologies (i.e., the different management and practices in a homogenous landscape). Table 4 summarizes the R^2 values obtained in the individual years 2017 and 2018 along with a combination of two years for σ_{VV}^0, σ_{VH}^0 and their linear combination ($\sigma_{VV}^0 + \sigma_{VH}^0$). The individual and combined backscatter coefficients in the two VV and VH polarizations over 2017 and 2018 pointed out a clear relationship with the in situ measurements of soil moisture.

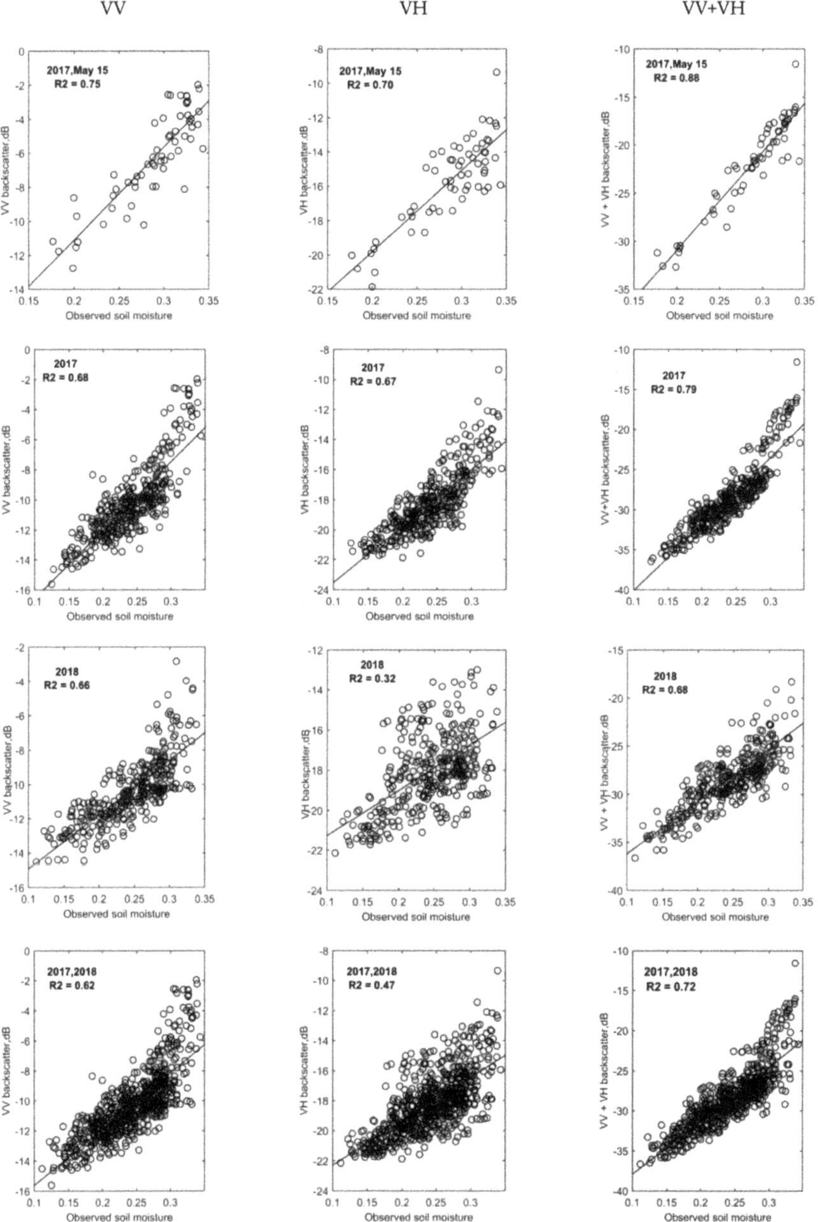

Figure 5. Localized and generalized linear models between soil moisture and backscatter. (**Top**) Examples of localized models refer to the Sentinel-1 acquisition of 15 May 2017. The remaining rows refer to the generalized models obtained using all Sentinel-1 images acquired in 2017, 2018, and in the total study from 2017 to 2018. The images from left to right represent Sentinel-1 images VV, VH, and VV + VH backscattering coefficients.

Table 3. Localized relationship between soil moisture and backscatter.

Sentinel-1 Acquisition Date	R^2 (σ^0_{VV})	R^2 (σ^0_{VH})	R^2 ($\sigma^0_{VV}+\sigma^0_{VH}$)
4 March 2017	0.63	0.66	0.83
28 March 2017	0.65	0.65	0.87
21 April 2017	0.68	0.67	0.84
3 May 2017	0.69	0.56	0.80
15 May 2017	0.75	0.70	0.88
27 May 2017	0.62	0.43	0.71
11 March 2018	0.69	0.32	0.75
23 March 2018	0.63	0.34	0.78
4 April 2018	0.56	0.47	0.60
16 April 2018	0.64	0.33	0.72
28 April 2018	0.63	0.31	0.66
10 May 2018	0.65	0.62	0.86
22 May 2018	0.57	0.54	0.78

Table 4. Generalized relationship between soil moisture and backscatter.

Year	σ°	R^2
2017	VV	0.68
2017	VH	0.67
2017	VV + VH	0.79
2018	VV	0.66
2018	VH	0.32
2018	VV + VH	0.62
2017, 2018	VV	0.62
2017, 2018	VH	0.47
2017, 2018	VV + VH	0.72

3.3. Soil Moisture Evaluation

Multi-linear regression and linear regression were applied to determine the value of empirical constants (A, B, and T) in both the localized and generalized models. Tables 5 and 6 summarize the results. Each localized model comprises images from one date of pass over of the study area, totaling 39 equations from March 2017 to May 2018, and considering individual σ^0_{VV}, σ^0_{VH}, and their linear combination $\sigma^0_{VV} + \sigma^0_{VH}$. Each generalized model combines all images acquired during a year for individual σ^0_{VV}, σ^0_{VH} and linear combination $\sigma^0_{VV} + \sigma^0_{VH}$. Three models were obtained for each year 2017 and 2018 and three additional models using all the images used in the study, making it a total of nine generalized models. For the localized model, 40 samples for calibration and 22 samples for validation were used during 2017 and 2018 (N = 62). The model calibration for individual dates (localized models) with combined backscatter coefficient ($\sigma^0_{VV} + \sigma^0_{VH}$) during the study years of 2017 and 2018 estimated an R^2 ranging from 0.91 to 0.70 and RSE ranging from 0.03 to 0.01.

As far as generalized models are concerned, N = 368 points were used in 2017, 258 for calibration, and 110 for validation, and N = 427 in 2018, with 299 for calibration and 128 for validation. The total number of points for both years was N = 795, with 557 used for calibration and 238 for validation. The nine linear equations modeled each backscatter coefficient σ^0_{VV}, σ^0_{VH} and a linear combination of both ($\sigma^0_{VV} + \sigma^0_{VH}$) for each year (2017, 2018) and 2017 and 2018 put together. Table 6 summarizes the R^2 values modeled from σ^0_{VV} and σ^0_{VH} as a function of (ϑ_v) during 2017 and 2018. RSE was 0.03 for 2017, 2018 from σ^0_{VV}, and 0.03 and 0.04 from σ^0_{VH} for 2017 and 2018, respectively. A linear combination of both backscatter coefficients during 2017 and 2018 showed an R^2 of 0.80 and 0.70, respectively. The RSE values were 0.02. The backscatter coefficient from each polarization and a linear combination of both polarizations put together for two years were also attempted. The R^2 from σ_{VV}^0 was 0.60 with a RSE of 0.03 and from σ_{VH}^0 it was 0.50 with a RSE 0.03. The R^2 from a linear combination of both polarizations ($\sigma_{VV}^0 + \sigma_{VH}^0$) was 0.70 with a RSE of 0.02.

Model validation was done for individual dates from a linear combination of the two polarizations. Table 5 and Figure 6 summarize the RMSE values for 2017 and 2018. The nine generalized models were also validated and the RMSE values are reported in Table 6 and Figure 7.

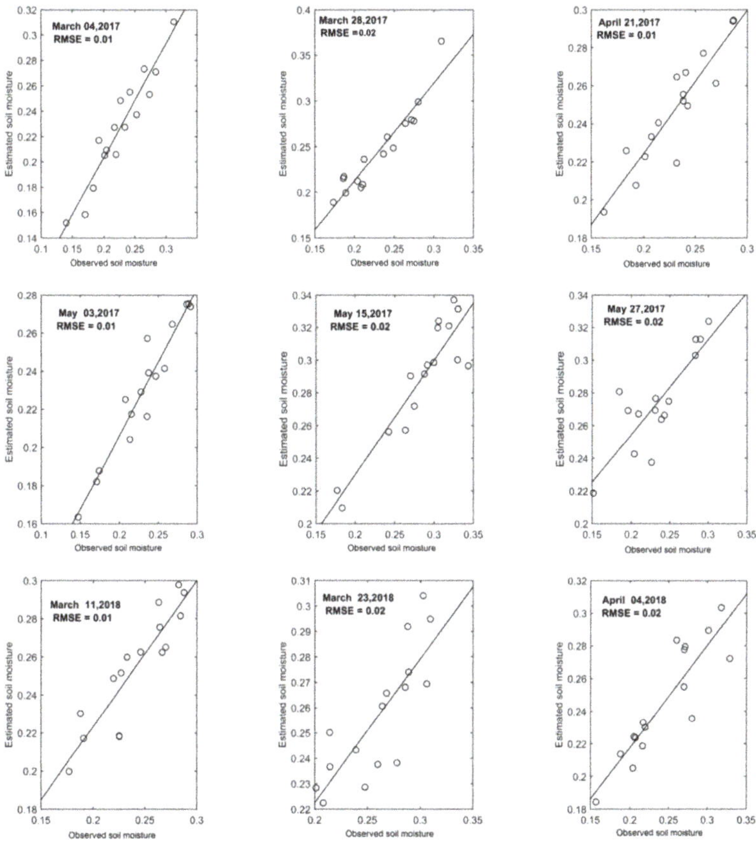

Figure 6. *Cont.*

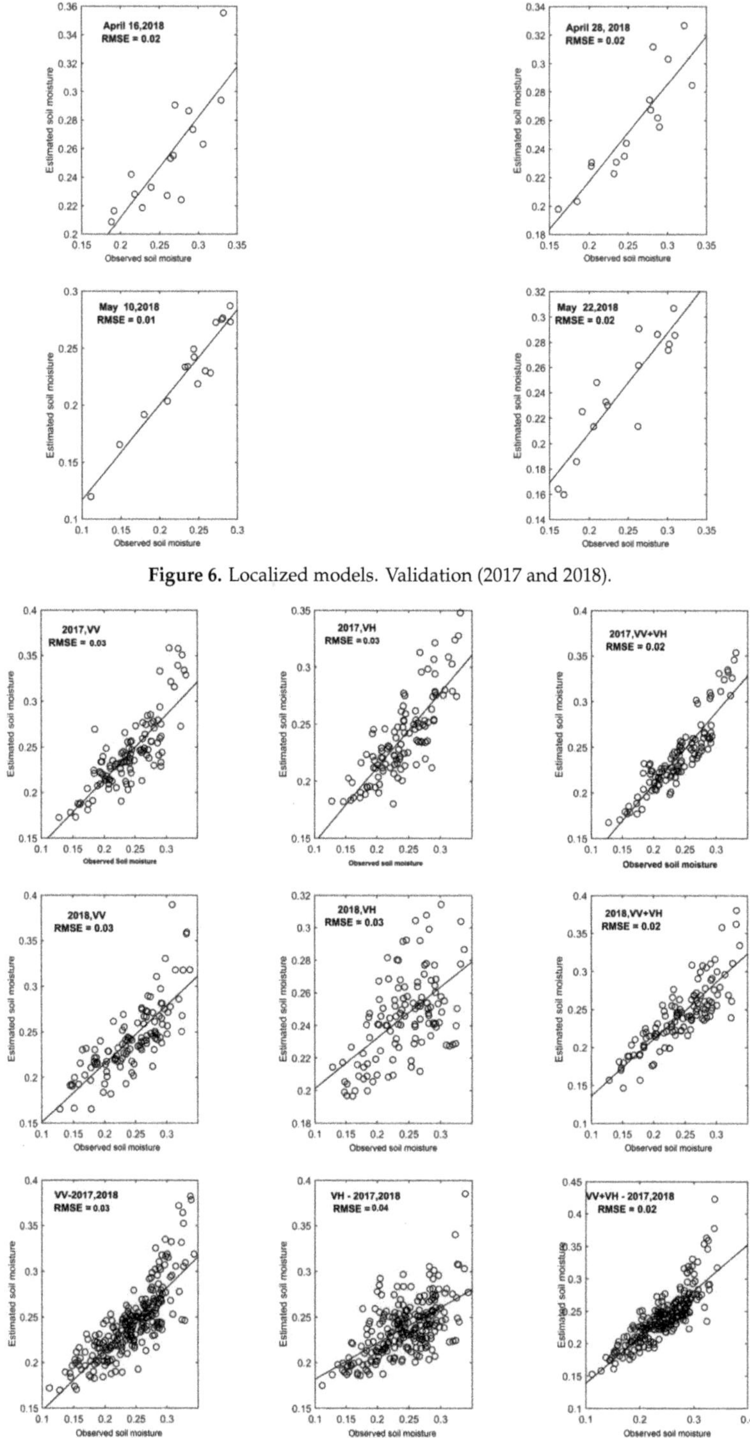

Figure 6. Localized models. Validation (2017 and 2018).

Figure 7. Generalized models. Validation between the estimated and observed soil moisture.

4. Discussion

Accurate estimation of ϑ_v was envisaged using a linear equation of σ^0_{VV} and σ^0_{VH} radar cross section from bare agricultural soils. A thorough data collection campaign was undertaken during 2017 and 2018, synchronizing with the pass of the satellite. Bare soil areas were mostly post-harvest cropped areas with little or no crop residue, depending on the crop sown. In the study area, 50% of the agricultural land comprises rice cropped and irrigated from a seasonal stream. Sentinel-1 SAR, dual polarized imagery was used to estimate soil moisture over bare soils using a semi-empirical model. Model parameters were estimated using linear and multi-linear regression. Performance evaluation was conducted based on a 70:30 ratio of sampled points and low RMSE was found between the observed and estimated soil moisture, when a linear relationship between σ^0_{VV} and σ^0_{VH} was combined for 2017 and 2018.

Table 5. Empirical constants (A, B, and T) of the localized model. The total number of samples for each date was 62. The linear equation was derived using 70% of the total population.

Sentinel-1 Image	Acquisition Date	Model	A	B	T	R^2	RMSE
							Validation
1	4 March 2017	$(\vartheta_v) = A \times \sigma^0_{VV} + B \times \sigma^0_{VH} + T$	0.014	0.011	0.60	0.82	0.01
2	28 March 2017	$(\vartheta_v) = A \times \sigma^0_{VV} + B \times \sigma^0_{VH} + T$	0.013	0.014	0.65	0.88	0.02
3	21 April 2017	$(\vartheta_v) = A \times \sigma^0_{VV} + B \times \sigma^0_{VH} + T$	0.011	0.015	0.65	0.84	0.01
4	3 May 2017	$(\vartheta_v) = A \times \sigma^0_{VV} + B \times \sigma^0_{VH} + T$	0.014	0.012	0.62	0.76	0.01
5	15 May 2017	$(\vartheta_v) = A \times \sigma^0_{VV} + B \times \sigma^0_{VH} + T$	0.008	0.008	0.47	0.90	0.02
6	27 May 2017	$(\vartheta_v) = A \times \sigma^0_{VV} + B \times \sigma^0_{VH} + T$	0.017	0.012	0.63	0.75	0.03
7	11 March 2018	$(\vartheta_v) = A \times \sigma^0_{VV} + B \times \sigma^0_{VH} + T$	0.016	0.008	0.57	0.82	0.02
8	23 March 2018	$(\vartheta_v) = A \times \sigma^0_{VV} + B \times \sigma^0_{VH} + T$	0.013	0.008	0.52	0.84	0.02
9	4 April 2018	$(\vartheta_v) = A \times \sigma^0_{VV} + B \times \sigma^0_{VH} + T$	0.022	0.004	0.55	0.76	0.02
10	16 April 2018	$(\vartheta_v) = A \times \sigma^0_{VV} + B \times \sigma^0_{VH} + T$	0.015	0.008	0.53	0.77	0.03
11	28 April 2018	$(\vartheta_v) = A \times \sigma^0_{VV} + B \times \sigma^0_{VH} + T$	0.022	0.010	0.67	0.70	0.02
12	10 May 2018	$(\vartheta_v) = A \times \sigma^0_{VV} + B \times \sigma^0_{VH} + T$	0.016	0.017	0.74	0.90	0.02
13	22 May 2018	$(\vartheta_v) = A \times \sigma^0_{VV} + B \times \sigma^0_{VH} + T$	0.014	0.019	0.76	0.78	0.02

Table 6. Empirical constants (A, B, and T) of the generalized model. The total number of samples for 2017 was 368, for 2018 it was 427, and for a combination of 2017 and 2018, it was 795. The linear equations were derived using 70% of the total sample of each year and combined year.

Year	σ	Model	A	B	T	R^2	RMSE
							Validation
2017	VV	$(\vartheta_v) = A \cdot \sigma^0_{VV} + T$	0.015	-	0.39	0.70	0.03
2017	VH	$(\vartheta_v) = B \cdot \sigma^0_{VH} + T$	-	0.017	0.56	0.67	0.03
2017	VV + VH	$(\vartheta_v) = A \cdot \sigma^0_{VV} + B \cdot \sigma^0_{VH} + T$	0.009	0.009	0.51	0.80	0.02
2018	VV	$(\vartheta_v) = A \cdot \sigma^0_{VV} + T$	0.019	-	0.44	0.60	0.03
2018	VH	$(\vartheta_v) = B \cdot \sigma^0_{VH} + T$	-	0.014	0.50	0.32	0.03
2018	VV + VH	$(\vartheta_v) = A \cdot \sigma^0_{VV} + B \cdot \sigma^0_{VH} + T$	0.016	0.009	0.58	0.70	0.02
2017, 2018	VV	$(\vartheta_v) = A \cdot \sigma^0_{VV} + T$	0.016	-	0.41	0.60	0.03
2017, 2018	VH	$(\vartheta_v) = B \cdot \sigma^0_{VH} + T$		0.016	0.59	0.50	0.04
2017, 2018	VV + VH	$(\vartheta_v) = A \cdot \sigma^0_{VV} + B \cdot \sigma^0_{VH} + T$	0.011	0.009	0.59	0.70	0.02

4.1. Relationship between σ^0_{VV} and σ^0_{VH} with Observed Data

Soil moisture data were collected over two years (2017 to 2018) during the dry summer season from March to May from the 62 plots on the dates of the satellite passes. The relationship between σ^0_{VV}, σ^0_{VH}, and observed soil moisture by individual dates of the satellite pass (13 images) showed that in both years, the backscatter and observed soil moisture had a significant positive correlation [2,10,27,28]. In both years, VV polarization had a higher backscatter dB value than VH polarization. In cross-polarization (VH), signal attenuation occurs due to volumetric scattering [29]. In 2017, soil moisture constantly increased from 4 March to 27 April. The R^2 between radar backscattering coefficient and in situ measurements of soil moisture is reported in Table 3. A sudden increase in R^2 (VV) can be observed on 15 May, corresponding to the consecutive rainfall events that occurred during the three days before the date of the satellite pass (Figure 8). This means that there is a better correlation for high values of soil moisture, probably because under this condition, the radar backscattering coefficient's dependence on soil moisture is more important than it is on surface roughness.

Similarly, an unexpected increase in σ^0_{VH} was observed (Figure 5). 27 May 2017 (Table 3) had a low R^2 value from σ^0_{VH} compared to the rest of the dates due to the rainfall event (Figure 8), weeds or crop residue moisture [24]. In 2018, R^2 for the relationship between σ^0_{VV} and observed soil moisture was significant during March because of residual soil moisture (i.e., the crop residual moisture influenced the radar backscattering coefficient, Table 3). Residual soil moisture was low on 4 April and 22 May due to evaporative demand and higher between 16 April and 10 May due to consecutive rainfall events (Figure 8). R^2 did not decrease from March to May, probably due to irregular changes in crop residue moisture, since σ^0_{VH} is sensitive to it [24]. The R^2 values from σ^0_{VH} during March were relatively low despite no rainfall in the month because of residual soil moisture from the previous crop. The cumulative moisture due to rainfall during April is reflected in the low R^2 of 16 April and 28 April (Figure 8). During May, high R^2 values were due to bare soils. A linear combination of σ^0_{VV} and σ^0_{VH} during each date in 2017 produced higher R^2 compared to R^2 from individual polarization. This shows that the addition of σ^0_{VV} and σ^0_{VH} or vice versa improves the backscatter and soil moisture relationship rather than a single relationship with different polarization (Table 3). A similar relationship existed during 2018 from a linear combination of σ^0_{VV} and, which improved R^2 significantly (Table 3).

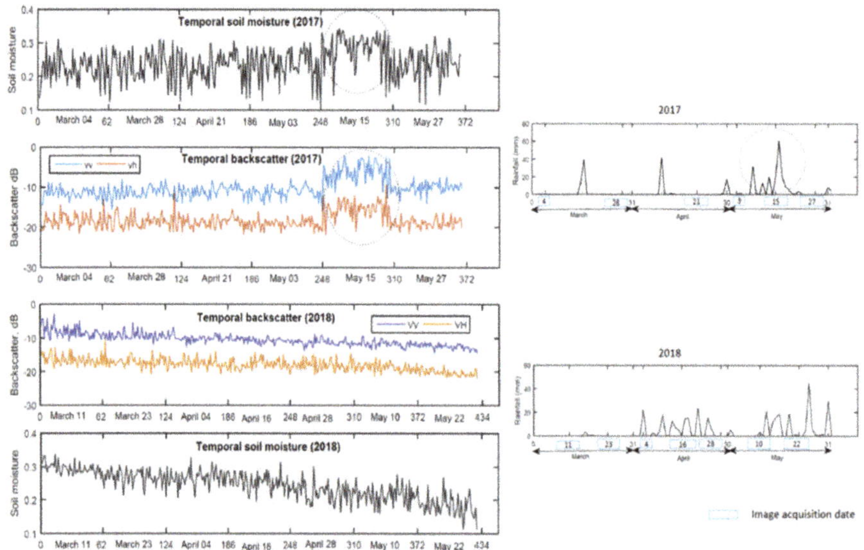

Figure 8. Temporal soil moisture, backscatter, and rainfall, 2017–2018 (March to May).

4.2. Localized and Generalized Relationships

To operationalize the accurate estimation of soil moisture for decision making, a global relationship was envisaged considering all dates during the dry season. The R^2 value of global relationship from VV polarization during 2017 was 0.68, which was higher than the mean of the local relationships.

The generalized relationship was found to be more useful for an accurate soil moisture estimate. In addition, R^2 for the generalized relationship performed better than the mean of the localized relationship (0.67) with VH polarization. The scenario during 2018 from VV polarization was more influenced by rainfall events in the dry season. The R^2 values ranged from 0.56 to 0.69 with a mean of 0.62 from localized relationships and 0.66 from the generalized relationship, which was more than the local mean. R^2 was very low from VH polarization due to cumulative moisture from rainfall events. However, the generalized relationship produced a lower R^2 than the mean localized relationship for VH in 2018 (see Tables 3 and 4). The usefulness of a generalized relationship was exhibited with a consistent increase in the accuracy of the soil moisture estimates over two years. The relationship from VV and VH polarization during 2017 showed significantly lower R^2 than the linear combination of VV and VH during the same year. Similarly, also during 2018, R^2 was significantly higher than the individual polarization. Finally, the best relationship was obtained when the linear combination of two polarizations was combined (appended) for the two years 2017 and 2018, than from single polarizations combined for the two years. It was inferred that generalized relationships are more promising in terms of building a model compared to localized relationships, which may not relate to the entire population.

4.3. Modeling the Relationships

The relationships of localized and generalized modeling were explored and tested for multicollinearity, especially linear combination models $\sigma_{VV}^0 + \sigma_{VH}^0$. Multicollinearity is a statistical phenomenon in which two or more predictor variables in a multiple regression model are highly correlated [30]. To detect multicollinearity, we used an indicator called variance inflation factor (VIF), which is a tool to measure and quantify how much the variance is inflated [30]. If any of the model's VIF values exceed 5 or 10, it is an indication that the associated regression coefficient is poorly estimated because of multicollinearity [31]. The P value indicates statistical significance for independent variable contribution in the model, which is explained in Section 2.4.3.

For generalized models, nine different types of linear relationships were explored with σ_{VV}^0 and σ_{VH}^0 data (Table 6) during 2017 and 2018. In 2017, among the three possible models σ_{VV}^0, σ_{VH}^0 and, $\sigma_{VV}^0 + \sigma_{VH}^0$, individual models σ_{VV}^0 and σ_{VH}^0 showed a low RSE and the p value was statistically significant. When σ_{VH}^0 was added to σ_{VV}^0, the RSE was lower than in the individual models. This indicates a very good relationship with a low VIF as well as p value for both backscatter coefficients. Similar results were observed in 2018. Generalized models derived by combining two years (2017 and 2018) of data showed similar results as individual year models. This study also showed that the linear combination equations from the localized models also performed well with low VIF (<2) and a p value statistically significant for both backscatter coefficients (Tables 7 and 8).

A collinearity test on the generalized and localized models showed that the VIF for a linear combination of both backscatter coefficients (VV + VH) was <3. Hence, these models are non-collinear. All models showed low p value, indicating that both backscatter coefficients made meaningful addition to the models. During modeling relationships with a linear combination of individual backscatter coefficient, it was inferred that the individual backscatter coefficients were non-collinear, contributing to R^2 independently. It was found that the localized models from individual dates varied over time, and any one equation with a low RSE and VIF may not represent the whole season. In addition, the generalized models produced lower RSE representing the whole season, and were hence better than each localized model.

Table 7. Analysis of variance in the generalized model.

Year	(σ°) Variable	t Value	$Pr(>|t|)$	VIF
2017	VV	24.24	$<2 \times 10^{-16}$ ***	-
2017	VH	23.24	$<2 \times 10^{-16}$ ***	-
2017	VV, VH	12.17, 11.20	$<2 \times 10^{-16}$ ***	2.11
2018	VV	21.23	$<2 \times 10^{-16}$ ***	-
2018	VH	11.98	$<2 \times 10^{-16}$ ***	-
2018	VV, VH	20.76, 11.50	$<2 \times 10^{-16}$ ***	1.10
2017, 2018	VV	29.49	$<2 \times 10^{-16}$ ***	-
2017, 2018	VH	24.31	$<2 \times 10^{-16}$ ***	-
2017, 2018	VV, VH	21.57, 16.32	$<2 \times 10^{-16}$ ***	1.40

*** p value < 0.001.

Table 8. Analysis of variance in the localized model.

Date	(σ°) Variable	t Value	$Pr(>|t|)$	VIF
4 March 2017	VV, VH	4.63, 7.54	$<2 \times 10^{-16}$ ***	1.36
28 March 2017	VV, VH	8.24, 8.04	$<2 \times 10^{-16}$ ***	1.30
21 April 2017	VV, VH	6.93,9.02	$<2 \times 10^{-16}$ ***	1.54
3 May 2017	VV, VH	7.39, 4.34	$<2 \times 10^{-16}$ ***	1.63
15 May 2017	VV, VH	6.93, 5.02	$<2 \times 10^{-16}$ ***	1.71
27 May 2017	VV, VH	5.03, 5.17	$<2 \times 10^{-16}$ ***	1.12
11 March 2018	VV, VH	10.45, 6.07	$<2 \times 10^{-16}$ ***	1.09
23 March 2018	VV, VH	11.25, 6.00	$<2 \times 10^{-16}$ ***	1.10
4 April 2018	VV, VH	10.19, 1.94	$<2 \times 10^{-16}$ ***	1.14
16 April 2018	VV, VH	9.17, 4.47	$<2 \times 10^{-16}$ ***	1.11
28 April 2018	VV, VH	9.10, 4.12	$<2 \times 10^{-16}$ ***	1.0
10 May 2018	VV, VH	10.69, 11.09	$<2 \times 10^{-16}$ ***	1.27
22 May 2018	VV, VH	6.66, 6024	$<2 \times 10^{-16}$ ***	1.26

*** p value < 0.001.

4.4. Validation of Models

Models were validated using 30% of the sampled points. Results for the localized models are summarized in Table 5. In 2017, the lowest RMSE (0.01) was found on 21 April. Figure 8 shows that no rainfall or very weak rainfall was observed on this day. An increase in RMSE was observed on 15 May. Similarly, in 2018, the lowest RMSE was observed on 23 March and the highest (0.03) on 16 April 2018, probably due to the increase in rainfall. The results seem to show that the RMSE of the models is related to the amount of rainfall. Localized models performed better in drier soils.

As far as the generalized models are concerned, the validation results showed that generalized models obtained using co-polar σ^0_{VV} data provided a lower RMSE than those based on cross-polar σ^0_{VH} data for both 2017 and 2018 and taking all data acquired from 2017 to 2018. We also found that the linear combination of both co-polar and cross-polar backscattering coefficients always provided a lower RMSE than the models using only one polarization. The best results came when using the linear combination of polarizations and all the data acquired along the two years, resulting in an RMSE of 0.02 (Table 6). This globalized model was used to produce maps of soil moisture and its spatial variability (Figures 9–11). This is probably the most important result, as a simple multi-linear model using both co-polar and cross-polar Sentinel-1 data acquired over long time periods can reproduce the spatial variability of soil moisture.

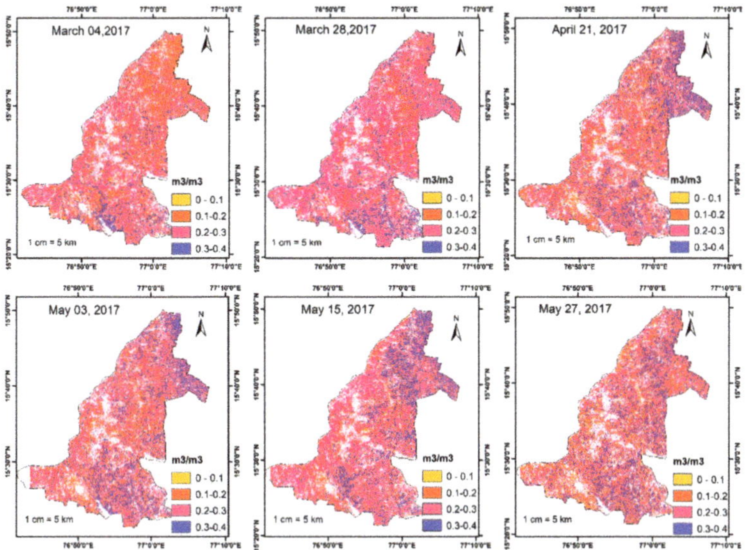

Figure 9. Spatial variability in the soil moisture in Siruguppa *taluk* during 2017.

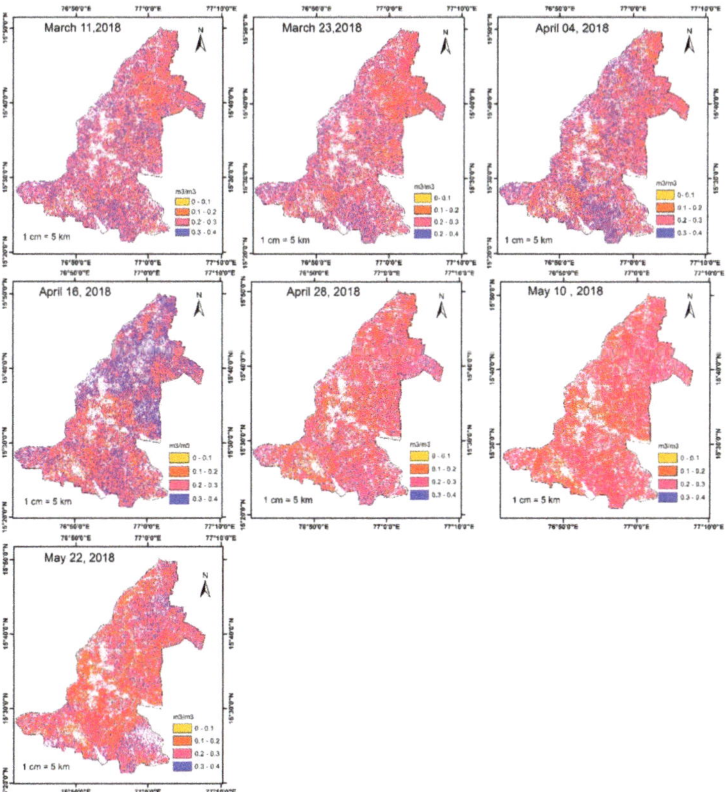

Figure 10. Spatial variability in the soil moisture in Siruguppa *taluk* during 2018.

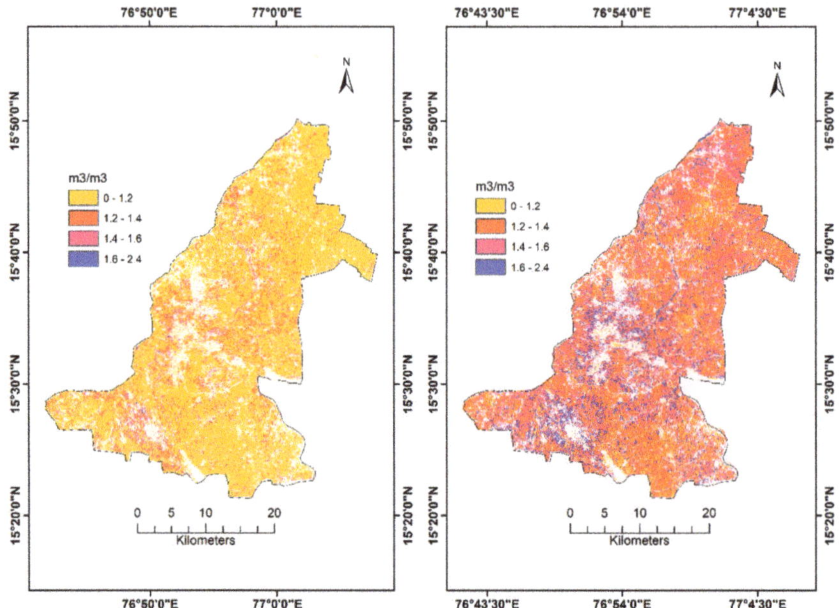

Figure 11. Spatial variability in the estimated soil moisture from combining all the dates in 2017 and 2018.

5. Conclusions

This study aimed to accurately estimate the soil moisture of bare, post-harvest agricultural areas collected from Siruguppa *taluk* (sub-district) in the Karnataka state of India. Fifty percent of this agricultural area is grown with rice that is irrigated by seasonal canal irrigation. An accurate estimate of volumetric soil moisture (ϑ_v) was envisaged using a semi-empirical model based on a linear equation of co-polarized and cross-polarized radar cross section obtained by Sentinel-1 images. A thorough data collection campaign was undertaken during 2017 and 2018 during the pass of the satellite.

Both localized and generalized models were developed using Sentinel-1 image independently and all images together, respectively. Results indicate that the accuracy of the soil moisture estimates increased when using both co-polar and cross-polar images instead of only σ^0_{VV} or σ^0_{VH}, independently.

The use of localized models revealed that the RMSE of soil moisture estimates decreased corresponding to dry periods, with little or no rainfall. This indicates that better estimates of soil moisture can be obtained for drier soils. Coming to globalized models, soil moisture estimates with lower RMSE were observed when merging all data acquired in 2017 and 2018, and co-polar and cross-polar images, with a R^2 of 0.7 and RMSE of 0.02. The availability of a large amount of in situ data collected over a large area demonstrated that a generalized linear model based on the joint use of co-polar and cross-polar C-band SAR images acquired for a long time period, with a short revisiting time of twelve days, could capture spatial variability in soil moisture. This is an important result as the availability of Sentinel-1 data can provide farmers with timely and accurate estimates of soil moisture and enable the mapping of its spatial variability by using simple semi-empirical models. This information, when provided in the immediate weeks and months preceding the cropping season, could be very crucial in determining planting dates and assessing early season plant growth, thereby playing a key role in influencing productivity.

Author Contributions: Conceptualization, methodology, and software: A.K.H.; Investigation: M.I.A., A.K.H., G.N., and A.W.; Original draft preparation: M.I.A., A.K.H., and G.N.; Review and editing: M.I.A., A.K.H., A.W., and G.N.; Funding acquisition: A.W. All authors have read and agree to the published version of the manuscript.

Funding: The authors gratefully acknowledge the financial support given by the Earth System Science Organization, Ministry of Earth Sciences, Government of India (IITM/MM-II/ICRISAT/2018/IND-11 and IITM/MM-II/CRIDA-ICRISAT-IIPR /2018/IND-9) to conduct this research under Monsoon Mission. This research was supported by the CGIAR Research Program on Climate Change, Agriculture and Food Security (CCAFS) carried out with support from the CGIAR Trust Fund and through bilateral funding agreements. For details, visit https://ccafs.cgiar.org/donors.

Acknowledgments: The authors are thankful to Prasad S. Thenkabail, Western Geographic Center, USGS, for his valuable insights into data analysis, and João Catalão, Department of Geographic Engineering, Geophysics and Energy, Faculty of Sciences, University of Lisbon, for engaging and fruitful discussions. The authors are also thankful to Smitha Sitaraman for English language editing.

Conflicts of Interest: The authors declare no conflict of interest. The funders had no role in the design of the study, in the collection, analyses, or interpretation of data, and in the writing of the manuscript, or in the decision to publish the results.

References

1. Barrett, B.W.; Dwyer, E.; Whelan, P. Soil Moisture Retrieval from Active Spaceborne Microwave Observations: An Evaluation of Current Techniques. *Remote Sens.* **2009**, *1*, 210–242. [CrossRef]
2. Huang, S.; Ding, J.; Zou, J.; Liu, B.; Zhang, J.; Chen, W. Soil Moisture Retrival Based on Sentinel-1 Imagery under Sparse Vegetation Coverage. *Sensors* **2019**, *19*, 589. [CrossRef] [PubMed]
3. Thoma, D.P.; Moran, M.S.; Bryant, R.; Rahman, M.; Holifield-Collins, C.D.; Skirvin, S.; Sano, E.E.; Slocum, K. Comparison of four models to determine surface soil moisture from C-band radar imagery in a sparsely vegetated semiarid landscape. *Water Resour. Res.* **2006**, *42*, W01418. [CrossRef]
4. Romshoo, S.; Taikan, O.; Katumi, M. C-band radar for soil moisture estimation under agricultural conditions. *IEEE Int. Geosci. Remote Sens. Symp.* **2002**, *4*, 2217–2219. [CrossRef]
5. Suni, T.; Guenther, A.; Hansson, H.; Kulmala, M.; Andreae, M.; Arneth, A.; Artaxo, P.; Blyth, E.; Brus, M.; Ganzeveld, L.; et al. The significance of land-atmosphere interactions in the Earth system—iLEAPS achievements and perspectives. *Anthropocene* **2015**, *12*, 69–84. [CrossRef]
6. Verhoest, N.E.C.; Lievens, H.; Wagner, W.; Álvarez-Mozos, J.; Moran, M.S.; Mattia, F. On the Soil Roughness Parameterization Problem in Soil Moisture Retrieval of Bare Surfaces from Synthetic Aperture Radar. *Sensors* **2008**, *8*, 4213–4248. [CrossRef]
7. Dubois, P.C.; van Zyl, J.; Engman, T. Measuring soil moisture with imaging radars. *IEEE Trans. Geosci. Remote Sens.* **1995**, *33*, 915–926. [CrossRef]
8. Fung, A.; Li, Z.; Chen, K. Backscattering from a randomly rough dielectric surface. *IEEE Trans. Geosci. Remote Sens.* **1992**, *30*, 356–369. [CrossRef]
9. Haider, S.S.; Said, S.; Kothyari, U.C.; Arora, M.K. Soil moisture estimation using ERS 2 SAR data: A case study in the Solani River catchment/Estimation de l'humidité du sol grâce à des données ERS-2 SAR: Étude de cas dans le bassin de la rivière Solani. *Hydrol. Sci. J.* **2004**, *49*, 334. [CrossRef]
10. Baghdadi, N.; Aubert, M.; Cerdan, O.; Franchisteguy, L.; Viel, C.; Martin, E.; Zribi, M.; Desprats, J.-F.; Eric, M. Operational Mapping of Soil Moisture Using Synthetic Aperture Radar Data: Application to the Touch Basin (France). *Sensors* **2007**, *7*, 2458–2483. [CrossRef]
11. Rao, S.S.; Pandey, A.K.; Das, S.N.; Nagaraju, M.S.S.; Venugopal, M.V.; Rajankar, P.; Laghate, P.; Reddy, M.S.; Joshi, A.K.; Sharma, J.R. Modified Dubois Model for Estimating Soil Moisture with Dual Polarized SAR Data. *J. Indian Soc. Remote Sens.* **2013**, *41*, 865–872. [CrossRef]
12. Alexakis, D.D.; Mexis, F.-D.K.; Vozinaki, A.-E.K.; Daliakopoulos, I.N.; Tsanis, I. Soil Moisture Content Estimation Based on Sentinel-1 and Auxiliary Earth Observation Products. A Hydrological Approach. *Sensors* **2017**, *17*, 1455. [CrossRef] [PubMed]
13. Conde, V.; Nico, G.; Catalao, J. Comparison of In-Field Measurements and INSAR Estimates of Soil Moisture: Inversion Strategies of Interferometric Data. In Proceedings of the IGARSS 2019—2019 IEEE International Geoscience and Remote Sensing Symposium, Yokohama, Japan, 28 July–2 August 2019; pp. 6186–6189.
14. Alvino, A.; Marino, S. Remote Sensing for Irrigation of Horticultural Crops. *Horticulturae* **2017**, *3*, 40. [CrossRef]
15. Lopez-Sanchez, J.M.; Cloude, S.R.; Ballester-Berman, J.D. Rice Phenology Monitoring by Means of SAR Polarimetry at X-Band. *IEEE Trans. Geosci. Remote Sens.* **2012**, *50*, 2695–2709. [CrossRef]

16. McNairn, H.; Ellis, J.; Van Der Sanden, J.; Hirose, T.; Brown, R. Providing crop information using RADARSAT-1 and satellite optical imagery. *Int. J. Remote Sens.* **2002**, *23*, 851–870. [CrossRef]

17. Holmes, M.G. Applications of Radar in Agriculture. In *Applications of Remote Sensing in Agriculture*; Clark, J.A., Steven, M.D., Eds.; Butterworths: Sevenoaks, UK, 1990; pp. 307–330.

18. Baghdadi, N.; El Hajj, M.; Zribi, M.; Bousbih, S. Calibration of the Water Cloud Model at C-Band for Winter Crop Fields and Grasslands. *Remote Sens.* **2017**, *9*, 969. [CrossRef]

19. Oh, Y. Effect of standing stubble on radar backscatter from harvested rice fields. *Electron. Lett.* **2008**, *44*, 1423. [CrossRef]

20. Petersen, R.G.; Calvin, L.D. Sampling. In *Methods of Soil Analysis Part 3*; Soil Science Society of America Book Series; American Society of Agronomy, Inc.: Madison, WI, USA, 1996; pp. 1–18.

21. Cline, M.G. Principles of soil Sampling. *Soil Sci.* **1944**, *58*, 275–288. [CrossRef]

22. Ozerdem, M.S.; Acar, E.; Ekinci, R. Soil Moisture Estimation over Vegetated Agricultural Areas: Tigris Basin, Turkey from Radarsat-2 Data by Polarimetric Decomposition Models and a Generalized Regression Neural Network. *Remote Sens.* **2017**, *9*, 395. [CrossRef]

23. Small, D. Flattening Gamma: Radiometric Terrain Correction for SAR Imagery. *IEEE Trans. Geosci. Remote Sens.* **2011**, *49*, 3081–3093. [CrossRef]

24. Smith, A.; Major, D. Radar Backscatter and Crop Residues. *Can. J. Remote Sens.* **1996**, *22*, 243–247. [CrossRef]

25. Yin, H.; Udelhoven, T.; Fensholt, R.; Pflugmacher, D.; Hostert, P. How Normalized Difference Vegetation Index (NDVI) Trends from Advanced very High Resolution Radiometer (AVHRR) and Système Probatoire d'Observation de la Terre Vegetation (SPOT VGT) Time Series Differ in Agricultural Areas: An Inner Mongolian Case Study. *Remote Sens.* **2012**, *4*, 3364–3389. [CrossRef]

26. Jin, Y.; Liu, X.; Chen, Y.; Liang, X. Land-cover mapping using Random Forest classification and incorporating NDVI time-series and texture: A case study of central Shandong. *Int. J. Remote Sens.* **2018**, *39*, 8703–8723. [CrossRef]

27. Oldak, A.; Jackson, T.; Starks, P.; Elliott, R. Mapping near-surface soil moisture on regional scale using ERS-2 SAR data. *Int. J. Remote Sens.* **2003**, *24*, 4579–4598. [CrossRef]

28. Puri, S.; Stephen, H.; Ahmad, S. Relating TRMM precipitation radar land surface backscatter response to soil moisture in the Southern United States. *J. Hydrol.* **2011**, *402*, 115–125. [CrossRef]

29. Bousbih, S.; Zribi, M.; Chabaane, Z.L.; Baghdadi, N.; El Hajj, M.; Gao, Q.; Mougenot, B. Potential of Sentinel-1 Radar Data for the Assessment of Soil and Cereal Cover Parameters. *Sensors* **2017**, *17*, 2617. [CrossRef]

30. Jamal, I. Multicollinearity and Regression Analysis. *J. Phys. Conf. Ser.* **2017**, *949*, 012009.

31. Montgomery, D.C.; Peck, E.A.; Vinning, G.G. *Introduction to Linear Regression Analysis*, 5th ed.; John Wiley & Sons: Hoboken, NJ, USA, 2012.

 *remote sensing*

Article

Integration of Time Series Sentinel-1 and Sentinel-2 Imagery for Crop Type Mapping over Oasis Agricultural Areas

Luyi Sun, Jinsong Chen *, Shanxin Guo, Xinping Deng and Yu Han

Shenzhen Institutes of Advanced Technology, Chinese Academy of Sciences, 1068 Xueyuan Avenue, Shenzhen University Town, Shenzhen 518055, China; ly.sun@siat.ac.cn (L.S.); sx.guo@siat.ac.cn (S.G.); xp.deng1@siat.ac.cn (X.D.); yu.han@siat.ac.cn (Y.H.)
* Correspondence: js.chen@siat.ac.cn; Tel.: +86-755-86392331

Received: 5 November 2019; Accepted: 28 December 2019; Published: 2 January 2020

Abstract: Timely and accurate crop type mapping is a critical prerequisite for the estimation of water availability and environmental carrying capacity. This research proposed a method to integrate time series Sentinel-1 (S1) and Sentinel-2 (S2) data for crop type mapping over oasis agricultural areas through a case study in Northwest China. Previous studies using synthetic aperture radar (SAR) data alone often yield quite limited accuracy in crop type identification due to speckles. To improve the quality of SAR features, we adopted a statistically homogeneous pixel (SHP) distributed scatterer interferometry (DSI) algorithm, originally proposed in the interferometric SAR (InSAR) community for distributed scatters (DSs) extraction, to identify statistically homogeneous pixel subsets (SHPs). On the basis of this algorithm, the SAR backscatter intensity was de-speckled, and the bias of coherence was mitigated. In addition to backscatter intensity, several InSAR products were extracted for crop type classification, including the interferometric coherence, master versus slave intensity ratio, and amplitude dispersion derived from SAR data. To explore the role of red-edge wavelengths in oasis crop type discrimination, we derived 11 red-edge indices and three red-edge bands from Sentinel-2 images, together with the conventional optical features, to serve as input features for classification. To deal with the high dimension of combined SAR and optical features, an automated feature selection method, i.e., recursive feature increment, was developed to obtain the optimal combination of S1 and S2 features to achieve the highest mapping accuracy. Using a random forest classifier, a distribution map of five major crop types was produced with an overall accuracy of 83.22% and kappa coefficient of 0.77. The contribution of SAR and optical features were investigated. SAR intensity in VH polarization was proved to be most important for crop type identification among all the microwave and optical features employed in this study. Some of the InSAR products, i.e., the amplitude dispersion, master versus slave intensity ratio, and coherence, were found to be beneficial for oasis crop type mapping. It was proved the inclusion of red-edge wavelengths improved the overall accuracy (OA) of crop type mapping by 1.84% compared with only using conventional optical features. In comparison, it was demonstrated that the synergistic use of time series Sentinel-1 and Sentinel-2 data achieved the best performance in the oasis crop type discrimination.

Keywords: oasis crop type mapping; Sentinel-1 and 2 integration; statistically homogeneous pixels (SHPs); red-edge spectral bands and indices; recursive feature increment (RFI); random forest (RF)

---

1. Introduction

The Xinjiang Uygur Autonomous Region is a major agricultural region in the arid and semi-arid areas of Northwest China. Due to the dry climate, almost all agriculture in Xinjiang depends on irrigation, leading to water shortage. This region relies on large-area cotton cultivation for profit, with cotton production accounting for over 70% of the national total. Cotton planting consumes large amounts of water, exacerbating the problem of water scarcity. Some areas in Xinjiang have undergone structural adjustments of agriculture, reducing the cultivation area of cotton and expanding the planting scale of two other cash crops, i.e., chili pepper and tomato. These adjustments resulted in a more complex cropping structure, requiring the timely and accurate mapping of crop distribution. Crop type distribution is vital information for estimating water availability and environmental carrying capacity. This is especially important in the arid and semi-arid areas in Northwest China, where oasis agriculture is the economic pillar, while the ecological environment is relatively fragile.

Optical remote sensing has been widely used in agricultural area mapping and crop classification in recent years. Approaches utilizing MODIS (Moderate Resolution Imaging Spectroradiometer) vegetation indices for crop type discrimination only suit for large-open fields, due to the low resolution (250–500 m) of MODIS data [1–3]. A number of studies used Landsat spectrum and vegetation indices for crop mapping, but the data availability is heavily limited by cloud cover due to Landsat's 16-day revisit interval [4]. Landsat data also encounters mixed-pixel problems in heterogeneous smallholder farming areas. In addition, crop type discrimination places a higher demand on the spectral resolution. The increased temporal, spatial, and spectral resolution of Sentinel-2 A/B imagery provides new opportunities for improving crop type classification over heterogeneous cultivated land compared with other optical sensors [5,6].

As pointed out by previous studies, due to cloud cover, the optical data discontinuity in key growth stages of crops can still happen [7]. Furthermore, for crop types with similar phenological cycles, only using spectral information is still challenging for reliable discrimination of crop types. As synthetic aperture radar (SAR) can reflect the structure of vegetation, and optical imagery captures the multi-spectral information of crops, it has been indicated that the synergetic use of SAR and optical data can be complementary to each other [8,9].

Space-borne SAR, due to its all-day, all-weather capability, wide coverage, and strong penetrating ability, has been increasingly used in crop classification, to complement with the use of optical imagery. It was found that considerable improvement can be achieved by increasing polarization channels [10,11]. Some studies suggested that using multi-temporal acquisitions can improve the accuracy of crop type mapping, and cross-polarized backscatter outperforms other polarization modes [12,13]. The launches of Sentinel-1 A and B satellites dramatically increased the volume of freely available SAR data, with dual-polarization modes, a 12-day revisit time, and 20 m spatial resolution. Thus, Sentinel-1 data is more desirable for medium to high-resolution crop mapping. However, affected by speckles inherent in SAR imaging systems, crop type mapping using SAR data alone yields quite limited accuracy [14]. The work by Ban et al. suggested that apart from speckles, the single parameter, high incidence angle SAR system used in their study did not provide sufficient differences to differentiate some crop species [15]. Thus, due to the limited viewing angles and orbits of available SAR data in most study cases, the sole use of SAR data may not be sufficient for crop type classification, especially in complex cropping systems.

By the synergic use of microwave Sentinel-1 features and optical Sentinel-2 features, the accuracy of crop discrimination can be potentially improved [15–18]. However, despite the existing studies to combine SAR and optical images for crop classification, few studies (1) explored the performance of individual InSAR products (such as coherence, amplitude dispersion, and master versus slave intensity ratio) in crop type identification. Regarding the de-speckling of SAR intensity, the conventional procedure presented in previous studies uses a regular-shaped window (e.g., boxcar filter, Lee filter, refined Lee filter, etc.) to reduce speckle effects, but in the meantime blurs the image especially over textural areas [19].

The combination of SAR and optical imagery resulted in hundreds of input features (also known as input variables) for the classification model. We adopted a supervised random forest classifier in the crop classification due to its high capacity to deal with a large number of input features. Nevertheless, it has been reported that the classification accuracy can be considerably increased by removing redundant features [20]. Feature selection is a crucial step to improve the performance of a classifier. From an operational perspective, the manual selection of such high-dimensional features is not desirable. Many approaches used separability criterion or hypothetical tests to select features based on assumptions of the sample distribution. In some cases, these assumptions were not satisfied, particularly when using SAR features. Moreover, crop samples can even break the assumption of unimodal distribution. For example, the sowing and harvesting dates are usually farmer customized; crop growth stages are affected by local weather conditions and soil conditions; therefore, they are site-specific. Thus, we prefer not to use separability criterion or hypothetical tests to select features. In the literature, several methods based on machine learning algorithms have been proposed for feature selection [21,22]. Some studies indicated the built-in attribute of random forest, the feature importance score, can be utilized as a ranking criterion to aid feature selection.

The objective of this research is to develop a method to integrate time series Sentinel-1 and Sentinel-2 features for the mapping of typical oasis crop distribution in heterogeneous smallholder farming areas. Firstly, in addition to SAR backscatter intensity, a number of InSAR products were extracted from time series Sentinel-1 data, such as the interferometric coherence, amplitude dispersion, and master versus slave intensity ratio. A statistically homogeneous pixel (SHP) distributed scatterer interferometry (DSI) [23,24] algorithm, originally proposed in the interferometric SAR (InSAR) community to identify distributed scatters (DSs), was adopted for the de-speckling of backscatter intensity and bias mitigation of coherence coefficient, so as to improve the quality of SAR features. To the best of our knowledge, this is the first time the use of amplitude dispersion and bias mitigated coherence is explored in crop type discrimination. Secondly, optical features were extracted from multi-temporal Sentinel-2 images. In particular, red-edge spectral bands and 11 indices were derived and included as input features for the oasis crop classification. Thirdly, a recursive feature increment (RFI) approach, on the basis of random forest feature importance, was proposed to obtain the optimal combination of S1 and S2 features for crop type discrimination. Finally, a random forest classifier was applied to the optimal feature set to produce a crop type distribution map. This study aims to answer the following questions: (1) Does the integration of Sentinel-1 and Sentinel-2 features achieve better performance than using SAR or optical features alone in the oasis crop type mapping? (2) If yes, which SAR feature has the most significant contribution? Are there any InSAR products that are capable of distinguishing oasis crop types? (3) To what extent can the inclusion of red-edge spectral bands and indices improve the accuracy of the oasis crop type identification? Which red-edge band or indices contribute most?

2. Materials and Methods

2.1. Study Area

The study area is located in oasis agricultural areas in Bayingolin Mongol Autonomous Prefecture, Xinjiang Uyghur Autonomous Region, China, encompassing the area 40.61°–42.44° N, 85.82°–87.14° E (Figure 1). This area is situated in the southern foothills of the Tianshan Mountains, on the north-eastern edge of the Tarim Basin, adjacent to the west bank of Bosten Lake, and to the north of the Taklimakan Desert. This region features a warm temperate continental climate, with much more evaporation than precipitation. It has a representative planting pattern in the arid and semi-arid regions of Northwest China.

The major crop types cultivated from spring to autumn include cotton, spring corn, summer corn, pear, chili pepper, tomato, etc. Cotton is sown from early April to early May and harvested by the end of September. Spring corn is sown in mid-April to early June and harvested in mid-August to mid-September. Summer corn is sown in mid-June to mid-July and harvested by early October. Pear starts budding in late March, flowering from late March to the end of April, fruiting from May to early September, and is harvested in September. Chili pepper and tomato seedlings are nurtured in greenhouses from February to March, transplanted to open fields from mid-April to early May. Tomato is harvested in August, and chili pepper is harvested in October. Phenological calendars of the five crop types are summarized in Table 1.

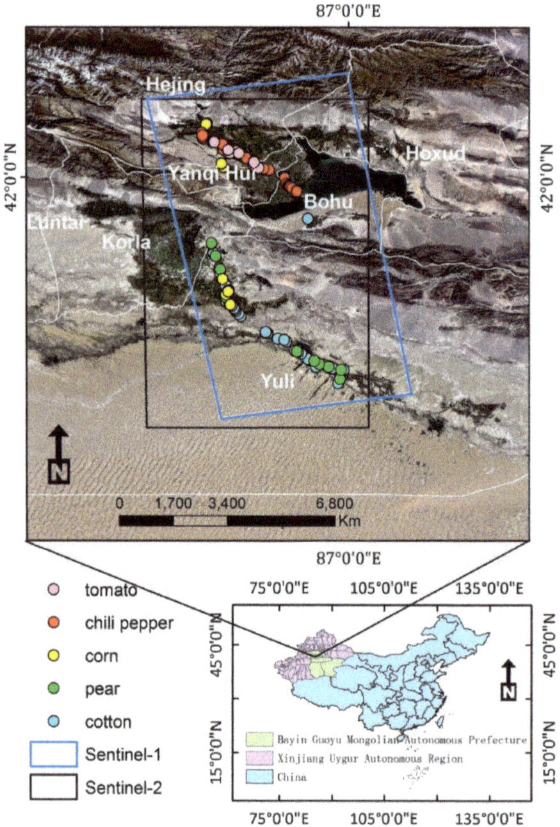

Figure 1. The study area across Yuli, Korla, Bohu, Yanqi, Hejing, and Hoxud counties in Bayin Guoyu Mongolian Autonomous Prefecture, Xinjiang Uygur Autonomous Region, China. The in-situ collected sample points of major crop types are highlighted. The overlapped area of Sentinel-1 IW2 sub-swath (in the blue rectangle) and mosaicked coverage of Sentinel-2 scenes (in the black rectangle) is the area of interest in this study. The background image is a true color Google Earth high-resolution image. Map data: Google Earth, Image © 2020 Europa Technologies.

Table 1. Phenological calendars of the major crop types in the study site.

	Jan	Feb	Mar	Apr	May	Jun	Jul	Aug	Sep	Oct	Nov	Dec
Cotton				△	△☆	☆	☆	☆	☆√			
Corn (Spring)				△	△	△☆	☆	☆√	√			
Corn (Summer)						△	△☆	☆	√	√		
Pear			♀	☆	☆	☆	☆	☆	√			
Chili pepper		♠	♠	★	★☆	☆	☆	☆	☆√	√		
Tomato		♠	♠	★	★☆	☆	☆	√				

Legend: △ Sowing; ♀ Budding; ♠ seedling nurturing; ★ Transplanting; ☆ Growing; √ Harvesting.

2.2. Data

2.2.1. In-Situ Reference Data

Ground samples of six land cover types (cropland, forest, urban area, desert, waterbody, and wetlands) were visually identified from Google Earth high-resolution images in terms of polygons. These sample polygons were resampled to generate randomly placed points for each cover type (Table 2). Individual fields of five crop species were collected in a field campaign in July 2018. Crop sample points were randomly extracted from each field under the condition that the minimum distance between any two points must be no less than 20 m. This was done in ArcGIS Data Management toolbox. Stratified K fold was applied in the units of crop fields; that is, sample points from the same field can only be used for training or testing. Thus, training samples are not spatially correlated to the samples used in validation. In the crop type classification, five folds were used to verify the accuracy, and the details are shown in Table 3.

Table 2. Ground samples of the six land cover types in the study area.

Land Cover Type	Sample Points
Cropland	3817
Forest	1468
Desert	1468
Urban area	2349
Waterbody	3257
Wetlands	3125

Table 3. Ground samples collected for major crop types in the study area.

Crop Type	Fold	Training Samples		Testing Samples	
		No. of Fields	No. of Points Points	No. of Fields	No. of Points
Chili pepper	1st	57	148	18	37
	2nd	50	147	25	38
	3rd	68	147	7	38
	4th	63	151	12	34
	5th	71	147	4	38
Corn	1st	23	89	9	22
	2nd	25	88	7	23
	3rd	30	90	2	21
	4th	29	87	3	24
	5th	21	90	11	21
Cotton	1st	37	150	13	38
	2nd	43	151	7	37
	3rd	43	152	7	36
	4th	34	151	16	37
	5th	43	152	7	36
Pear	1st	22	76	2	19
	2nd	20	78	4	17
	3rd	15	76	9	19
	4th	18	77	6	18
	5th	21	73	3	22
Tomato	1st	19	104	2	31
	2nd	16	108	5	27
	3rd	19	108	2	27
	4th	15	110	6	25
	5th	15	110	6	25

2.2.2. Satellite Data

To monitor the growth stages of major crop types cultivated from spring to autumn in the study area, we examined Sentinel-1 and Sentinel-2 data acquired in the time period from late March to early October. After discarding Sentinel-2 acquisitions with too much cloud cover (more than 20%), there were in total eight Sentinel-2 acquisitions left. The available number of the Sentinel-1 and Sentinel-2 acquisitions for each month is summarized in Table 4.

Table 4. The employed Sentinel-1 and Sentinel-2 acquisitions of each month.

Acquisition Time	Mar	Apr	May	Jun	Jul	Aug	Sep	Oct
Sentinel-1A	1	2	3	3	3	3	2	2
Sentinel-2A	0	0	1	0	0	1	1	1
Sentinel-2B	0	0	0	1	1	0	1	1

The employed Sentinel-1 data is the level-1 interferometric wide swath (IW) single look complex (SLC) data. In total, 18 acquisitions were used (Table 5).

Table 5. Metadata of the data stack of Sentinel-1 (S1) interferometric wide swath (IW) single look complex (SLC) data using the parameters from the first image. These values remain very close for all subsequent acquisitions.

Sentinel-1 IW SLC Data	
First acquisition	26 March 2018
Last acquisition	16 October 2018
Pass direction	Ascending
Polarization mode	VV + VH
Incidence angle (°)	36.12–41.84
Wavelength (cm)	5.5 (C-band)
Range spacing (m)	2.33
Azimuth spacing (m)	13.92

The Sentinel-2 data used in this study is the MSI Level-1C data (Table 6). In total, eight acquisitions were used, spanning the time from 9 May 2018 to 11 October 2018.

Table 6. Central wavelength and bandwidth of different spectral bands of Sentinel-2 data used in this research.

Spatial Resolution (m)		Spectral Bands	S2A	S2B
			Central Wavelength (nm)	Central Wavelength (nm)
10	B2	Blue	496.6	492.1
	B3	Green	560	559
	B4	Red	664.5	665
	B8	Near-infrared (NIR)	835.1	833
20	B5	Red-edge 1	703.9	703.8
	B6	Red-edge 2	740.2	739.1
	B7	Red-edge 3	782.5	779.7
	B8A	NIR narrow	864.8	864
	B11	Short-wave infrared (SWIR) 1	1613.7	1610.4
	B12	SWIR 2	2202.4	2185.7

2.3. Methods

The workflow used in this research is provided in Figure 2.

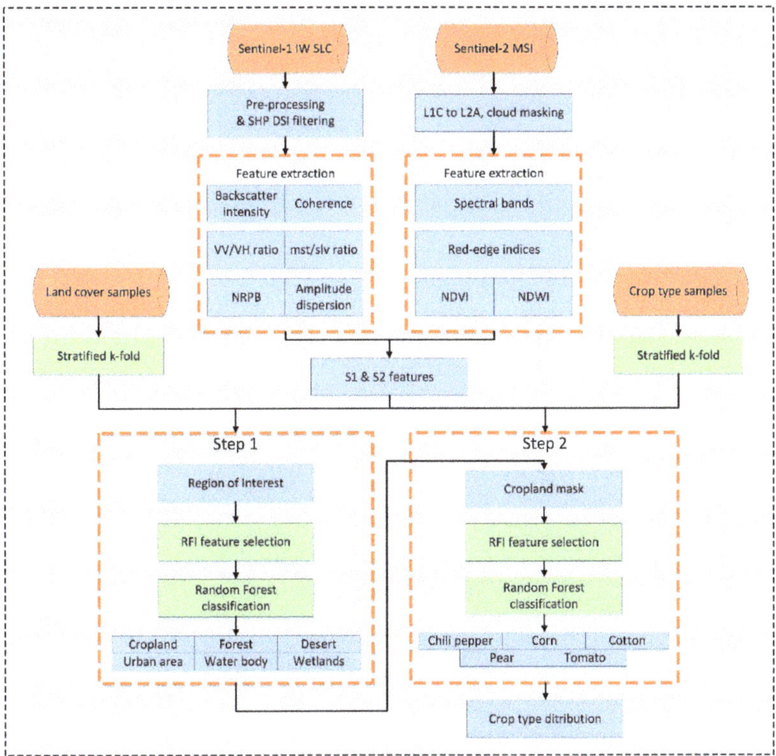

Figure 2. The workflow of crop type mapping by the integration of time series Sentinel-1 and Sentinel-2 features.

Firstly, a number of SAR features were extracted from time series Sentinel-1 data. It has been reported intensity and derived ratios (VV versus VH ratio, and the normalized ratio procedure between bands (NRPB)) can monitor the vegetation changes. Some studies indicated that SAR interferometric coherence can potentially improve the discrimination capability of different land cover types [25–28]. Thus, the backscatter intensity, interferometric coherence, and their derivative products were derived as SAR features for crop mapping. A by-product from interferometric pairs was also included; that is, the master versus slave intensity ratio computed respectively for the VH and VV polarization modes. In addition, we derived a statistic of time series SAR intensity, i.e., the amplitude dispersion computed respectively for VH and VV polarization. The amplitude dispersion was originally used in multi-temporal interferometric SAR (InSAR) for the initial selection of persistent scatters, which is a good indicator to distinguish vegetated surface and man-made structures, bare rocks, etc.

Secondly, optical features were derived from Sentinel-2 images. In addition to the conventional features, such as the visible, NIR (Near-infrared), SWIR (Short-wave infrared) bands, normalized difference vegetation index (NDVI), and normalized difference water index (NDWI), the red-edge spectral bands were also extracted. A total of 11 red-edge indices (will be detailed in Table 7) were calculated for each S2 acquisition, in order to thoroughly explore the contribution of red-edge wavelengths in crop type discrimination.

To deal with the high dimensional input features for the classifier, we proposed a recursive feature increment approach to select the optimal combination of SAR and optical features (detailed in Section 2.3.2) based on the random forest feature importance ranking.

A two-step hierarchical cotton mapping strategy (Figure 2) was implemented in this research. In step 1, the entire field site was classified into six land cover types, i.e., cropland, forest, desert, urban area, water body, and wetlands, so as to create a cropland mask. In the second step, the cropland was mapped into five different crop types, including chili pepper, corn, cotton, pear, and tomato. The automated feature selection method and RF classification were used both in step 1 and step 2, as illustrated in Figure 2.

2.3.1. Data Processing and Feature Extraction

1. Sentinel-1 Data Processing and Feature Extraction

All Sentinel-1 IW SLC images were pre-processed by the SNAP—ESA Sentinel Application Platform v7.0.0 (http://step.esa.int). Processing steps include co-registration, de-burst, and subset. To derived InSAR products, in total, 17 interferometric pairs were formed for each polarization mode using the acquisition on 26 March 2018 as the common master image, and all the subsequent acquisitions as the slave images. The next step is the de-speckling of intensity and accurate estimation of coherence. As pointed out by [29], using regular-shaped windows, conventional de-speckling methods average the values of neighboring pixels indiscriminately, leading to degraded image resolution and blurred edges between objects of different scattering characteristics. Furthermore, conventionally, the coherence is calculated by a regular-shaped sliding window, which unavoidably averages pixel values from different distributions (e.g., belonging to different land cover types), resulting in a biased estimation. In this study, a SHP DSI algorithm [23,24] was applied to the single look complex SAR data prior to SAR feature extraction. Here, we briefly summarize the principles and processing steps of the SHP DSI algorithm:

In SAR images, there are large numbers of distributed scatters (DSs) that exist in a resolution cell in cropland, forest, desert, etc. DSs cannot dominate the backscatter characteristics of a resolution cell, and the pixel appears as a random variable along the time dimension. When the SAR time series is sufficiently long, according to the central limit theorem, the vector sum of all the distributed scatters from a pixel can be defined as a complex Gaussian random variable. In a SAR image, we can identify if a pixel has the same statistical distribution with the other using the confidence interval. In this way, pixels can be divided into different statistically homogeneous pixel (SHP) subsets. The SHP DSI algorithm applies a modified Lee filtering method to the diagonal elements of the complex covariance matrix of an arbitrary pixel, on the basis of the SHP subset that the pixel belongs to, to obtain filtered time series intensity.

As for the interferometric coherence, for an arbitrary pixel i, the coherence coefficient is calculated as:

$$\gamma = \frac{\left| \sum_{i=1}^{K} S_1(i) S_2^*(i) \right|}{\sqrt{\sum_{i=1}^{K} |S_1(i)|^2 \sum_{i=1}^{K} |S_2(i)|^2}} \tag{1}$$

where $S_1(i)$ and $S_2(i)$ represent the corresponding complex values of the pixel i in the master and slave images, and K is the number of pixels in an SHP subset where the pixel i is located. Firstly, the coherence coefficient is calculated on the basis of each SHP subset for the purpose of removing the errors caused by the image texture. Secondly, a maximum-likelihood fringe rate algorithm [24] is applied to compensate for the interferometric fringe pattern. Finally, the bias in the coherence estimation is further mitigated using the second kind of statistical estimator proposed by [30].

After the implementation of the SHP DSI algorithm, we obtained the de-speckled backscatter intensity and bias-corrected coherence coefficient for each acquisition date. The units of VH and VV intensity were both converted decibels (dB). On this basis, a number of other Sentinel-1 features were generated as follows.

As one of the InSAR products, the master versus slave intensity ratio of the *VH* and *VV* polarization modes was calculated as

$$Intensity\_VH\_mst/slv\_ratio = \frac{int_{VH,mst}}{int_{VH,slv}}$$
$$Intensity\_VV\_mst/slv\_ratio = \frac{int_{VV,mst}}{int_{VV,slv}}$$
(2)

where $int_{VH,mst}$, $int_{VV,mst}$ and $int_{VH,slv}$, $int_{VV,slv}$ are respectively the backscatter intensity (in the unit of dB) of the master and slave images of each interferometric pair with the *VH*, *VV* polarization mode.

A statistic previously used in multi-temporal InSAR, i.e., amplitude dispersion is calculated as

$$amp\_disp_{VH} = \frac{std(amp_{VH})}{mean(amp_{VH})}, \quad amp\_disp_{VV} = \frac{std(amp_{VV})}{mean(amp_{VV})}$$
(3)

where amp_{VH} and amp_{VV} represent the amplitude of each Sentinel-1 acquisition of the *VH* and *VV* polarization mode.

A by-product of backscatter intensity, i.e., *VV* versus *VH* intensity ratio is calculated as

$$Intensity\_VV/VH\_ratio = \frac{int_{VV}}{int_{VH}}$$
(4)

where int_{VV} and int_{VH} are respectively the backscatter intensity (in the unit of dB) of each Sentinel-1 acquisition.

Another intensity-based ratio, the normalized ratio procedure between bands (NRPB) [31] is calculated as

$$NRPB = \frac{int_{VH} - int_{VV}}{int_{VH} + int_{VV}}$$
(5)

Table 7. Spectral indices calculated from Sentinel-2 data.

	Reference Spectral Indices	Formula
NDVI	Normalized Difference Vegetation Index	$NDVI = \frac{(B8-B4)}{(B8+B4)}$
NDWI	Normalized Difference Water Index	$NDWI = \frac{(B3-B8)}{(B3+B8)}$
Red-edge spectral indices		**Formula**
NDVIre1	Normalized Difference Vegetation Index red-edge 1 [32]	$NDVIre1 = \frac{(B8-B5)}{(B8+B5)}$
NDVIre1n	Normalized Difference Vegetation Index red-edge 1 narrow [6]	$NDVIre1n = \frac{(B8a-B5)}{(B8a+B5)}$
NDVIre2	Normalized Difference Vegetation Index red-edge 2 [6]	$NDVIre2 = \frac{(B8-B6)}{(B8+B6)}$
NDVIre2n	Normalized Difference Vegetation Index red-edge 2 narrow [6]	$NDVIre2n = \frac{(B8a-B6)}{(B8a+B6)}$
NDVIre3	Normalized Difference Vegetation Index red-edge 3 [32]	$NDVIre3 = \frac{(B8-B7)}{(B8+B7)}$
NDVIre3n	Normalized Difference Vegetation Index red-edge 3 narrow [6]	$NDVIre3n = \frac{(B8a-B7)}{(B8a+B7)}$
CIre	Chlorophyll Index red-edge [33]	$CIre = \frac{B7}{B5} - 1$
NDre1	Normalized Difference red-edge 1 [32]	$NDre1 = \frac{(B6-B5)}{(B6+B5)}$
NDre2	Normalized Difference red-edge 2 [34]	$NDre2 = \frac{(B7-B5)}{(B7+B5)}$
MSRre	Modified Simple Ratio red-edge [35]	$MSRre = \frac{(B8/B5)-1}{\sqrt{(B8/B5)+1}}$
MSRren	Modified Simple Ratio red-edge narrow [6]	$MSRren = \frac{(B8a/B5)-1}{\sqrt{(B8a/B5)+1}}$

In summary, we derived 10 subcategories of features from Sentinel-1 data. Adding features of different acquisition dates together, 142 Sentinel-1 features were produced, geocoded, and resampled to 10 m resolution.

2. Sentinel-2 Data Processing and Feature Extraction

Sentinel-2 data were firstly processed from Level-1C to Level-2A surface reflectance using a Sen2Cor—ESA Sentinel-2 Level 2A processor (https://step.esa.int/main/third-party-plugins-2/sen2cor/ (accessed on 30 October 2019)). Cirrus cloud correction was done during the Level-1C to Level-2A processing by Sen2Cor. Pixels covered by thick clouds were masked out using the opaque clouds band in the Level-2A product metadata. All spectral bands of 10 m and 20 m spatial resolution, including the red-edge bands (Band 5, Band 6, and Band 7) were included in the analysis. The 20 m resolution bands were resampled to 10 m resolution for further processing. Apart from the normalized difference vegetation index (NDVI) and normalized difference water index (NDWI), 11 red-edge indices (Table 7) were generated for each Sentinel-2 acquisition.

All geocoded Sentinel-1 features were aligned with the Sentinel-2 features. In summary, 33 subcategories, a total of 326 Sentinel-1 and Sentinel-2 features, were produced for subsequent analysis.

2.3.2. Optimal Feature Selection and Classification

1. Random Forest Classification

The random forest (RF) classifier [36] was deployed in this study for both cropland extraction and crop type classification. Random forest is a machine learning algorithm for classification, which consists of an ensemble of decision trees where each tree has a unit vote for the most popular class to classify. RF has been widely used for remote sensing classification, as it runs efficiently on large databases and performs well with high-dimensional features.

In this study, we used the RF classifier from the Scikit-learn Python module [37] for classification. The two key parameters, i.e., the number of trees and the number of features in each node, were chosen by analyzing the out-of-bag (OOB) errors. As a result, the number of trees was set as 450 in step 1 and 550 in step 2, and the number of maximum features in each tree was set as the square root of the total number of input features in both steps.

2. Optimal S1 and S2 Input Variable Selection Using Recursive Feature Increment (RFI) Method

Inspired by the backward feature elimination method [22], we proposed a forward feature selection approach to obtain the optimal combination of S1 and S2 features, which is referred to as the recursive feature increment (RFI) method.

Firstly, a feature importance ranking was obtained by the permutation importance of random forest calculated for each feature. The permutation importance assesses the impact of each feature on the accuracy of the random forest model. The general idea is to randomly permute the feature values and measure the decrease of the accuracy due to the permutation. In this study, we utilized the permutation importance function from the ELI5 package to estimate the feature importance.

Secondly, using the feature ranking with the most important feature placed at the top, RF classification was implemented by recursively considering bigger and bigger feature sets, starting from one feature and adding one at a time. For the feature set at each iteration, an RF classifier is parameterized and assessed using stratified k-fold cross-validation, with the mean overall accuracy (OA), mean kappa coefficient, and mean f1 score of all classes, as well as the f1 score of individual class recorded.

Finally, by analyzing the time series of the accuracy metrics, for example, the kappa coefficient, the feature set corresponding to the iteration achieving the highest accuracy is selected as the optimal combination of S1 and S2 features.

3. Accuracy Assessment

The performance of classification was assessed using stratified k-fold cross-validation. The entire sample dataset was split into k stratified "folds". The folds were made by preserving the percentage of samples for each class. In the case of unbalanced samples, this is to ensure that each fold is a good representative of the whole. For each fold in the sample dataset, the classification model was trained on k-1 folds and tested on the kth fold. This was repeated until each fold had served as the test set. Three accuracy metrics were used in this study, i.e., the overall accuracy (OA), Cohen's kappa coefficient, and F1 score of each class. All these metrics were averaged over the k folds.

The OA is calculated from the confusion matrix by adding the number of correctly classified sites together and dividing it by the total number of the reference sites. It is to test what proportion were mapped correctly out of all of the reference sites. From the confusion matrix, the producer's accuracy (PA, also referred to as recall) and the user's accuracy (UA, also referred to as precision) can also be calculated. UA (precision) is the ratio of correctly predicted positive observations to the total predicted positive observations; PA (recall) is the ratio of correctly predicted positive observations to all the observations in the actual class. The kappa coefficient is to evaluate if the classification does better than randomly assigning values by taking into account the possibility of the agreement occurring by chance. In the case of uneven class distribution, the F1 score is a more useful metric than OA, as it is a weighted average of precision and recall. The F1 score for each class is calculated as follows:

$$F_1 = 2\frac{precision \times recall}{precision + recall} \tag{6}$$

3. Results

3.1. Step 1: Cropland Extraction

A random forest (RF) classifier was applied to the Sentinel-1 and Sentinel-2 features for land cover classification, so as to generate a cropland mask. As the purpose of this step is to extract a cropland mask, we use the F1 score of cropland (Figure 3) as the accuracy metric for the RFI feature selection (Section 2.3.2). The top 114 features in the importance ranking were selected. The feature importance scores of the top six features are displayed in Figure 4, with corresponding boxplots of different land cover types shown in Figure 5.

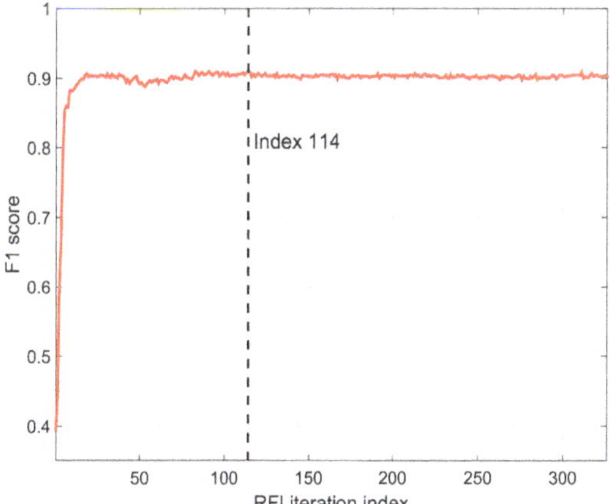

Figure 3. The F1 score of cropland recorded in each iteration of the recursive feature increment (RFI) process, reaching the maximum at the 114th iteration. Therefore, the top 114 features in the importance ranking were chosen as the optimal feature set for land cover classification in step 1.

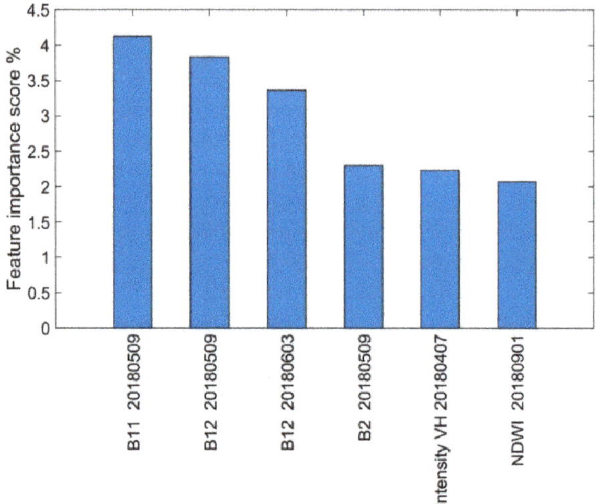

Figure 4. The feature importance scores of the top six features selected by the RFI method for cropland extraction, in descending order. Each importance score was normalized and converted to percentage. Each feature is named by its subcategory and acquisition time in the format of "yyyymmdd".

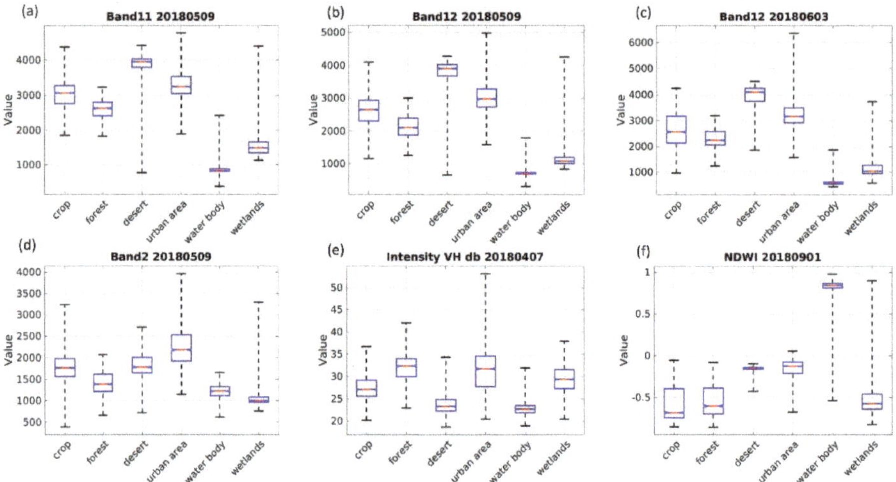

Figure 5. Boxplots of the top six most important features of different land cover types. Each feature is named by its subcategory and acquisition time in the format of "yyyymmdd".

In Figure 5, the top six features exhibit a good ability to distinguish different cover types, which reflects the reliability of the RFI feature selection method.

The random forest classifier was parameterized using the selected 114 features to generate a land cover map (Figure 6a). Then, a cropland mask (Figure 6b) was produced from the land cover classification results. The land cover classification accuracy was assessed using 10-fold cross-validation (Table 8).

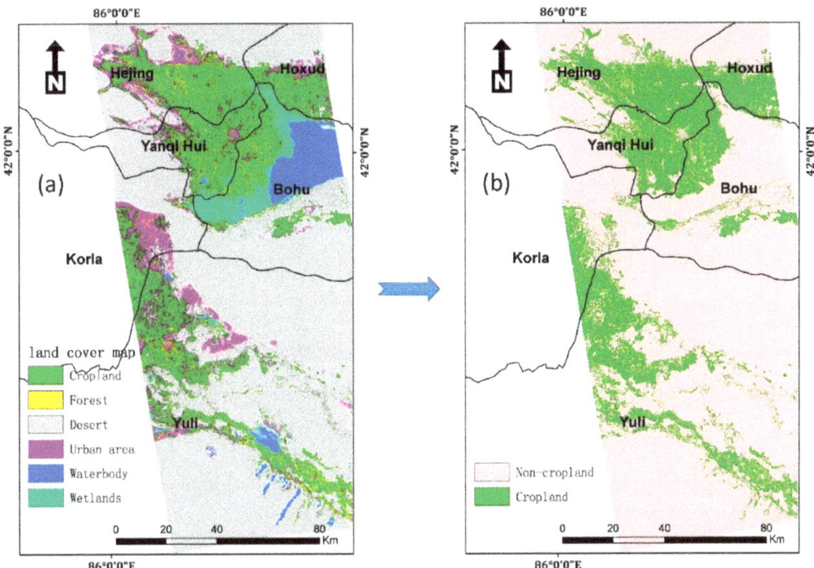

Figure 6. (**a**) Land cover classification map; (**b**) cropland mask extracted from the land cover classification results.

Table 8. Accuracy of step 1 land cover classification, assessed by stratified 10-fold cross-validation using the ground samples. OA: overall accuracy.

Mean OA (%)	Kappa Coefficient	F1 Score	
		Cropland	0.942
		Forest	0.835
		Desert	0.908
94.05%	0.927	Urban area	0.932
		Waterbody	0.995
		Wetlands	0.945

3.2. Step 2: Crop Type Mapping

3.2.1. Optimal S1 and S2 Feature Combination and Crop Type Mapping

Using the RFI feature selection method, we obtained the optimal combination of Sentinel-1 and 2 features for crop type discrimination. In this case, the kappa coefficient was used as the main accuracy metric to determine the final feature set. The mean OA achieved its maximum at the same time as the kappa coefficient, at the 113th iteration. Thus, in total, 113 features were selected. The mean OA and mean kappa coefficient averaged over the k-fold cross-validation are plotted in Figure 7. The feature importance scores of the top six features selected for crop type classification are shown in Figure 8.

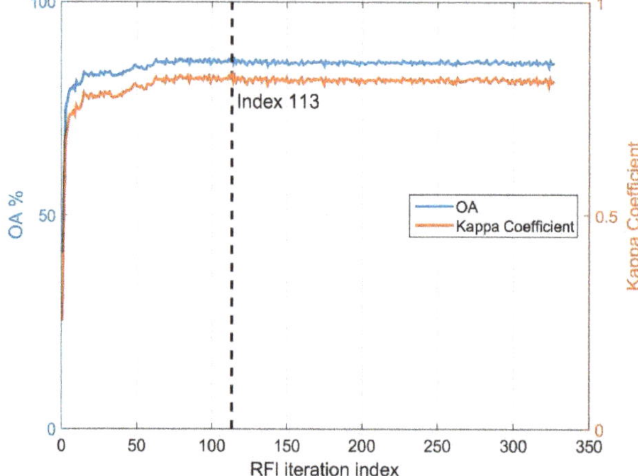

Figure 7. The mean overall accuracy (OA) and mean kappa coefficient of fivefold cross-validation recorded for each iteration during the RFI feature selection process.

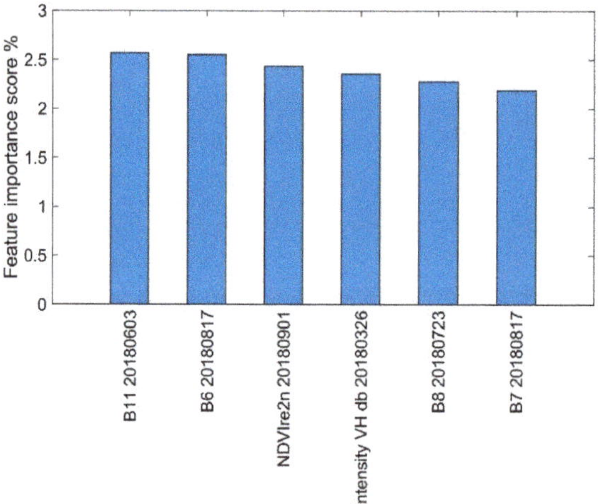

Figure 8. The feature importance score of the top six features selected by the RFI method for crop type classification, in descending order. Each importance score was normalized and converted to a percentage. Each feature is named by its subcategory and acquisition time in the format of "yyyymmdd".

The boxplots of the top six features (Figure 9) reveal the good separability of different crop types, which proves the rationality of the RFI feature selection results. For example, the intensity of the VH mode reflects a good ability to distinguish pear from other crop types in March. Pear values in Band 8 (NIR) in July show a clear distinction from other crops. Corn can be easily separated in the red-edge index NDVIre2n in September. Cotton is clearly distinguished in Band 11 (SWIR 1) in June.

Based on the optimal combination of S1 and S2 features, a random forest classification model was trained using the ground samples and applied to the whole study area to produce a crop distribution map (Figure 10). The classification accuracy was assessed by fivefold cross-validation (Table 9). Besides, the accuracy was also verified by a "one sample per field" method; that is, from an individual validation field (as listed in Table 3), only one sample was extracted. In this method, the validation samples are independent from each other. The corresponding accuracy metrics are listed in Table 10.

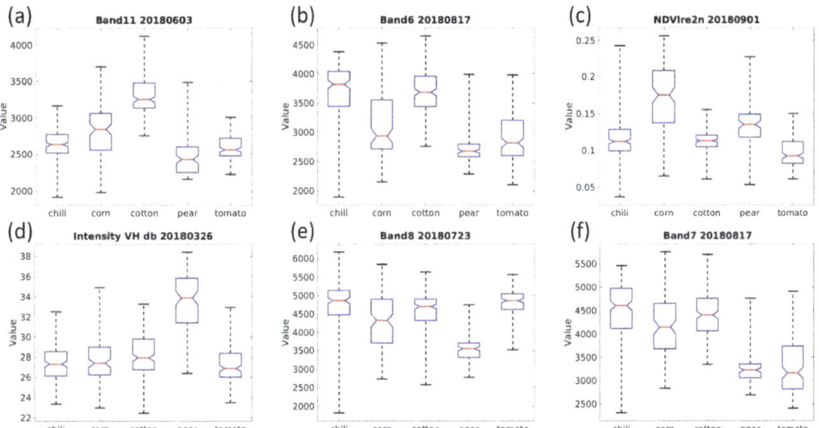

Figure 9. Boxplots of different crop types of the top six important features selected by the RFI approach. Each feature is named by its subcategory and acquisition time in the format of "yyyymmdd". (**a**) Boxplots of the band 11 values from the 3 June 2018 acquisition of Sentinel-2; (**b**) boxplots of the band 6 values from the 17 August 2018 acquisition of Sentinel-2; (**c**) boxplots of the NDVIre2n indices derived from the 1 September 2018 acquisition of Sentinel-2; (**d**) boxplots of the VH intensity values from the 26 March 2018 acquisition of Sentinel-1; (**e**) boxplots of the band 8 values from the 23 July 2018 acquisition of Sentinel-2; (**f**) boxplots of the band 7 values from the 17 August 2018 acquisition of Sentinel-2.

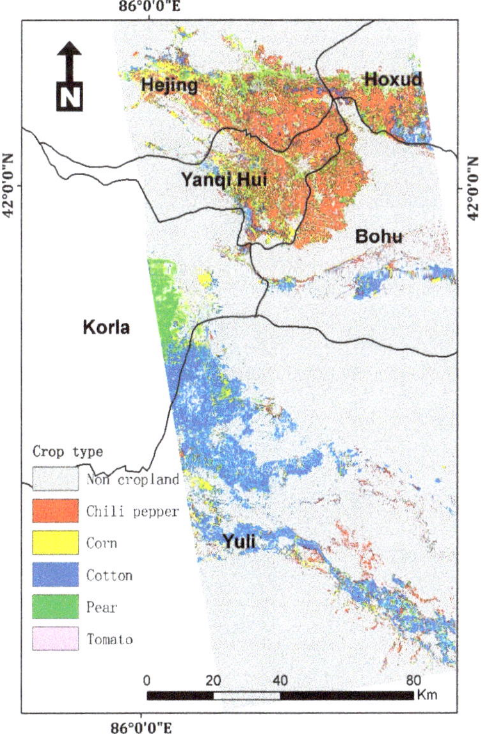

Figure 10. Crop distribution map of the study area derived from random forest (RF) classification using the combined Sentinel-1 and Sentinel-2 feature set.

Table 9. Accuracy of step 2 crop type classification using the optimal combination of S1 and S2 features, assessed by stratified fivefold cross-validation using the ground samples.

Mean OA (%)	Kappa Coefficient	F1 Score	
		Chili pepper	0.84
		Corn	0.71
86.98%	0.83	Cotton	0.97
		Pear	0.94
		Tomato	0.79

Table 10. Accuracy of step 2 crop type classification using the optimal combination of S1 and S2 features, assessed by stratified fivefold cross-validation using one sample from each validation field.

Mean OA (%)	Kappa Coefficient	F1 Score	
		Chili pepper	0.87
		Corn	0.69
83.22%	0.77	Cotton	0.91
		Pear	0.89
		Tomato	0.71

From Figure 10, we can see that pears are mainly planted in Korla County; the four counties near Bosten Lake (Bohu, Yanqi, Hejing, and Hoxud) are the main production areas of chili peppers and tomatoes; cotton is mostly cultivated in Yuli County on the edge of the Tarim Basin. Corn cultivation spreads across the whole study area.

3.2.2. Performance Comparison of SAR Features Filtered by Different Methods in Crop Classification

In this section, we compare the crop classification results obtained using SAR features processed by conventional de-speckling methods and the SHP DSI method. The input variables include all the SAR features mentioned in Section 2.3.1. In the conventional procedure, the intensity and intensity-derived features were de-speckled using a 7×7 refined Lee filter, and the coherence was estimated using a 7×7 sliding window. In the SHP DSI method, statistically homogeneous pixels (SHPs) were identified based on a 5×9 window using the intensity stack formed by 18 acquisitions. Then, the de-speckled intensity and accurately estimated coherence were obtained by the procedure described in Section 2.3.1. An example intensity of a subset area extracted from the original SAR intensity, intensity filtered by refined Lee method, and intensity filtered by the SHP DSI method are compared in Figure 11. Figure 12 shows the example coherence of the same subset area extracted from the interferometric coherence estimated respectively using a 7×7 sliding window and the SHP DSI method. The crop classification accuracy obtained by using SAR features processed by different filtering algorithms is listed in Table 11.

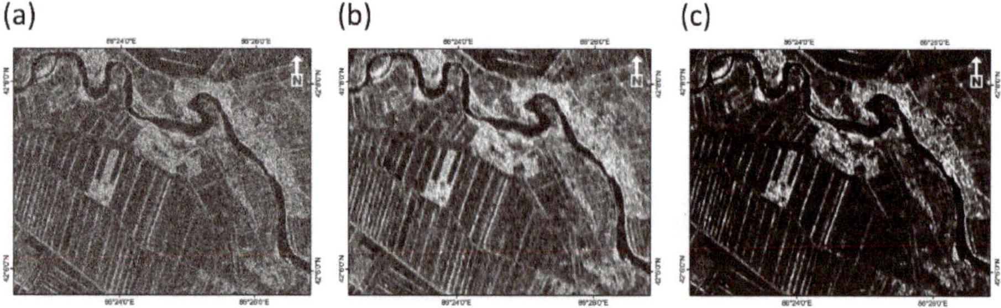

Figure 11. A subset of VH intensity on 26 March 2018 extracted from (**a**) Original synthetic aperture radar (SAR) intensity; (**b**) SAR intensity filtered by the refined Lee method; (**c**) SAR intensity filtered by the SHP distributed scatterer interferometry (DSI) method.

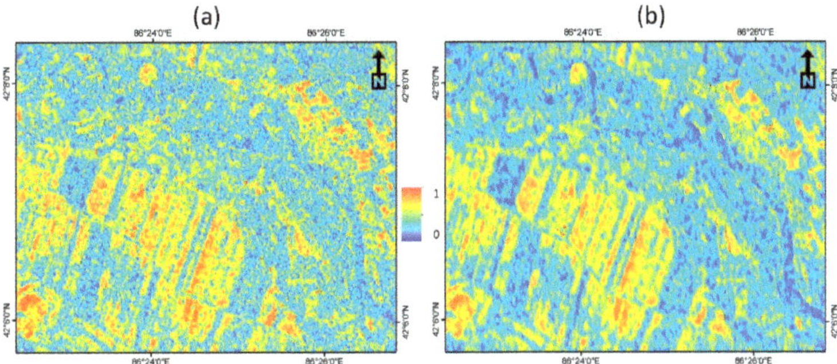

Figure 12. A subset of VV coherence on 7 April 2019 extracted from (**a**) Coherence coefficient estimated by a 7×7 sliding window; (**b**) Coherence coefficient estimated by the SHP DSI method with bias mitigation.

Table 11. Accuracy metrics of crop type classification using SAR features processed by different filters. SHP: statistically homogeneous pixel.

	Mean OA	Kappa Coefficient	F1-Score Chili	F1-Score Corn	F1-Score Cotton	F1-Score Pear	F1-Score Tomato
Original	60.20%	0.48	0.55	0.33	0.67	0.72	0.58
Refined Lee	73.21%	0.65	0.66	0.52	0.80	0.83	0.74
SHP DSI	79.46%	0.73	0.75	0.60	0.88	0.86	0.77

In Figure 11, it is evident that the speckles in original SAR intensity are both reduced by a refined Lee filter and SHP DSI filter, but the sharpness and details of each image is better preserved in (c). In Figure 12, it is found the coherence over a water body is overestimated in (a). This bias is reduced in (b) by using the SHP DSI method. It is clear the coherence in (b) exhibits less salt and pepper-like noise.

In Table 11, it is demonstrated that the SAR features filtered by the SHP DSI method yield the best accuracy in crop type classification. Compared with the results of using a refined Lee algorithm, the overall accuracy is increased by 6.25%, the kappa coefficient is improved by 0.08, and the F1-scores of each crop type are all improved.

3.2.3. Performance Comparison of Features from Different Sources in Crop Type Mapping

A comparison was conducted on the performance of crop type classification using four groups of features. The first group includes only Sentinel-1 features; the second group contains only conventional optical features exclusive of red-edge features; the third group comprises all Sentinel-2 features inclusive of red-edge contribution; the last group consists of the both SAR and optical features. The same feature importance ranking obtained in Section 3.2.1 was used for all groups. For each feature group, the optimal feature set was individually determined by the RFI feature selection method, using the kappa coefficient of each group as the accuracy metric. The mean OA and kappa coefficient reached the maximum at the same iteration index for each test, as summarized in Table 12. The corresponding F1 scores of different crop types are compared in Figure 13.

Table 12. Accuracy assessment of crop type discrimination using different groups of features. The mean overall accuracy (OA) and kappa coefficient were averaged over fivefold cross-validation in both multi-sample per field validation and one sample per field validation.

	Optimal Number of Features	Mean OA		Kappa Coefficient	
		Multi Samples per Field	One Sample per Field	Multi Samples per Field	One Sample per Field
Sentinel-1	133	79.46%	76.91%	0.73	0.69
Sentinel-2 without red-edge features	58	82.37%	79.80%	0.77	0.72
Sentinel-2	104	85.43%	81.64%	0.81	0.75
Sentinel-1 and Sentinel-2	113	86.98%	83.22%	0.83	0.77

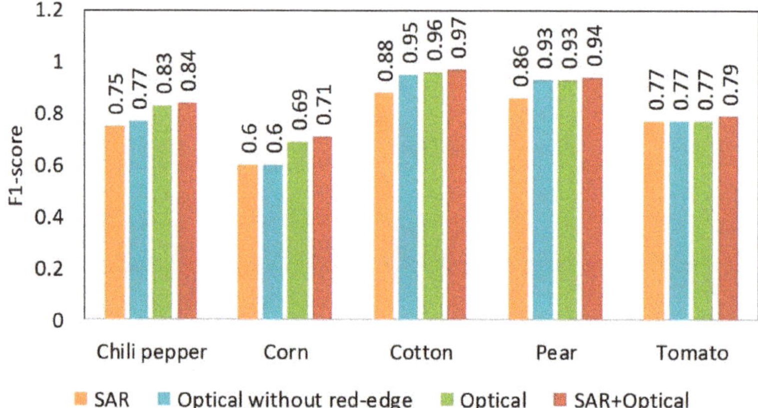

Figure 13. Comparison of crop type mapping results using different feature combinations. Four groups of features were tested, the first group contains only SAR (Sentinel-1) features, the second group contains only optical (Sentinel-2) features without the red-edge contribution, the third group has all of the Sentinel-2 features, the fourth group includes both SAR and optical (Sentinel-1 and 2) features.

In both Table 12 and Figure 13, it is evident that the combination of SAR and optical features achieved the best accuracy in crop type discrimination. The inclusion of red-edge bands and indices improved the mean OA and kappa coefficient respectively by 1.84% and 0.03 (3.06% and 0.04% in multi-samples per field validation), compared with the results using other Sentinel-2 features. By integrating the optical and SAR features, the mean OA and kappa coefficient were improved by 1.58% (1.55% in multi-samples per field validation) and 0.02% compared with the results only using Sentinel-2 features. In Figure 13, among the five crop types, the F1 score of chili pepper and corn were significantly improved by the inclusion of red-edge features.

4. Discussion

4.1. Importance of Features from Different Sources

The accumulated importance scores were calculated by subcategories (Figure 14), different data sources (Figures 15a and 16a), and the month of acquisition (Figures 15b and 16b).

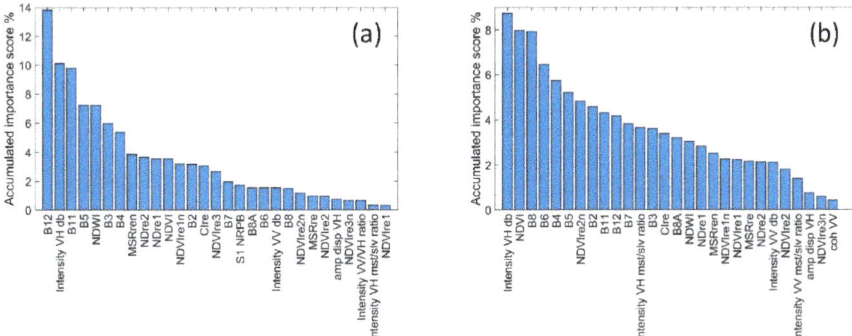

Figure 14. Accumulated feature importance scores of features selected for (**a**) step 1: cropland extraction and (**b**) step 2: crop type discrimination, calculated for each subcategory regardless of acquisition time. Each importance score was normalized and converted to a percentage. Each feature is named by its subcategory and acquisition time in the format of "yyyymmdd".

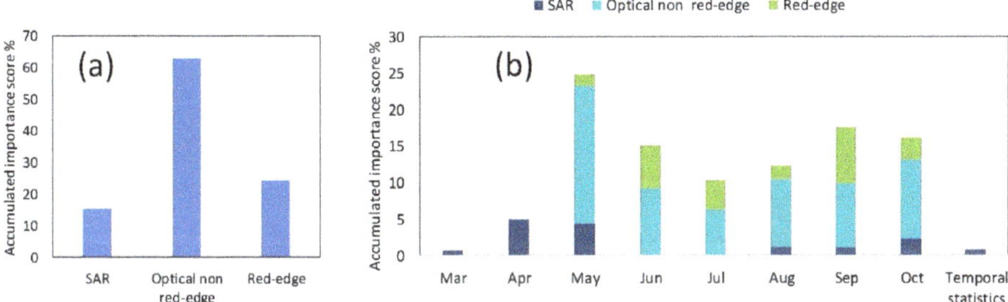

Figure 15. The accumulated importance scores of features selected in step 1: cropland extraction in three groups. The first group contains only Sentinel-1 features; the second group comprises only Sentinel-2 features exclusive of the red-edge features; the third group includes only the red-edge features. (**a**) Feature scores in the three groups calculated regardless of the acquisition time; (**b**) Feature scores in the three groups calculated for each month. The VH amplitude dispersion, as a single-phase feature, is plotted on the rightmost bar. Each importance score was normalized and converted to a percentage.

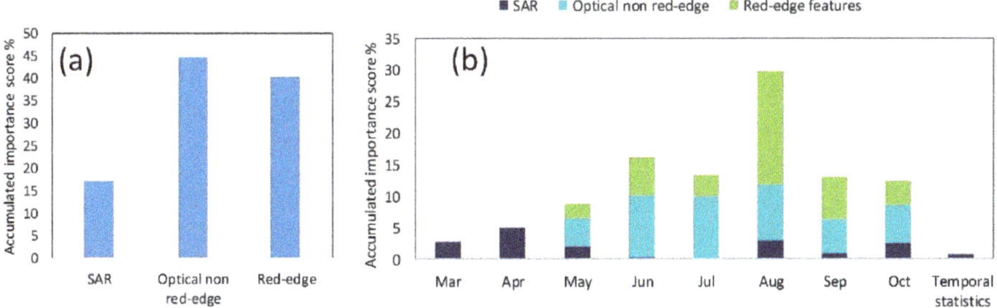

Figure 16. The accumulated importance scores of features selected in step 2: crop type discrimination in three groups. The first group contains only Sentinel-1 features; the second group comprises only Sentinel-2 features exclusive of the red-edge features; the third group includes only the red-edge features. (**a**) Feature scores in the three groups calculated regardless of the acquisition time; (**b**) Feature scores in the three groups calculated for each month. The VH amplitude dispersion, as a single-phase feature, is plotted on the rightmost bar. Each importance score was normalized and converted to a percentage.

Among the features selected for cropland extraction (Figure 14a), it is found that the SWIR 2 band (Band 12) contributed the most in the cropland extraction, followed by the Sentinel-1 VH intensity and Band 11 (SWIR 1), again the red-edge 1 (Band 5), and NDWI.

In the features selected for crop type discrimination (Figure 14b), the VH intensity exhibited the highest accumulated score for crop type classification, followed by the NDVI, Band 8 (NIR), Band 6 (red-edge 2), Band 4 (red), and Band 5 (red-edge 1). The optimal feature set comprises six subcategories of SAR features, 13 subcategories of red-edge features, and nine subcategories of conventional optical features. The contribution of features from different data sources will be explored in detail later.

It should be noted that the VH intensity held a significant share in both land cover classification and crop discrimination. The intensity ratio between bands, i.e., S1 NRPB and VV versus VH intensity ratio were only used in land cover classification. As for InSAR products, it is found that the VV coherence was selected in step 2. No coherence was selected in step 1. The coherence feature with the highest score in its subcategory, i.e., the VH coherence on 17 August 2018, ranked 118 in the RF permutation importance results in step 1. Its importance score was 0.27%, which is comparable to the score of 0.29% of the last chosen feature, NDre2 (red-edge) index on 23 July 2018. The amplitude dispersion of VH polarization, the master to slave intensity ratio in both polarization modes, were used in both step 1 and step 2.

All conventional optical features and the three red-edge spectral bands contributed to both the land cover classification. Regarding the red-edge indices, "NDVIre3" was only used in step 1; all the other 10 indices were selected for both cropland extraction and crop type identification.

As shown in Figures 15a and 16a, the three groups of features held similar proportions in both steps. The conventional optical features had the largest proportion in step 1 and step 2; the SAR features accounted for −15% in step 1 and −16% in step 2; the red-edge features accounted for −24% in step 1 and −22% in step 2.

Comparing Figures 15b and 16b, the features selected in step 1 and step 2 had significantly different temporal distribution. For land cover classification, the features in May had the largest proportion, followed by September, October, and June. Red-edge features contributed more in September, June, and July. In crop type classification, it is clear that features in August held the biggest share, of which red-edge features had a significant proportion. Compared with other months, the red-edge wavelengths performed the best in August, when most crops were ripening. This is likely associated with the sensitivity of the red-edge wavelengths to differences in the leaf structure and chlorophyll content of crops. Conventional optical features (visible, NIR, and SWIR wavelengths) were dominant from May to July and in October, and held a significant share in August and September.

In both steps, there were no optical features available in March and April, but some SAR features show good separability in the early season. SAR features in June and July were not selected for either of the steps. The proportion of SAR features re-increased in August and at the end of harvesting season in October. The amplitude dispersion was used in both steps, as a single-phase feature, plotted on the rightmost bar in both Figures 15b and 16b.

In addition, we examine the correlation between selected features in both step 1 and step 2, as shown in Figure 17. In both step 1 and step 2, most of the selected features showed a low correlation, as revealed by the histograms in (c) and (d).

4.2. The Contribution of SAR Features in Crop Type Discrimination

The electromagnetic radiation of the C-band cannot penetrate through vegetation cover. Therefore, the radiation from the Sentinel-1 C-band SAR sensor reflects the interaction between the radar signal and the ground surface cover, which explains the sensitivity of Sentinel-1 SAR to the vegetation cover.

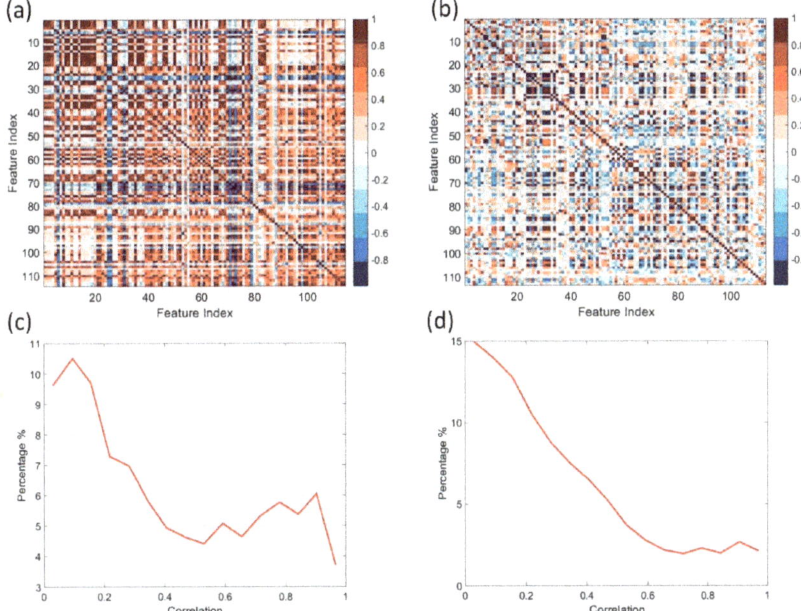

Figure 17. Heat maps showing correlation between selected features. (**a**) Correlation heat map of selected features in step 1; (**b**) correlation heat map of selected features in step 2. In both (**a**,**b**), the feature indexes follow the feature rankings obtained through the RF feature importance score. Correlation close to 1 or −1 indicates high positive or negative correlation between features. The diagonal elements in both correlation matrices are self-correlation coefficients, so they constantly equal one. (**c**) The histogram of correlation between selected features in step 1; (**d**) the histogram of correlation between selected features in step 2. In both (**c**,**d**), the histograms were calculated after removing the diagonal elements of the correlation matrices and converting a negative correlation coefficient to corresponding positive values. Thus, the '0' in the histogram indicates low correlation, while '1' indicates high correlation.

It has been reported that VH polarization is more sensitive to the structural and geometrical arrangements of plants [11,38]. This is in accordance with the result that VH intensity bands held the biggest proportion in the selected SAR features (also in all features) for crop type discrimination (Figure 14b). We examined the selected SAR features and found that 13 out of 22 features were in VH polarization mode, including six features of the VH intensity (Figure 18) and six features of the master versus slave VH intensity ratio (Figure 19), as well as the VH amplitude dispersion (Figure 20a). Pear is the most distinguishable crop in most of these features. This is highly likely because SAR intensity is sensitive to target structure, as the planting density and vegetation height in pear fields are significantly different from those in other crop fields. It is interesting to notice that the VV coherence on 7 April 2018 revealed a good ability to differentiate cotton from other crop types (Figure 20b). In Figure 20b, we can see the VV coherence is significantly low in the cotton field. According to local crop phenological calendars (Section 2.1), cotton is the crop that is most likely to be sown in early April. Other annual crops are usually sown or transplanted in mid-April or later. Thus, this is possibly related to the changes of the cotton field surface in the sowing season (early April) from bare soil to plastic-mulched land, which leads to a fast decrease of coherence in 12 days, while the other crop fields almost remain as before. In addition, chili pepper can be easier recognized from the VH intensity on 16 October 2018 (Figure 18f).

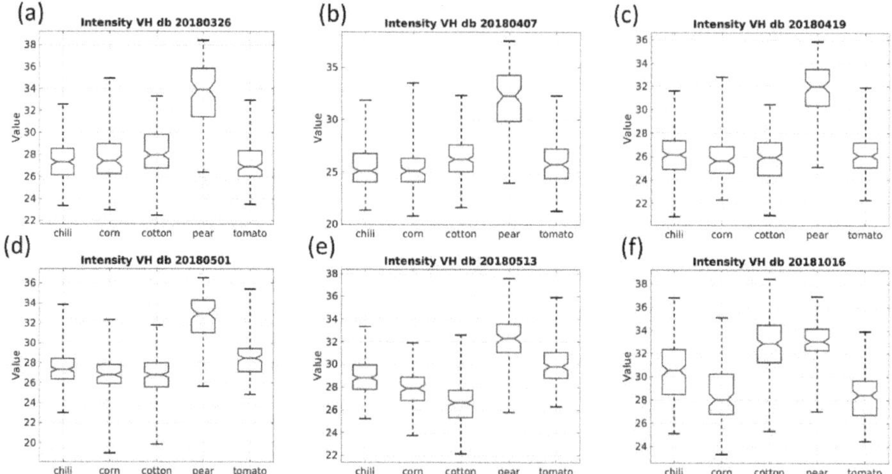

Figure 18. Boxplots of selected VH intensity features of different crop types, spanning the time from 26 March 2018 to 13 May 2018, as well as a feature on 16 October 2018. Each feature is named by its subcategory and acquisition time in the format of "yyyymmdd". In (**a–e**), pear values show an apparent distinction from other crops, while in (**f**), chili can be easier recognized.

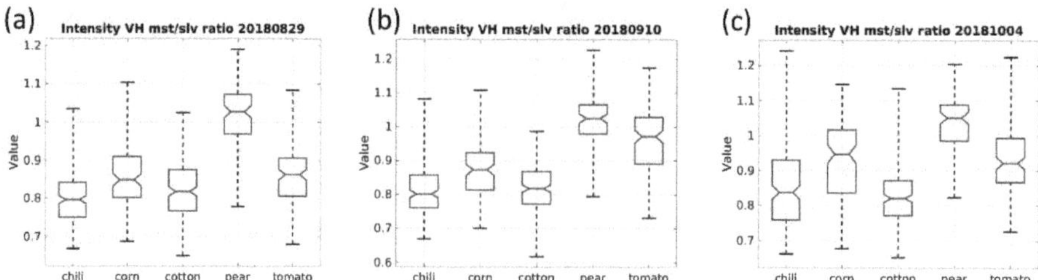

Figure 19. Boxplots of the top three scoring master versus slave VH intensity ratio of different crop types. Each feature is named by its subcategory and acquisition time in the format of "yyyymmdd". (**a**) Boxplots of the master vs. slave VH intensity ratio derived from the 29 August 2018 acquisition of Sentinel-1; (**b**) boxplots of the master vs. slave VH intensity ratio derived from 10 September 2018 acquisition of Sentinel-1; (**c**) boxplots of the master vs. slave VH intensity ratio derived from 4 October 2018 acquisition of Sentinel-1.

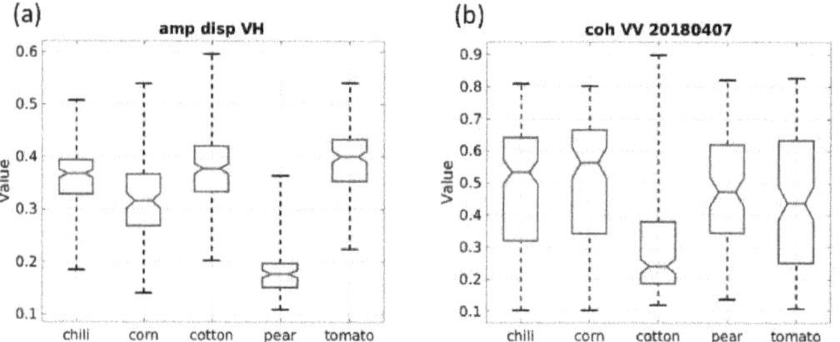

Figure 20. Boxplots of selected SAR features of different crop types. (**a**) Coherence coefficient in VV polarization mode; (**b**) amplitude dispersion in VH polarization mode. Each feature is named by its subcategory and acquisition time in the format of "yyyymmdd".

4.3. The Sensitivity of Red-Edge Wavelengths to Different Crop Types

Red-edge refers to the region of the sharp rise in vegetation reflectance between the red and NIR part of the electromagnetic spectrum. It is an important wavelength range that is sensitive to vegetation conditions. This research demonstrated the advantage of adding red-edge spectral bands and indices in the crop type discrimination, which increased the overall accuracy by 3.06% compared with the results only using conventional optical features. In Figure 14, among the selected red-edge features, Band 6 (red-edge 2), Band 5 (red-edge 1), and the NDVIre2n (red-edge index based on Band 6 and Band 8A) exhibited greater importance. As for red-edge indices, 10 out of 11 indices were selected for crop type discrimination (NDVIre3n was discarded by RFI method), with a number of indices showing a good capability to differentiate corn from other crops (Figure 21). This explains why the inclusion of red-edge features significantly improved the mapping accuracy of corn (Figure 13). Apart from corn, some re-edge indices also revealed good separability of tomato (Figure 21c–f,h) and pear (Figure 22). Several red-edge features show good capability to identify chili pepper (Figure 23). These findings reinforce the benefit of using red-edge wavelengths in crop type mapping.

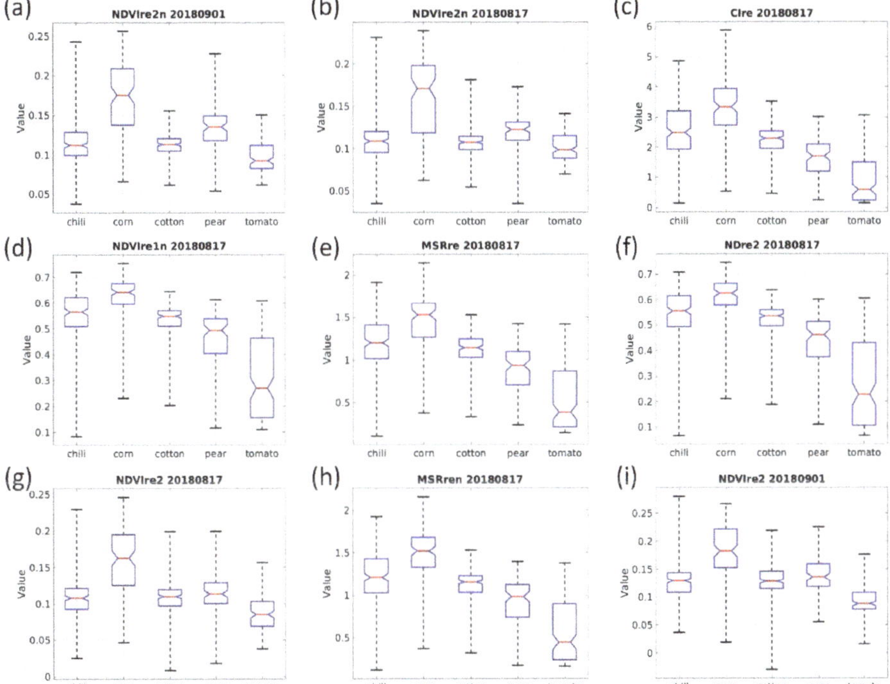

Figure 21. Boxplots of selected red-edge indices showing a good capability to distinguish corn from other crops. Each feature is named by its subcategory and acquisition time in the format of "yyyymmdd". (**a**) Boxplots of the NDVIre2n indices derived from the 1 September 2018 acquisition of Sentinel-2; (**b**) boxplots of the NDVIre2n indices derived from the 17 August 2018 acquisition of Sentinel-2; (**c**) boxplots of the CIre indices derived from the 17 August 2018 acquisition of Sentinel-2; (**d**) boxplots of the NDVIre1n indices derived from the 17 August 2018 acquisition of Sentinel-2; (**e**) boxplots of the MSRre indices derived from the 17 August 2018 acquisition of Sentinel-2; (**f**) boxplots of the NDre2 indices derived from the 17 August 2018 acquisition of Sentinel-2; (**g**) boxplots of the NDVIre2 indices derived from the 17 August 2018 acquisition of Sentinel-2; (**h**) boxplots of the MSRren indices derived from the 17 August 2018 acquisition of Sentinel-2; (**i**) boxplots of the NDVIre2 indices derived from the 01 September 2018 acquisition of Sentinel-2.

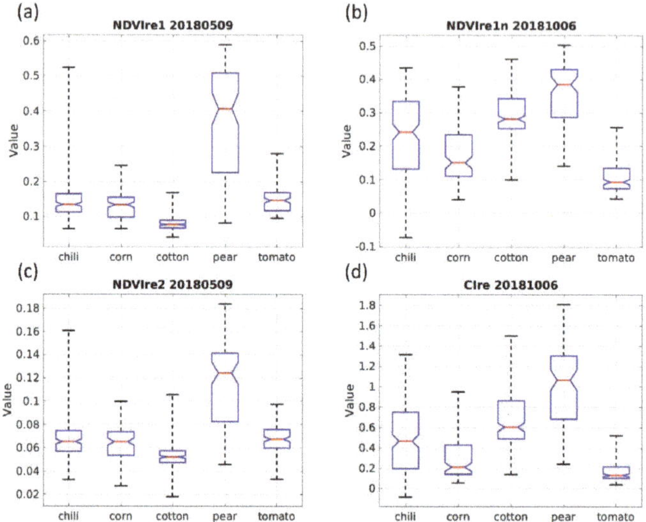

Figure 22. Boxplots of selected red-edge indices revealing a clear distinction between the pear and other crops. Each feature is named by its subcategory and acquisition time in the format of "yyyymmdd". (**a**) Boxplots of the NDVIre1 indices derived from the 9 May 2018 acquisition of Sentinel-2; (**b**) boxplots of the NDVIre1n indices derived from the 6 October 2018 acquisition of Sentinel-2; (**c**) boxplots of the NDVIre2 indices derived from the 9 May 2018 acquisition of Sentinel-2; (**d**) boxplots of the CIre indices derived from the 06 October 2018 acquisition of Sentinel-2.

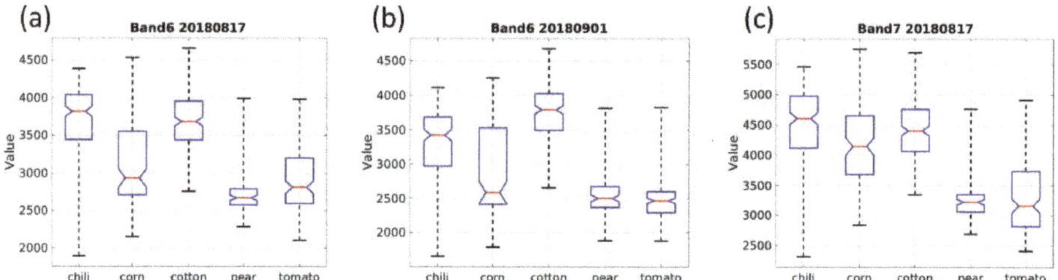

Figure 23. Boxplots of selected red-edge features showing good separability of chili pepper. Each feature is named by its subcategory and acquisition time in the format of "yyyymmdd". (**a**) Boxplots of the band 6 values from the 17 August 2018 acquisition of Sentinel-2; (**b**) boxplots of the band 6 values from the 1 September 2018 acquisition of Sentinel-2; (**c**) boxplots of the band 7 values from the 17 August 2018 acquisition of Sentinel-2.

From the top six scoring features for crop type mapping (Figure 8) and the accumulated importance score of each subcategory (Figure 14b), it can be inferred that the most important red-edge band is Band 6, and the most useful red-edge indices is the NDVIre2n (calculated using Band 6 and Band 8A (NIR narrow)). The top six features emphasized the significance of three red-edge features, i.e., Band 6 (red-edge 2), Band 7 (red-edge 3), and NDVIre2n (red-edge index based on Band 6). The accumulated importance ranking (Figure 14b) indicated the contribution of Band 6, Band 5, and NDVIre2n. It has been reported that the red-edge close to red wavelengths (Band 5) is mainly related to the difference in chlorophyll content, while the red-edge close to NIR (Band 7) is usually correlated to variations in the leaf structure [6]. The above results suggest that the separability of red-edge features lies in both the leaf structure and chlorophyll content of different crop species.

In a nutshell, SAR features distinguish crops based on the structural and geometrical arrangements of plants or changes of the crop field surface, while optical features rely on the multi-spectral information to differentiate crops. In experiments, the red-edge wavelengths exhibited good separability between different crop types, due to their sensitivity to the variations in chlorophyll content and leaf structure. Therefore, the combination of SAR and optical integrated the physical and spectral characteristics of crops, improving the performance of crop classification.

With the development of deep learning technology, powerful convolutional networks, such as U-net [39], can be explored for crop type mapping in future work. U-net has the potential to support multi-dimensional input variables. By implementing U-net with multi-temporal, multi-sensor features, it is expected that a higher level of accuracy can be achieved in crop type identification.

5. Conclusions

This research proposed a method for the synergistic use of Sentinel-1 and Sentinel-2 features for oasis crop type mapping through a case study in a smallholder farming area in Northwest China. First of all, a SHP DSI algorithm was introduced for the de-speckling of SAR intensity and accurate estimation of interferometry coherence to improve the quality of SAR features. It was demonstrated that the use of the SHP DSI method improved the crop classification accuracy by 6.25% when only using SAR features. A variety of SAR features and optical features were derived from multi-temporal Sentinel-1 and Sentinel-2 images, including several InSAR products and red-edge spectral bands and indices. Secondly, based on the permutation importance of the random forest classifier, a recursive feature increment feature selection method was proposed to obtain the optimal combination of Sentinel-1 and Sentinel-2 features for cropland extraction and crop type classification. Finally, a crop distribution map was generated with an overall accuracy of 83.22% and kappa coefficient of 0.77. The contribution of SAR and optical features were explored thoroughly. Among all the Sentinel-1 features, the VH intensity held the biggest proportion, indicating the better sensitivity of VH polarization to vegetation changes. It was also noted that some of the InSAR products, such as the VH amplitude dispersion, the master versus slave intensity ratio, and the VV coherence in early April revealed good separability of certain crop types. As for Sentinel-2 features, we demonstrated the merits of using red-edge spectral bands and indices in oasis crop type mapping. The inclusion of red-edge features improved the crop classification OA by 1.84% compared with only using conventional optical features. This proves the superiority of Sentinel-2 data due to the increased spectral resolution. A comparison was conducted on the performance of oasis crop classification using four combinations of features. The results indicated that the integration of SAR and optical features achieved the best performance. We concluded that the integration of time series S1 and S2 imagery is advantageous, and thanks to the free, full, and open data policy, it can be further explored in the vast majority of regions for the monitoring of crop status.

Author Contributions: Methodology & investigation & writing—original draft preparation, L.S.; supervision, J.C.; Writing—review and editing, J.C., S.G. and X.D.; Project administration, S.G.; Validation, Y.H. All authors have read and agreed to the published version of the manuscript.

Funding: This research has been supported in part by the National Key Research and Development Program of China [Grant No. 2017YFB0504203], the Strategic Priority Research Program of Chinese Academy of Sciences [Grant No. XDA19030301], and the National Natural Science Foundation of China [Grant No. 41801360, 41601212].

Acknowledgments: The authors thank J. Liu, Q. Liu, X. Zheng, Y. Shen, and Y. Xiong from SIAT, and Wang and her team from Fuzhou University for their work in the field investigation.

Conflicts of Interest: The authors declare no conflict of interest.

References

1. Wardlow, B.D.; Egbert, S.L. Large-area crop mapping using time-series MODIS 250 m NDVI data: An assessment for the US Central Great Plains. *Remote Sens. Environ.* **2008**, *112*, 1096–1116. [CrossRef]
2. Arvor, D.; Jonathan, M.; Meirelles, M.S.P.; Dubreuil, V.; Durieux, L. Classification of MODIS EVI time series for crop mapping in the state of Mato Grosso, Brazil. *Int. J. Remote Sens.* **2011**, *32*, 7847–7871. [CrossRef]

3. Pan, Y.; Li, L.; Zhang, J.; Liang, S.; Zhu, X.; Sulla-Menashe, D. Winter wheat area estimation from MODIS-EVI time series data using the Crop Proportion Phenology Index. *Remote Sens. Environ.* **2012**, *119*, 232–242. [CrossRef]

4. Azzari, G.; Lobell, D.B. Landsat-based classification in the cloud: An opportunity for a paradigm shift in land cover monitoring. *Remote Sens. Environ.* **2017**, *202*, 64–74. [CrossRef]

5. Korets, M.A.; Ryzhkova, V.A.; Danilova, I.V.; Sukhinin, A.I.; Bartalev, S.A. Forest Disturbance Assessment Using Satellite Data of Moderate and Low Resolution. In *Environmental Change in Siberia*; Springer Press: Dordrecht, The Netherlands, 2010; pp. 3–19.

6. Fernández-Manso, A.; Fernández-Manso, O.; Quintano, C. SENTINEL-2A red-edge spectral indices suitability for discriminating burn severity. *Int. J. Appl. Earth Obs. Geoinf.* **2016**, *50*, 170–175. [CrossRef]

7. Lambert, M.-J.; Traoré, P.C.S.; Blaes, X.; Baret, P.; Defourny, P. Estimating smallholder crops production at village level from Sentinel-2 time series in Mali's cotton belt. *Remote Sens. Environ.* **2018**, *216*, 647–657. [CrossRef]

8. Chatziantoniou, A.; Psomiadis, E.; Petropoulos, G. Co-Orbital Sentinel 1 and 2 for LULC mapping with emphasis on wetlands in a mediterranean setting based on machine learning. *Remote Sens.* **2017**, *9*, 1259. [CrossRef]

9. Ferrant, S.; Selles, A.; Le Page, M.; Herrault, P.-A.; Pelletier, C.; Al-Bitar, A.; Mermoz, S.; Gascoin, S.; Bouvet, A.; Saqalli, M. Detection of irrigated crops from Sentinel-1 and Sentinel-2 data to estimate seasonal groundwater use in South India. *Remote Sens.* **2017**, *9*, 1119. [CrossRef]

10. Silva, W.F.; Rudorff, B.F.T.; Formaggio, A.R.; Paradella, W.R.; Mura, J.C. Discrimination of agricultural crops in a tropical semi-arid region of Brazil based on L-band polarimetric airborne SAR data. *ISPRS J. Photogramm. Remote Sens.* **2009**, *64*, 458–463. [CrossRef]

11. Zeyada, H.H.; Ezz, M.M.; Nasr, A.H.; Shokr, M.; Harb, H.M. Evaluation of the discrimination capability of full polarimetric SAR data for crop classification. *Int. J. Remote Sens.* **2016**, *37*, 2585–2603. [CrossRef]

12. Skriver, H.; Mattia, F.; Satalino, G.; Balenzano, A.; Pauwels, V.R.N.; Verhoest, N.E.C.; Davidson, M. Crop classification using short-revisit multitemporal SAR data. *IEEE J. Sel. Top. Appl. Earth Obs. Remote Sens.* **2011**, *4*, 423–431. [CrossRef]

13. McNairn, H.; Kross, A.; Lapen, D.; Caves, R.; Shang, J. Early season monitoring of corn and soybeans with TerraSAR-X and RADARSAT-2. *Int. J. Appl. Earth Obs. Geoinf.* **2014**, *28*, 252–259. [CrossRef]

14. Jia, K.; Li, Q.; Tian, Y.; Wu, B.; Zhang, F.; Meng, J. Crop classification using multi-configuration SAR data in the North China Plain. *Int. J. Remote Sens.* **2012**, *33*, 170–183. [CrossRef]

15. Ban, Y. Synergy of multitemporal ERS-1 SAR and Landsat TM data for classification of agricultural crops. *Can. J. Remote Sens.* **2003**, *29*, 518–526. [CrossRef]

16. Sun, C.; Bian, Y.; Zhou, T.; Pan, J. Using of multi-source and multi-temporal remote sensing data improves crop-type mapping in the subtropical agriculture region. *Sensors* **2019**, *19*, 2401. [CrossRef]

17. Torbick, N.; Chowdhury, D.; Salas, W.; Qi, J. Monitoring Rice Agriculture across Myanmar Using Time Series Sentinel-1 Assisted by Landsat-8 and PALSAR-2. *Remote Sens.* **2017**, *9*, 119. [CrossRef]

18. Van Tricht, K.; Gobin, A.; Gilliams, S.; Piccard, I. Synergistic use of radar sentinel-1 and optical sentinel-2 imagery for crop mapping: A case study for Belgium. *Remote Sens.* **2018**, *10*, 1642. [CrossRef]

19. Lee, J. Sen Polarimetric SAR speckle filtering and its implication for classification. *IEEE Trans. Geosci. Remote Sens.* **1999**, *37*, 2363–2373.

20. Millard, K.; Richardson, M. On the importance of training data sample selection in random forest image classification: A case study in peatland ecosystem mapping. *Remote Sens.* **2015**, *7*, 8489–8515. [CrossRef]

21. Mahesh, P.; Foody, G.M. Feature selection for classification of hyperspectral data by SVM. *IEEE Trans. Geosci. Remote Sens.* **2010**, *48*, 2297–2307.

22. Guyon, I.; Weston, J.; Barnhill, S.; Vapnik, V. Gene selection for cancer classification using support vector machines. *Mach. Learn.* **2002**, *46*, 389–422. [CrossRef]

23. Jiang, M.; Ding, X.; Hanssen, R.F.; Malhotra, R.; Chang, L. Fast Statistically Homogeneous Pixel Selection for Covariance Matrix Estimation for Multitemporal InSAR. *IEEE Trans. Geosci. Remote Sens.* **2014**, *53*, 1213–1224. [CrossRef]

24. Jiang, M.; Ding, X.; Li, Z. Hybrid Approach for Unbiased Coherence Estimation for Multitemporal InSAR. *IEEE Trans. Geosci. Remote Sens.* **2014**, *52*, 2459–2473. [CrossRef]

25. Ramsey, E.W., III; Lu, Z.; Rangoonwala, A.; Rykhus, R. Multiple baseline radar interferometry applied to coastal land cover classification and change analyses. *GIScience Remote Sens.* **2006**, *43*, 283–309. [CrossRef]
26. Thiel, C.J.; Thiel, C.; Schmullius, C.C. Operational large-area forest monitoring in Siberia using ALOS PALSAR summer intensities and winter coherence. *IEEE Trans. Geosci. Remote Sens.* **2009**, *47*, 3993–4000. [CrossRef]
27. Jin, H.; Mountrakis, G.; Stehman, S.V. Assessing integration of intensity, polarimetric scattering, interferometric coherence and spatial texture metrics in PALSAR-derived land cover classification. *ISPRS J. Photogramm. Remote Sens.* **2014**, *98*, 70–84. [CrossRef]
28. Zhang, M.; Li, Z.; Tian, B.; Zhou, J.; Zeng, J. A method for monitoring hydrological conditions beneath herbaceous wetlands using multi-temporal ALOS PALSAR coherence data. *Remote Sens. Lett.* **2015**, *6*, 618–627. [CrossRef]
29. Lee, J.-S.; Pottier, E. *Polarimetric Radar Imaging: From Basics to Applications*; CRC Press: Cleveland, OH, USA, 2009.
30. Abdelfattah, R.; Nicolas, J.-M. Interferometric SAR coherence magnitude estimation using second kind statistics. *IEEE Trans. Geosci. Remote Sens.* **2006**, *44*, 1942–1953. [CrossRef]
31. Filgueiras, R.; Mantovani, E.C.; Althoff, D.; Fernandes Filho, E.I.; da Cunha, F.F. Crop NDVI Monitoring Based on Sentinel 1. *Remote Sens.* **2019**, *11*, 1441. [CrossRef]
32. Gitelson, A.; Merzlyak, M.N. Spectral reflectance changes associated with autumn senescence of *Aesculus hippocastanum* L. and Acer platanoides L. leaves. Spectral features and relation to chlorophyll estimation. *J. Plant Physiol.* **1994**, *143*, 286–292. [CrossRef]
33. Gitelson, A.A.; Gritz, Y.; Merzlyak, M.N. Relationships between leaf chlorophyll content and spectral reflectance and algorithms for non-destructive chlorophyll assessment in higher plant leaves. *J. Plant Physiol.* **2003**, *160*, 271–282. [CrossRef]
34. Barnes, E.M.; Clarke, T.R.; Richards, S.E.; Colaizzi, P.D.; Haberland, J.; Kostrzewski, M.; Waller, P.; Choi, C.; Riley, E.; Thompson, T. Coincident detection of crop water stress, nitrogen status and canopy density using ground based multispectral data. In Proceedings of the Fifth International Conference on Precision Agriculture, Bloomington, MN, USA, 16–19 July 2000; Volume 1619.
35. Chen, J.M. Evaluation of vegetation indices and a modified simple ratio for boreal applications. *Can. J. Remote Sens.* **1996**, *22*, 229–242. [CrossRef]
36. Breiman, L. Random forests. *Mach. Learn.* **2001**, *45*, 5–32. [CrossRef]
37. Pedregosa, F.; Varoquaux, G.; Gramfort, A.; Michel, V.; Thirion, B.; Grisel, O.; Blondel, M.; Prettenhofer, P.; Weiss, R.; Dubourg, V. Scikit-learn: Machine learning in Python. *J. Mach. Learn. Res.* **2011**, *12*, 2825–2830.
38. Moran, M.S.; Hymer, D.C.; Qi, J.; Kerr, Y. Comparison of ERS-2 SAR and Landsat TM imagery for monitoring agricultural crop and soil conditions. *Remote Sens. Environ.* **2002**, *79*, 243–252. [CrossRef]
39. Ronneberger, O.; Fischer, P.; Brox, T. U-net: Convolutional Networks for Biomedical Image Segmentation. In Proceedings of the International Conference on Medical Image Computing and Computer-Assisted Intervention, Munich, Germany, 5–9 October 2015; Springer Press: Cham, Switzerland, 2015; pp. 234–241.

Review

Applications of UAV Thermal Imagery in Precision Agriculture: State of the Art and Future Research Outlook

Gaetano Messina and Giuseppe Modica *

Dipartimento di Agraria, Università degli Studi Mediterranea di Reggio Calabria, Località Feo di Vito, I-89122 Reggio Calabria, Italy; gaetano.messina@unirc.it
* Correspondence: giuseppe.modica@unirc.it; Tel.: +39-0965-1694261

Received: 16 March 2020; Accepted: 6 May 2020; Published: 8 May 2020

Abstract: Low-altitude remote sensing (RS) using unmanned aerial vehicles (UAVs) is a powerful tool in precision agriculture (PA). In that context, thermal RS has many potential uses. The surface temperature of plants changes rapidly under stress conditions, which makes thermal RS a useful tool for real-time detection of plant stress conditions. Current applications of UAV thermal RS include monitoring plant water stress, detecting plant diseases, assessing crop yield estimation, and plant phenotyping. However, the correct use and interpretation of thermal data are based on basic knowledge of the nature of thermal radiation. Therefore, aspects that are related to calibration and ground data collection, in which the use of reference panels is highly recommended, as well as data processing, must be carefully considered. This paper aims to review the state of the art of UAV thermal RS in agriculture, outlining an overview of the latest applications and providing a future research outlook.

Keywords: unmanned aerial vehicles (UAVs); remote sensing (RS); thermal UAV RS; thermal infrared (TIR); precision agriculture (PA); crop water stress monitoring; plant disease detection; yield estimation; vegetation status monitoring

---

1. Introduction

Remote sensing (RS) is the practice of obtaining information regarding an object, an area, or a phenomenon through the analysis of images acquired by means of a device that does not make physical contact with them [1]. RS investigations are mostly based on the development of a deterministic relationship between the amount of reflected emitted or backscattered electromagnetic energy, in specific bands or frequencies, and the chemical, biological, and physical characteristics of the studied phenomena. Conventionally, RS has been associated with satellites and manned aircrafts equipped with different sensors [2]. In recent years, considerable technological developments—mainly concerned with the use of unmanned aerial vehicle (UAV) platforms—have been registered [3]. Born for military applications, UAVs have become a common tool for use in geomatics for data acquisition in various research and operational fields: regional security, monitoring of structures and infrastructures, monitoring of archeological sites, environmental monitoring, application in agriculture, etc. [2,4,5]. Indeed, low-altitude RS by means of UAVs is one of the most powerful tools in precision agriculture (PA).

The International Society of Precision Agriculture (ISPAG) defines PA as "a management that gathers, processes and analyzes temporal, spatial and individual data and combines it with other information to support management decisions according to estimated variability for improved resource use efficiency, productivity, quality, profitability and sustainability of agricultural production" (www.ispag.org, last access 18 April 2020). The general stages of PA practice involve data collection, field variability mapping through the use and development of algorithms, decision making, and management practice [6].

In this framework, thermal RS has been found to be a promising tool by measuring surface temperature [7]. In the last years, thermal sensors have also gained popularity because of the improvements in sensor technology and the reduction in costs. The surface temperature detected by thermal sensors has been found to be a rapid response variable to monitor plant growth and stress [7,8]. Indeed, the temperature is a fundamental environmental variable that plays an essential role in plant physiological processes, such as transpiration, leaf water potential, and photosynthesis [9].

The potentialities of the application of thermal UAV RS concern several mapping and monitoring issues, such as yield estimation, plant phenotyping [10,11], plant water stress detection [12,13], and plant disease detection [14], as explained in the further paragraphs of this review. When considering these last two applications, the capability to identify, using sensors mounted on UAVs, crop stress, and health before said crops are significantly damaged could be a crucial goal.

Concerning water stress, it has been found that thermal images show a correlation between minor changes in water stress that are undetectable by the normalized difference vegetation index (NDVI) [15]. In this regard, temperature-based indices represent a fast and practical way to evaluate and estimate crop water status, indicating plants' water content [16]. Among these indices, the crop water stress index (CWSI) [17], which is the most used, is often exploited to monitor plants' water status and, therefore, for the management of irrigation [18]. Such indices have been applied to different crops, both tree and herbaceous, such as olives [19], grapevines [20], sugar-beet [21], maize [22], rice [23], wheat and cotton [24].

However, the correct interpretation and use of a thermal image are based on basic knowledge of the nature of thermal radiation [1]. Thermal images can be influenced by various factors, including the characteristics of the thermal camera, meteorological conditions, and several sources of emitted and reflected thermal radiation [7]. For this reason, the aspects that are related to calibration, ground data collection, the step in which the use, and the measurement of the temperature of reference panels are recommended [25], and data processing must be carefully carried out for correct temperature retrieval.

The objective of this review is to cover the state of the art of thermal UAV RS in the framework of PA, outlining an overview of the latest applications. The structure of this paper is as follows. In Section 2, a brief description of the basic principles of thermography is provided. Section 3 deals with the characteristics of thermal cameras; the calibration, data collection, and data processing aspects are discussed. Section 4 is devoted to the main thermal RS in PA. Finally, in Section 5, we provide the present—as well as future—challenges in thermal RS.

2. Basic Principles of Thermography

Thermal RS uses the information at the emitted radiation in the thermal infrared (TIR) range (Figure 1) of the electromagnetic (EM) spectrum [26]. This information is converted into temperature [7]. Two categories can be distinguished within the IR region (0.7–100 µm), namely, the reflected-IR (0.7–3.0 µm) and TIR (3.0–100 µm). Generally, all of the elements of the landscape, such as vegetation, soil, water, and people, emit TIR radiation in the 3.0–14 µm portion of the EM spectrum [27].

In this range of the EM spectrum, part of the IR energy is transmitted to the Earth's surface through two so-called atmospheric windows that range from 3 to 5 µm and from 8 to 14 µm [28] (Figure 2). Almost all of the radiation between 5 and 8 µm is absorbed by atmospheric gasses (i.e., water, carbon dioxide (CO_2), and ozone (O_3) molecules), as indicated in Figure 2 [27] (Figure 2).

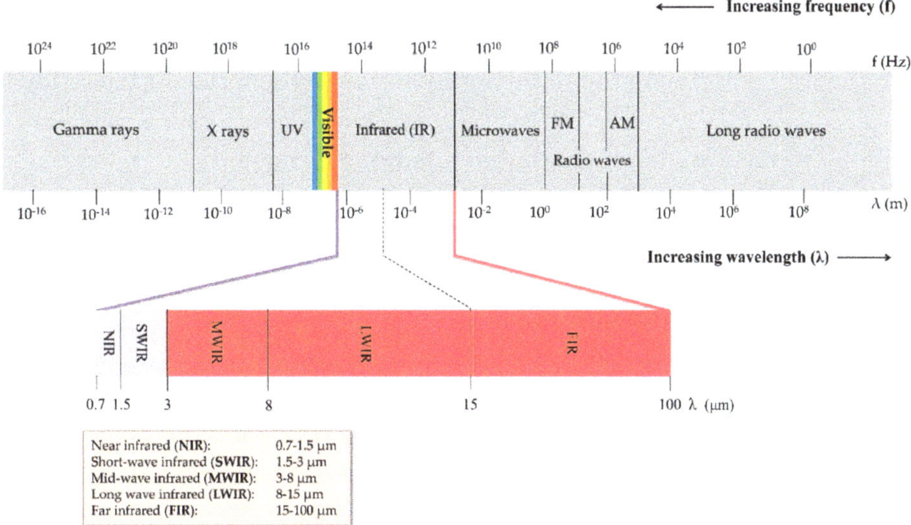

Figure 1. The electromagnetic (EM) spectrum. In evidence, the infrared (IR) region, in which the reflected-IR (0.7–3.0 µm) and the emitted-IR (3.0–100 µm) are further detailed.

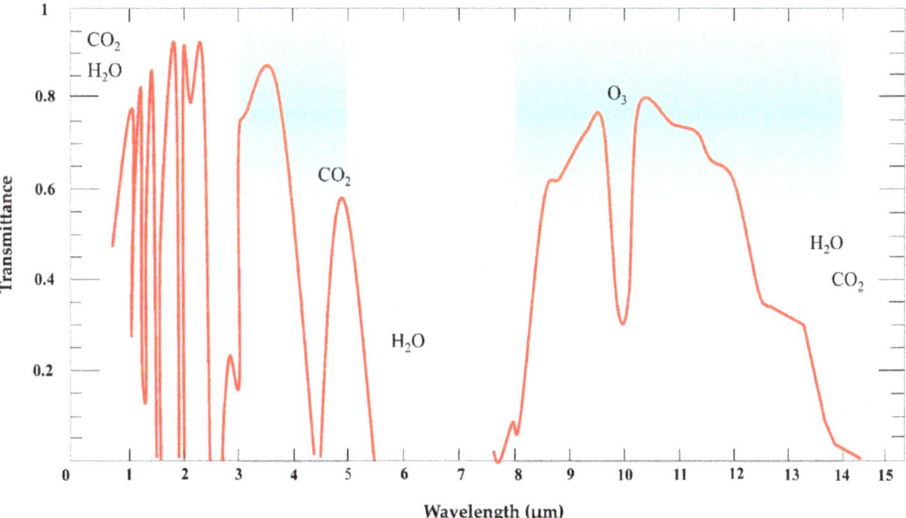

Figure 2. Atmospheric transmittance in the thermal region with typical absorption bands induced by gasses and water (modified from [29]).

The physical laws of Planck, Wien, Stefan–Boltzmann, and Kirchhoff make it possible to better understand the behavior of EM radiation. According to Planck, each energy element (Q) is proportional to its frequency (ν), while the Planck's constant h is used to adjust this relationship (Equation (1)):

$$Q = h\nu \tag{1}$$

When considering that the frequency of the wave (v) is directly proportional to the speed of light (c) and inversely proportional to its length (λ), Equation (1) can be rewritten as follows Equation (2):

$$Q = \frac{hc}{\lambda} \tag{2}$$

In other words, the energy of a quantum is inversely proportional to its wavelength. Thus, the longer the wave (larger wavelength), the lower its energy. Wien's and Boltzmann's laws describe the relationship of a black body's radiations (i.e., an ideal object that absorbs and reemits all of the incident energy) and the wavelength of the maximum emission with a black body's temperature [30].

Wien's displacement law explains the relationship between the true temperature of a black body, being expressed in degrees Kelvin, and its peak spectral exitance or dominant wavelength [27]. As the temperature increases, its maximum exitance shifts towards shorter wavelengths [31]:

$$\lambda_{\max} = \frac{b}{T} \tag{3}$$

Given that the Wien's constant b is equal to 2898 μm K, this formula indicates the wavelength at which the maximum radiant spectral exitance can be obtained. It is possible to observe such effects in nature. For example, a body with a very high absolute temperature like the sun (about 6000 K) has a λ_{max} and, thus, a peak of emission in the visible part of the spectrum [32]. The result of the above formula is useful to indicate what should be the measurement range of the sensor used to measure the radiation emitted by a given body [27].

The Stefan–Boltzmann law states that the emittance of a black body is proportional to the fourth power of its absolute temperature (Equation (4)):

$$E = \sigma T^4 \tag{4}$$

where E represents the spectral radiant exitance expressed in W·m^{-2}, σ is the Stefan–Boltzmann constant, and T is the absolute temperature [K]. The formula clearly shows that the total EM radiation that is emitted by a black body is a function of its absolute temperature. Therefore, the radiation emitted by a body increases as its temperature increases, as mentioned above.

Kirchhoff's law (1860) (Equation (5)) states that, at a given wavelength, the emittance of a body is equal to its absorption capacity, which is:

$$\varepsilon = \alpha \tag{5}$$

where ε represents the emittance and α the absorbance.

This is frequently formulated as "good absorbers are good emitters and good reflectors are poor emitters" [27]. The principle of energy conservation is defined by the following equation (Equation (6)):

$$\varepsilon + \rho + \tau = 1 \tag{6}$$

where ρ is the reflection and τ is the transmission. When considering that most objects are opaque to TIR radiation, the above equation becomes (Equation (7)):

$$\varepsilon + \rho = 1 \tag{7}$$

Materials with a high ε absorb a large quantity of incident energy and radiate large quantities of energy, while materials with low ε absorb and radiate less energy (Kirchhoff, 1860) [33]. All bodies with a temperature above absolute zero are characterized by random movement—i.e., the kinetic heat, whose measure is the kinetic temperature T_{kin} [27]. Besides, an object emits energy as a function of temperature, and the emitted energy is used in order to determine its radiant temperature T_{rad} [28]. Although there is a strong positive linear correlation between T_{kin} and T_{rad}, T_{rad} is lower than T_{kin}

due to emissivity (ε) [34]. For this reason, the temperature that is measured by a sensor (T_{rad}) will always be lower than the real temperature (T_{kin}) (Equation (8)) [31]. It follows from this, as explained by Kirchhoff's law (Equation (8)), that:

$$T_{rad} = \varepsilon^{\frac{1}{4}} T^{kin} \tag{8}$$

The emissivity is the ratio between the radiation that is emitted by the surface and the radiation emitted by a black body at the same temperature [35]. Because the radiance of any real body, at the same temperature, is always lower than that of a black body (equal to 1), its emissivity has a value between 0 and 1 [27]. Anybody absorbs and emits radiation less effectively at a given temperature when compared to a black body, as explained by Planck's law. Practically, the sensed temperatures of materials with low emissivity appear to be much lower than those of nearby objects with the same temperatures, making the T_{kin} assessment less precise [31]. Many factors influence emissivity: color, chemical composition, surface roughness, moisture content, field of view, viewing angle, spectral wavelength, etc. [27,35,36]. The emissivity of materials is difficult to measure, although it is more or less constant in the region of the EM that ranges from 8 to 14 µm. The emissivity of vegetation ranges from 0.96 to 0.99, while that of soil is around 0.89, and the emissivity of the water is 0.99 (Table 1). In particular, focusing on vegetation, emissivity values of the leaves are available in literature for many plant species [37,38] and some studies were dedicated to the determination of emissivity in herbaceous and tree species [39,40].

Table 1. The emissivity of different surfaces over the range of 8–14 µm [1,36].

Material	Average Emissivity (ε)
Healthy vegetation	0.96–0.99
Dry vegetation	0.88–0.94
Wood	0.93–0.94
Sand	0.90
Dry soil	0.92
Wet soil	0.95–0.98
Water	0.98–0.99
Snow	0.98–0.99

Regarding the intrinsic characteristics of a body, which affect its emissivity, thermal capacity (or heat capacity, measured in J kg^{-1}K^{-1}) measures the quantity of heat energy that is necessary for a body to increase its temperature by one degree—the lower the heat capacity, the less energy required. The thermal conductivity (measured in W m^{-1} K^{-1}), which measures the rate at which heat passes through a material, is another important parameter. This capacity is greater the higher the value of such a parameter in a material. These parameters can be integrated into the thermal inertia (expressed in J m^{-2} K^{-1} s$^{-\frac{1}{2}}$), which measures a body's tendency to change in temperature or the rate of heat transfer between two substances put in contact. This concept is of great importance in the field of TIR, because the capacity of a body to quickly change its temperature (if the thermal inertia is low) depends on this parameter.

3. Thermal Cameras and Unmanned Aerial Vehicles (UAVs)

Thermal cameras typically carry a sensor that detects the infrared radiation emitted by a body, displaying its temperature in a digital radiometric image. Two types of thermal cameras are currently available: scanning devices that allow for capturing a point or a line and those with a two-dimensional infrared focal plane array. The latter, which allows for capturing all of the elements of an image at once, is faster if combined with a better image resolution [41] and it is the most commonly used [42]. A further distinction can be made between thermal and photon (or quantum) detectors. The latter convert directly absorbed EM into a change in the distribution of electric energy in a semiconductor by varying the concentration of free charge carriers [42] and require a cooling system, usually made while using helium or liquid nitrogen at a temperature of −196 °C [27]. The cryocooler lowers the temperature of the sensor with the aim to reduce the thermally induced noise to a level that is lower than that of the signal coming from the image [43]. The more efficient the cooling system, the more accurate the measurements of the instrument, which makes cooled thermal cameras more precise and accurate [44], thus allowing the detection of the slightest temperature differences in the image. These cameras generally work in the mid-wavelength infrared (MWIR) region (3–8 μm), where the thermal contrast is high [42]. The thermal radiance of a target is easier to detect once it is distinguishable (higher or lower) from the background [43]. Unfortunately, cooled sensors, besides being large and expensive, have higher energy consumption and they are not suitable for UAVs because of their weight [45]. Thermal detectors are less sensitive (±0.1 °C) and they are slower than quantum detectors, but, instead, have the advantage of not requiring cooling elements [46]. Different types of uncooled detectors are available, all of which are made of different and unconventional materials; the three most common types are VOx (vanadium oxide), amorphous silicon (α-Si) microbolometers, and ferroelectrics [47]. The operating principle of thermal detectors is based on the conversion of absorbed EM radiation into thermal energy [48]. Ferroelectric detectors are based on the ferroelectric phase transition that can be detected in some dielectric materials. The microbolometer is a resistor organized in arrays—called focal plane arrays, which are made up of VOx and α -Si—which is composed of a thermometer, integrated on a micro-bridge, and an adsorber. Temperature increases, which are caused by absorption of IR radiation, determine large fluctuations of its electrical resistance, which can be converted into electrical signals and processed in order to generate an image [42,49–52], whose geometric resolution depends on the number of detectors. The low values of temperature differences equivalent to noise in uncooled thermal sensors, which reach 20 m °C, allows it to be used in applications where previously only cooled thermal sensors could be used [44]. Furthermore, especially due to the rapid development of micro- and nanotechnology, microbolometers have become cheaper and more efficient [53]. The lenses are made of germanium—a shiny semi-metal, chosen because of its transparency to infrared radiation—and they are reflective of visible radiation [42]. Several parameters characterize a thermal camera; among these, worthy of a mention, include the temperature range measured, usually between −20 and +120 °C, and the thermal sensitivity, which determines the minimum value of temperature difference (ΔT) detectable in an image and generally ranges from 40 to 20 m °C for uncooled and cooled devices, respectively [42]. Regarding the geometric resolution, to date, it is still very low when compared to RGB cameras (currently, higher image resolutions ranging from 320 × 240 to 640 × 512 pixels), while the spectral resolution normally ranges from 7 to 14 μm [12]. As for the price of cameras, their cost can vary from €1.000 to more than €10,000, depending on sensor resolution and radiometric calibration accuracy [54,55].

Cooled thermal infrared cameras have the widest use in satellite and aerial RS, thanks to their thermal sensitivity and precision [56]; on the other hand, these devices are larger and more expensive, also in terms of energy, than the uncooled ones [57]. In contrast, uncooled thermal cameras are normally mounted on UAVs (Figure 3a,b), because they are smaller, lighter, and have lower energy consumption [58].

The limited payload is perhaps the principal limit of UAVs, together with the limited battery life, which affects the duration of the flight. For example, it is not unlikely that 45 minutes or more pass between the first and the final image of the dataset when larger areas need to be covered and several flights are required [59].

Payload integration considerably changes between cooled and uncooled TIR cameras, where different ventilation modes are a key factor in image quality; thus, generally, an uncooled microbolometer is preferred for its weight benefits [60].

Figure 3. DJI Phantom 4 pro (**a**) and (**b**) DJI Inspire 1, both equipped with a FLIR uncooled thermal camera (photos taken by the authors).

3.1. Camera Calibration and Data Collection

Low-cost thermal cameras, which are generally not radiometrically calibrated, can only provide information regarding relative temperature differences [54]. The data provided by such instruments are represented in the form of raw digital numbers (DNs) that express radiance. Even when using radiometrically calibrated UAV cameras, it is not easy to derive accurate surface temperature measurements, because their low accuracy is due to the presence of uncooled microbolometers [54]. The sensitivity and, therefore, the accuracy of each microbolometer, is affected by both the temperature of the focal plane array [61] and that of the other components of the thermal camera, body, and lens, such as to create a weak signal-to-noise ratio [53].

Besides this, other causes make it necessary to calibrate a thermal camera, as shown in [53] and [57]. The atmosphere affects the quality of the thermal image as it absorbs and emits IR. The atmospheric effects on UAVs at low-altitude measurements are considered to be negligible when compared to airborne or satellite measurements [31].

The effect of relative humidity, air density, and altitude can only be avoided by making measurements within about 10 m or less of the target's surface [62]; under different conditions, their effects must be taken into account. Meteorological conditions can have an indirect effect on the temperature measurements of uncooled thermal cameras [59]. In the field, some recommendations are useful for reducing the effects; for instance, the critical factors for data acquisition are the time of day, weather, and the knowledge of the surrounding environment [60]. Regarding the applications in PA, midday has been identified as the best time for flying in terms of thermal accuracy [63] and the reduction of background effects [64].

UAV thermal surveys should always be done in the absence of clouds, rain, snow, smoke, dust, or any other darkening agents, because all of these reduce atmospheric transition, as well as change the temperature of the background [62]. Before the flight, after powering up, the temperature of the camera sensor is expected to stabilize from 20 minutes [25] up to one hour [65]. It is also essential to perform a test flight to allow the camera to acclimatize to local weather conditions, as well as to protect the thermal camera with a casing (to reduce air temperature effects on the sensor) when mounted on a quadcopter UAV [66].

Besides, temperature calibration in the field is an essential step. The temperature of the presumed hottest and coldest objects within the area of image acquisition should be measured for ground truth calibration [31]. Field calibration should be performed while using the thermal measurements of target surfaces during the flight [20,67]. These are placed in the area of the study where the temperature is measured using thermocouples [59] or infrared thermo-radiometers [25], combined with data-loggers capable of recording the temperature throughout the flight. Targets can be made using black and white polypropylene panels, which represent the thermal extremes of the study site, and whose size is such as to represent several homogeneous pixels in the infrared thermal image [25].

In our experience, as is the case in Messina et al.'s [68] work, wet and dry reference surfaces were used during the thermal UAV survey by taking images of them before and after the flights, as indicated by [59]. Reference surfaces were placed close to UAV take-off and landing points, and their temperature was measured while using a handheld infrared thermometer (FLIR E6) (Figure 6b).

The temperature was measured at three moments of the flight: At take-off, during, and end of every flight. Four reference surfaces were used: three dry panels (black, grey, and white) and one created delimiting a square piece of the ground while using circular targets and covered with aluminum (Figure 4).

Figure 4. The temperature reference targets used during thermal unmanned aerial vehicle (UAV) surveys in an onion crop field in Calabria (Italy) (photo taken by the authors).

Portions of dry and wet soil can also be considered as targets, as described in the following paragraph. In addition to these, as in the case of multispectral sensors, there are ground control points (GCP)s—points that are marked on the ground that have a known geographical position—which improve the positioning and accuracy of the mapping outputs.

To make GCPs more visible in thermal images, they should have a low emissivity when compared to the adjacent vegetation and other bodies [69].

In the case of thermal measurements, GCPs are made of aluminum, exploiting the low emissivity that makes it appear as a cold object in the images [70] (Figure 5b). This GCP was made using aluminum [71]. Our proposed GCPs were made using 50 cm × 50 cm white polypropylene panels and covering two quadrants while using aluminum sheets (Figure 5a). Black cardboard was used to partially cover the other two quadrants to locate the point and also to make the GCP clearly visible and usable in multispectral surveys (Figure 5c). Figure 5 shows the detectability of the target in thermal and spectral images.

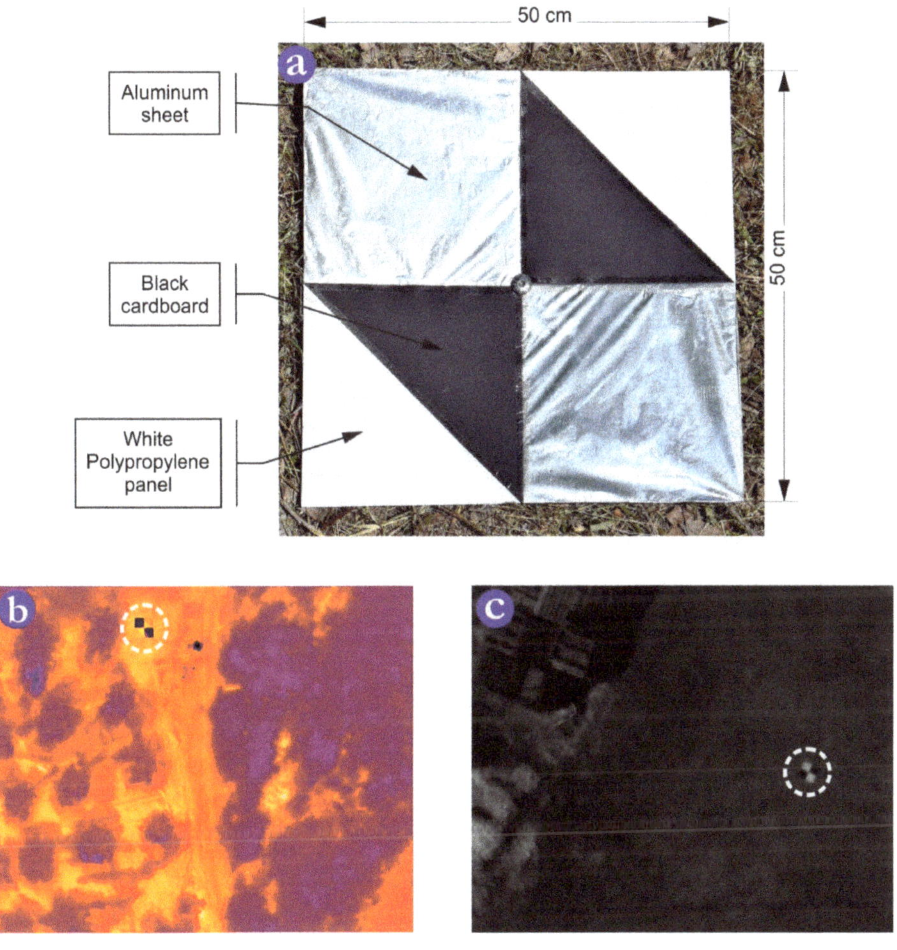

Figure 5. Example of a homemade target for ground control points (GCPs) designed to be easily detected in thermal as well as in multispectral UAV surveys. (**a**). The GCP target (white dashed circle) as it appears in (**b**) thermal and (**c**) near-infrared (NIR) multispectral images.

A further effect of the weather conditions to be taken into account is that of attenuating the thermal radiance by the atmosphere [59]. The sensor registers, for every pixel, an at-sensor radiance ($L_{at\text{-}sensor}$), being expressed in Wm^{-2}, which is determined by the following formula (Equation (9)) [72]:

$$L_{at\text{-}sensor} = \tau\, L_{surf} + L_{atm} \tag{9}$$

In Equation (9), τ represents the atmospheric transmittance, while L_{atm}, measured in Wm^{-2}, is the upwelling thermal radiation, created as a result of particles in the atmosphere, both depending on the distance of the sensor from the object and the water content in the atmosphere [59]. Parameters τ and L_{atm} can be obtained by exploiting one of the theoretical atmospheric models, such as MODTRAN

(MODerate resolution atmospheric TRANsmission) [73], which are widely used for thermal data that are acquired from satellites [7]. This model allows for the estimation of atmospheric emission, thermal scattering, and solar scattering by incorporating the effects of molecular absorbers and scatterers, aerosols, and clouds, when considering the wavelengths from the ultraviolet region to the far-infrared [36].

An example of MODTRAN application in UAV is shown by [65], which shows that it allows for obtaining the surface temperature by applying atmospheric correction methods that are based on entering the model as input data, i.e., data that are related to local atmospheric conditions. In the application on UAV data using MODTRAN, it is essential to use local measurements of temperature, relative humidity, and atmospheric pressure acquired from nearby meteorological stations (Figure 6a) placed in the field [39–43].

Figure 6. (**a**) The weather station used during the thermal surveys on an onion field. (**b**) Handheld infrared thermometer used to measure the temperature of reference surfaces (photos taken by the authors).

3.2. Data Processing

Several steps are needed to acquire georeferenced UAV imagery, starting from the flight planning that generates a navigation file guiding the UAV to capture images with the required overlaps and geometric resolution automatically. Thermal imagery can be processed with structure-from-motion (SfM) algorithm, which does not always work properly [59], as reported in [74,75]. Indeed, SfM can be unable to align thermal imagery, making it necessary to mosaic separate images and manually geo-reference them using GCPs. These issues are due to the reduced information contained in the thermal image, which complicates the identification of the common features that are needed for bundle adjustment [59]. Indeed, if compared to other types of imagery, such as RGB, thermal imagery is characterized by a lower geometric resolution and contrast and by a stronger optical distortion [69]. In addition to increasing vertical and horizontal overlaps, the presence of an incorporated multispectral (or RGB) camera, which is characterized by a higher geometric resolution, could help the alignment step of the photogrammetric process. A framework to process thermal imagery is shown in Turner et al.'s [76] study. Firstly, image pre-processing provided the removal of blurry imagery and the subsequent conversion of images to a 16-bit file format. In this way, all of the images have the same dynamic scale, and a specific temperature value corresponds to the same digital number (DN) in all images. Secondly, image alignment was executed using GPS log files and the imagery's time stamps. Finally, the spatial image was co-registered to RGB images. Some improvements of this framework were provided in [59] in order to optimize the alignment and processing of thermal images by exploiting the increased information contained in the RGB data. Although recent UAV models are provided with real-time kinematic global navigation satellite systems (RTK-GNSS) that are able to reach a centimeter planimetric accuracy, to ensure high positional precision, the placement of GCP within the surveyed site is generally expected. Imagery must be geometrically aligned (orthorectification), calibrated, and corrected, while taking atmospheric effects into account before final orthomosaics are obtained [7] (Figure 7).

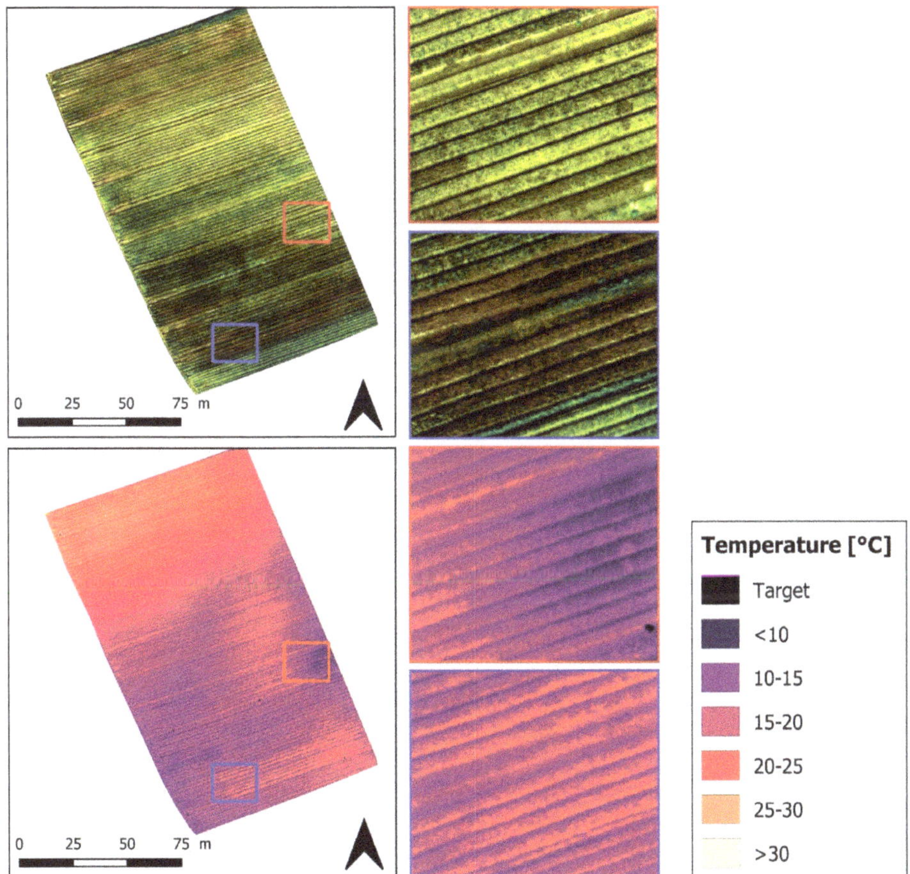

Figure 7. RGB (top) and thermal orthomosaic (bottom) images of an onion field (data collected by the authors).

4. Thermal UAV Imagery Analysis and Applications in Precision Agriculture (PA)

4.1. Crop water Stress Monitoring

Several detection systems have been developed for the production of specialty crops worldwide thanks to technological advances [77]. The collection of accurate information on the spatial variability of fields is essential in this context. A field's variability is affected by several factors, including crop yield and water content, and sensors that can be used to detect these factors include thermal cameras.

Plant water stress is one of the main critical factors of abiotic stress, as it limits the development of crops [78,79]. Therefore, among the applications of thermal UAV RS in agriculture, the identification of water stress from plant temperature data is of great importance, with irrigation resource management being a key issue for PA. The use of UAVs in the study of plant water status requires measurements of stem water potential (ψ_s) and stomatal conductance (g_s), which are the most common physiological water stress indicators that are used to determine crop water status [15,76,80,81]. Both indicators can be measured in the field while using a pressure chamber and porometers, respectively, although the interpolation of such local observations is not straightforward.

The use of thermal images allows for detecting the water stress conditions of a plant, because, at the foliar level, stomata closure reduces transpiration and evaporative cooling, which results in an increased temperature of the leaf [82,83]. This increase in temperature can be detected by thermal cameras; therefore, thermal images can provide spatially continuous information concerning the water status of plants in a wider area than that obtained by local measurements [84,85]. The temperature of the plant is not only regulated by the water supply, but also by the micrometeorological conditions [86]. Among the climatic factors, atmospheric humidity plays a crucial role and, in environments with humid climates, cloudiness also becomes a critical factor [87]. Several indices have been developed in the past decades with the aim of compensating for the variation of these conditions [86].

The CWSI was developed precisely for that purpose. Jackson and colleagues formulated the CWSI while using portable IR thermometers on herbaceous crops [84,88], developing a normalized index in order to overcome environmental effects that may affect the relationship between plant temperature and water stress. This index, which can assume values that range between 0 and 1, results in being directly proportional to the water stress level of many species of interest. The CWSI is based on the normalization of the canopy–air temperature difference with evaporative demand by means of the vapor pressure deficit (VPD) of the air [14]. The formula to calculate the CWSI is as follows, according to the methodology that was proposed by Idso et al. [17] (Equation (10)):

$$CWSI = \frac{(T_C - T_a) - (T_C - T_a)LL}{(T_C - T_a)UL - (T_C - T_a)LL} \tag{10}$$

Where $T_c - T_a$ is the canopy–air temperature difference, while LL refers to the $T_c - T_a$ values for the lower limit and UL for the upper limit. The normalization related to the VPD considers the $T_c - T_a$ difference of a canopy under two boundary conditions: (a) a lower limit when it transpires at its potential rate (i.e., under well-watered conditions), and (b) an upper limit under no transpiration. The lower limit is described by linear regression between $T_c - T_a$ and the VPD, which is known as the non-water-stressed baseline (NWSB). The NWSB is empirically derived by measuring the leaf–air temperature difference for a well-watered crop in the experimental environment and provides the lowest temperature difference likely in that environment. Once the NWSB, (Tc − Ta)LL can be calculated by solving the baseline equation for the current VPD. The upper limit (Tc − Ta)UL, which is a constant, is obtained by solving the same NWSB equation for a hypothetical slightly negative VPD. It represents the vapor pressure difference that is generated by the temperature differential (Tc − Ta) when VPD is 0 [17]. The upper limit value is close to the NWSB interception *a* (depending on the temperature) and only corresponds to *a* when *a* is equal to 0 [89]. This method is site-dependent. A second approach for obtaining the upper and lower limits of (Tc − Ta) is theoretical and it foresees the combination of energy balance and diffusion equations. This method requires knowledge of challenging to obtain variables, such as net radiation and aerodynamic resistance [19,89]. Another approach, defined statistical, was used with good results [15,63,90,91]. In brief, it foresees the use of the average temperature of the most cooling 5–10% of the canopy pixels as the wet reference to calculate the lower limit [92].

A further method that was proposed to obtain the two limits consists in the use of direct measurements over wet or dry reference surfaces, natural or artificial. The reference dry targets, as suggested by Jones [93,94], can be achieved, impeding leaf transpiration by covering the leaf surface (on one or both sides) with petroleum jelly. In this case, the temperature of the upper limit reference results from the measurement of the leaf temperature carried out about 30 minutes after the application of the petroleum jelly layer. The wet targets are obtained by spraying a thin layer of water on one or both leaf sides before taking the thermal image—this is valid on a single plant scale [94]. The advantage of using this method is that the stress levels are normalized to the actual response of the plants. However, the need to repeat the measurement for each test site after each flight of the UAV can be an obstacle to its applicability. In addition, another problem is related to the difficulty of identifying the leaves that are covered by petroleum jelly [95,96]. Maes proposed an example of a wet artificial reference target in [97] using a cloth and a steel wireframe. The target thus created, with an appearance, shape, and size similar to kiwi and grapevines leaves was kept wet for days by keeping the lower part of the cloth immersed inside a bottle filled with water. Artificial targets were also used in [98]. This study provides an approach in which the reference temperature of the upper limit is set at 5 °C above the air temperature, while that of the lower limit is derived from the temperature of the artificial target. Berni et al. [19] proposed an approach to monitor relatively large areas while using UAVs to detect water stress; the work, focused on the control of water stress in an olive grove, based on physical models for the estimation of input variables of energy balance equations, did not require the use of reference surfaces.

Two problems have prevented the widespread use of the CWSI: the first concerns its use under changing atmospheric conditions, and the second occurs when using lower resolution data from satellites or aircrafts platforms (as compared to UAVs), concerning, namely, the problem of "mixed-pixel value", which is where part of the temperature of the pixel comes from the background soil and not from the pure canopy, thus reducing the quality of the data [99].

As far as the first aspect is concerned, CWSI was found to work better in dry climates, while it showed considerable limitations in wet climates and in environments with substantial climatic variability [100]. In fact, it must be taken into account that the absolute in leaf-air temperature difference decreases as the atmospheric humidity deficit decreases, and the same goes for sensitivity to any measurements made. Furthermore, while taking the latter aspect into account in the calculation of the CWSI, as the humidity (and temperature) deficit decreases, the signal-to-noise ratio is reduced [101]. Finally, it must also be considered that the canopy temperature depends, in part, on variations in the roughness of the canopy, wind speed, and net radiation, all of which are more variable in humid climates [87].

As far as the problem of "mixed-pixel value" is concerned, while the resolution of satellite images is limited, the sensors mounted on UAVs, because of their sub-meter spatial resolution, make it easier to recover the pure canopy temperature while minimizing the thermal effects of the background soil [65,102,103]. Different approaches can be adopted, as shown in [99]. An approach to extracting the temperatures of canopies' sunlit leaves foresees distinguishing canopies' pixels from background pixels that are based on an analysis of the temperature distribution among the pixel population and then segment the image into two distinct classes. This solution is possible where there is a clear difference between the average temperature of the canopy and that of the soil/background. A valid solution also includes the use of additional information that is derived from RGB, multispectral and hyperspectral images (in cases where several sensors are used simultaneously), in order to identify plant's pixels [99]. In this case, the use of vegetation indices, like NDVI, to separate soil pixels and plant pixels can be useful.

In addition to the CWSI, several normalized thermal indices were developed between the 1970s and 1980s. Among these, there are the conductance index (IG) and the stomatal conductance index (I3), whose formulas (Equations (11) and (12)) are as follows:

$$IG = \frac{T_{dry} - T_C}{T_C - T_{wet}} \tag{11}$$

$$I3 = \frac{T_C - T_{wet}}{T_{dry} - T_C} \tag{12}$$

where T_c represents the surface temperature of the canopy, while T_{dry} and T_{wet} are entirely dry or wet reference surfaces to simulate leaf temperature under conditions of minimum and maximum transpiration.

The main aspects of these two indices are that IG increases with stomatal conductance, while I3 is positively correlated with stomatal resistance [94,96]. The three indices (CWSI, IG, and I3) need the knowledge of canopy temperature under both wet and dry conditions; however, the concept of the CWSI remains the most widely used, as it is best known [94].

Research carried out thus far has concerned the monitoring of water stress in both herbaceous and tree crops. In Sullivan et al.'s [24] and Bian et al.'s [104] research, thermal UAVs were used, respectively, to monitor *Gossypium hirsutum* L. crop residue management and the response to different irrigation treatments, calculating the CWSI. Martinez et al. [105] worked on monitoring sugar beet water stress by comparing thermal data, which were obtained using a conventional thermal camera, with those obtained using a low-cost infrared sensor.

Zhang et al. [106] monitored, at the farm scale, maize water stress using RGB and thermal images. The study of Crusiol et al. [107] dealt with the evaluation of the water status of soybean plants under different water conditions. Other works [108,109] focused on the monitoring of herbaceous crops while using different sensors (RGB, multispectral, and thermal) with the aim of producing orthomosaics and vigor maps.

The CWSI is also widely used for perennial crops. Bellvert et al. [110] calculated the index in peach trees (*Prunus Persica*), mapping the internal spatial variability of the orchard by thermal UAV and verifying the relationships between the index and the leaf water potential in different growing seasons. Gonzalez-Dugo et al. [91] studied the spatial variations in the water status of five different fruit tree species: almond (*Prunus dulcis*), apricot (*Prunus armeniaca*), peach (*Prunus persica*), lemon (*Citrus x limon*), and orange (*Citrus sinensis*). In this respect, it is important to remember that the stomatal response to environmental conditions can vary from species to species [111]. Olive and citrus show relevant stomatal closure at midday [89,91], while, in other species, such as almond trees, the stomatal behavior changes between the different cultivars [112].

Some studies have applied thermal UAV RS to citrus orchards (orange, *Citrus sinensis*, and mandarin, *Citrus reticulata*) in order to extract the temperature of the crowns for water stress detection [113,114]. Research has also been carried out on olive crops, a species of great importance in the countries of the Mediterranean basin [115], in order to verify the plant's behavior in response to various irrigation treatments. In particular, Poblete-Echeverría et al. [116] showed that the temperature difference between the canopy (T_c) and air (T_a) is related to the difference in water potential under different irrigation treatments when the plants are under water stress conditions. Berni et al. [19] used very high-resolution UAV thermal imagery to calculate and map the tree canopy conductance and the CWSI in a heterogeneous olive orchard.

Egea et al. [117] demonstrated the usefulness of the CWSI for monitoring water stress in a dense olive orchard by verifying sound relationships between the index and water stress indicators, such as stomatal conductance, stem water potential, and leaf transpiration rate. Ortega-Farías et al. [118], while using thermal and multispectral cameras that were mounted on UAVs, estimated the energy balance components on a drip irrigation olive grove, acquiring high-resolution images to evaluate intra-field spatial variability.

Several works have used thermal UAVs for the calculation of the CWSI in vineyards, such as that of Zarco-Tejada et al. [81]. In Baluja et al.'s [15] study, the relationships between the temperatures or indices that were derived from thermal and multispectral imagery and stomatal conductance and water potential were determined. In particular, air and leaf temperatures were recorded with a handheld thermometer, and stomatal conductance was measured with a leaf gas exchange system, while the stem water potential was measured using a Scholander pressure bomb; the CWSI, IG, and I3 indices were also calculated.

Similarly, the work by Bellvert et al. [20] related the CWSI with leaf water potential and that by Santesteban et al. [119] related the CWSI with stem water potential and seasonal leaf stomatal conductance to estimate the variability of plant water status in a vineyard. Matese et al. [120] and Pàdua et al. [16,121] exploited the use of different sensors (RGB, multispectral, and thermal) for several applications in precision viticulture: the production of vigor maps, multi-temporal analysis of vigor maps, and water stress detection. Gómez-Candón et al. [25], by testing in an apple orchard, proposed a methodology to derive thermal orthomosaics, including a method for the radiometric correction of thermal UAV imagery.

4.2. Plant Disease Detection, Phenotyping, Yield Estimation, and Vegetation Status Monitoring

Calderón et al. provides an interesting example of application in the framework of plant pathology of thermal UAV RS [14]. In particular, this research aimed to evaluate the use of thermal imagery and physiological indices derived from other types of sensors to detect the presence of infection that is caused by the soil-borne fungus *Verticillium dahliae*. The role of thermal RS in the diagnosis of the pathology is due to the water stress of plants, caused by the fungus, or by the stomatal closure, which determines the reduction of the transpiration rate. Therefore, decreasing evaporative cooling increases leaf temperature. By conducting field measurements, it was demonstrated that crown variations of temperature are higher and stomatal conductance lower as the severity level of the disease increased. Besides, crown temperature and the CWSI index were shown to be among the best indicators to detect *Verticillium dahliae* in the early stages of disease development.

With regard to the issue of yield estimation and related aspects, thermal UAV RS was exploited to research cotton, soybean, and rice crops [23,122,123]. Feng et al. [122] used three types of sensors (RGB, multispectral, and thermal) in order to analyze several features to see which one had the best performance for the yield prediction and at which stages of the crop cycle: Four vegetation indices, canopy cover, plant height, temperature, and a cotton fiber index.

The study of Maimaitijiang et al. [123] on soybean concerned testing the potentialities of using different sensors (including the thermal camera) in the framework of multimodal data fusion and deep neural network to derive useful information for the yield prediction model. Liu et al. [23] investigated rice lodging while using RGB and thermal sensor UAV images. The objectives of the research were to measure the daily temperature differences between non-lodged and lodged rice crops, as well as to detect the optimal time window. Because the differences in temperature between 10 a.m. and 4 p.m. were significant, thermal camera devices allowed for identifying lodged rice plants.

Some other works have focused on the monitoring of vineyards and related landscape elements, such as agricultural terraces [124–127]. In particular, Tucci et al. [125,126] used RGB and thermal cameras to investigate the thermal dynamics of a vineyard that was grown on dry-stone wall terraced land. These dry-stone terraces are part of the UNESCO's Representative List of the Intangible Cultural Heritage of Humanity and are typical and iconic features of the agricultural landscapes across Europe, including Italy [128]. However, the risk of abandonment and degradation threaten agricultural terraces, which can involve the increase of hazards that are linked to geo-hydrological processes triggered by rainfall events. In this framework, the use of RGB and thermal sensors mounted on UAVs represents an efficient and cost-effective monitoring methodology, also given the high resolution of the images and the reduced time for their acquisition. Tucci et al. [125], besides verifying the effectiveness of the used sensors in monitoring the terraced crops, investigated the influence that dry-stone walls can have on the microclimate of the vineyard and, consequently, on the quality of the obtained grapes. The results of the study showed differences in the temperature between the plants of the internal rows (lower temperatures) and the external rows (higher temperatures).

Phenotyping has an important role in crop science research. The acquisition of crop phenotypic information in different environments allows the association of genomic and phenotypic information useful for increasing yield [129]. In this framework, recently, the importance of the use of UAVs has increased, as they provide a rapid and non-destructive approach to phenotyping, also allowing the use of high spatial resolution images [129,130].

The use of UAVs has made it possible to overcome various limits that are linked to the use of traditional methods, such as the difficulty in making simultaneous measurements on different plots [131]. In this context, some works have been carried out involving the use of thermal UAV RS, thus proving its effectiveness, as in [9]. Natarajan et al. [132] employed different sensors (visible, multispectral, and thermal) that were mounted on UAVs for the phenotyping of indirect traits (including canopy temperature) for early-stage selection in sugarcane breeding.

Gracia-Romero et al.'s [133] work concerned the comparison of the performance of RGB, multispectral, and thermal data, which were derived from ground-based and UAV surveys, with the aim to assess genotypic differences in durum wheat's yield, under different growing conditions. In this case, the measurement of the canopies' temperature was an alternative valid to that of leaf stomatal conductance. Perich et al. [134] exploited thermal UAV RS to measure canopy temperature in wheat. Canopy temperature has a strong association with water status and stomatal conductance in wheat [135–137]; in particular, low canopy temperature can be associated with a 30% increase in yield, as well as an increase in water absorption by the deeper roots when measured during the grain filling phase [138]. Maimaitijiang et al.'s [139] study, which was performed on soybean, showed that the fusion of thermal and multispectral data could provide the best estimate of the biochemical traits of crops (chlorophyll content and N concentration) and biophysics (LAI, fresh and dry biomass).

The use of thermal UAV RS has proven to be very promising as a less-expensive way for mapping drainage systems. Subsurface tile drainage is of great importance in the Midwest of the United States [140]. Subsoil drainage allows for the level of groundwater to be lowered by removing excess water, as well as reducing soil erosion and increasing the aeration [141,142] and infiltration capacity of water derived from precipitation.

Knowing the precise layout of the drainage networks, as well as bringing benefits for their better functionality, favors soil and water conservation practices [143]. Additional benefits include increases in the soil productivity and yield of the crops, most of which do not tolerate excess water [143–145]. As the soil is directly above a drainage line, it is often drier than the soil between the drainage lines [146]. Differences in emissivity between dry and wet soil surfaces can be detected while using a thermal camera [147], which has been demonstrated in several studies using thermal cameras mounted on UAVs, even in combination with visible and multispectral cameras, showing the significant performance of thermal images and potentialities for their use in mapping agricultural drainage pipe systems [143,146,148–150].

Table 2 reports the dataset of all analyzed studies concerning the application of thermal UAV RS in precision agriculture (PA) and organized based on the aim of the study, the type of the used camera sensor, and the analyzed crops.

Table 2. Dataset compilation of studies concerning the application of thermal UAV remote sensing (RS) in precision agriculture (PA).

Aim of the Study	Camera Sensor	Analyzed Crops	References
Water stress detection and plant phenotyping	Thermal	Soybean and sorghum	[9]
Water stress and disease detection	Thermal / Multispectral / Hyperspectral	Olive	[14]
Water stress detection	Thermal / Multispectral	Grapevine	[15]
Water stress detection and monitoring	Thermal / Multispectral / RGB	Grapevine	[16]
	Thermal / Multispectral	Olive	[19]
	Thermal	Grapevine	[20]
	Thermal	Sugar beet	[21]
Monitoring lodging	Thermal / RGB	Rice	[23]
Irrigation and crop residue management	Thermal	Cotton	[24]
Water stress detection	Thermal / Multispectral / RGB	Apple	[25]
Monitoring and water stress detection	Thermal / Multispectral	Cotton, Corn, Olive and Peach	[65]
Monitoring	Thermal / Multispectral	Onion	[68]
	Thermal / Multispectral	Barley and Corn	[108]
	Thermal / RGB	Grapevine	[125,126]
Evapotranspiration estimation	Thermal	-	[74]
Water stress detection	Thermal	Nectarine and Peach	[70]
	Thermal / Multispectral	Grapevine	[81]
	Thermal	Almond, Apricot, Peach, Orange, Lemon	[91]
	Thermal / Multispectral	Cotton	[104]
	Thermal	Sugar beet	[105]
	Thermal / RGB	Corn	[106]
	Thermal / RGB	Soybean	[107]
	Thermal / Multispectral	Soybean	[109]
	Thermal	Nectarine and Peach	[110]
	Thermal / Hyperspectral	Mandarin and Orange	[113]
	Thermal / Hyperspectral	Grapevine	[114]
	Thermal	Grapevine and Olive	[116,117]
	Thermal	Grapevine	[119,120]
Estimation of energy balance components	Thermal / Multispectral	Olive	[118]
Monitoring and water stress detection	Thermal / Multispectral / RGB	Grapevine	[121]
Yield estimation	Thermal / Multispectral / RGB	Cotton	[122]
	Thermal / Multispectral / RGB	Soybean	[123]
	Thermal / Multispectral	Grapevine	[127]
Phenotyping	Thermal / Multispectral / RGB	Sugarcane	[132]
	Thermal / Multispectral / RGB	Wheat	[133]
	Thermal	Wheat	[134]
	Thermal / Multispectral	Soybean	[139]
Mapping drainage systems	Thermal / RGB	-	[143]
	Thermal / Multispectral / RGB	Corn and soybean	[146,148,149]

5. Conclusions and Future Research Outlook

This review covered the state-of-the-art thermal UAV RS technology and, to the best of our knowledge, is the first to deal with this topic. We also outlined an overview of the latest applications of thermal UAV RS in the framework of PA. Starting from a synthesis of the fundamental principles of thermography, necessary for the less experienced who approach this scientific field, this was followed by a brief description of the features of thermal cameras, from the operation of the sensor that allows the conversion of measurements into images to a hint about the cameras' cost, briefly covering the topics of field data acquisition, calibration, and data processing.

As far as the application of thermal UAV RS in agriculture is concerned, a literature review made it possible to verify the presence of numerous works devoted to the subject. To the best of our knowledge, most of the applications of thermal UAV RS have concerned the detection of crops water stress and the management of irrigation resources—both being crucial aspects for crop development and agriculture [12,83,151]. The other applications, in a smaller number, have dealt with the identification of symptoms that are caused by pathologies, phenotyping, monitoring, yield estimation, and the identification of drainage networks in the fields.

Certainly, the use of UAV thermal sensors is not as widespread as that of other sensors, such as optical and multispectral sensors, and this is probably due to the characteristics of thermal sensors and the type of data that are derived from them. Although the first aspect is important, an important limitation of thermal cameras is their geometric resolution, which is low when compared to, for example, RGB sensors. The second concerns the data that are derived from the sensor, i.e., the temperature, which has proved to be fundamental in the detection of water stress in plants, given the natural mechanisms that regulate the temperature of plants. Excluding, perhaps, this type of application, which sees thermal sensors as protagonists and advantageous over other sensors, especially for the possibility of detecting water stress connections in advance, other types of applications that have provided the exclusive use of thermal cameras are not many, especially when considering the field of plant pathology. The real potential of cameras on UAVs can be exploited at high levels and with maximum profit in terms of utility, focusing on the feature that makes UAVs unique: that of being able to simultaneously mount and use multiple sensors [152]. The thermal camera used side-by-side with RGB and multispectral sensors can increase the importance of UAVs in PA and expand their possibilities in terms of use. In particular, when considering, for example, plant pathology, in order to improve the ability to detect diseases or parasitic attacks at an early stage, the use of different sensors, including thermal sensors, and the fusion/combination of their derived data with optical and multispectral sensors, should be considered [153]. The possibility to perform surveys with a centimeter resolution, coupling different sensors at any time, and with more affordable costs, puts UAVs ahead of aircraft platforms. When compared to satellites, on the other hand, the introduction of platforms, including nanosatellites, equipped with sensors, capable of offering high-resolution images of less than 3 m or the ultra-high resolution of less than 1 m makes satellites increasingly competitive with regard to drones in PA applications [152,154–157]. It would be interesting to combine thermal (and optical) satellite data with UAV data, together with data that were collected on the ground, as shown, for example, in [109]. As things stand, when considering the different platforms and sensors of RS, no one is probably able to offer a high resolution in all spatial, spectral, and temporal dimensions [153]. Therefore, it would be desirable to synergize UAV images with high-resolution satellite images to improve the quality of the final products, including thermal RS, in order to overcome these limits.

Naturally, new developments are also expected in the framework of thermal RS. Likewise, it is also expected that the current trend of increasing user-friendliness for all types of users will continue in the technological development of UAVs and sensors. Greater automation, where possible, of aspects that are related to both the field data acquisition phase (preparation of the optimal flight plan, configuration, and calibration of the sensors before and during flight) and the data processing phase (together with the reduction of the time that is needed for data processing) is necessary as the next steps to implement the use of UAVs in agriculture. As far as ground data acquisition is concerned, as explained in the previous paragraphs, it remains an essential step for the moment. Specifically for thermal RS, a simpler and more easy and immediate combination of the data that were collected by weather stations with data derived from UAVs would be useful. Indeed, there are still important practical difficulties in the correct collection of data, when considering the mitigation of atmospheric effects, calibration, climatic conditions, and the complex interactions between soil and plants [7]—particularly true in the case of thermal RS, whose raw data before the processing steps are far from offering true and accurate temperature measurements. In this respect, thermal RS requires accurate knowledge of thermography [152] in all its application phases, from the preparation of the surveys to the final product. If this aspect, on the one hand, does not fail to stimulate the world of research to explore all aspects of thermal RS, then it might, on the other hand, constitute a limit for use outside this field.

However, it is fair to say that, given the important progress in the use of RS sensors in agriculture, in the short term, new solutions should also be able to simplify and expand the use of thermal RS in agriculture and PA, increasing its integration in decision making [7]. PA needs high-intensity procedures for the use of acquired images and it requires the presence of experienced and qualified personnel [126], which results in higher costs for companies. Therefore, the use of advanced technologies, including the use of UAVs, remains confined to those farmers with large agricultural areas available [158]. This aspect is more evident in the case of thermal UAV surveys, when considering that their operational costs per hectare are higher than those of multispectral surveys [159].

Author Contributions: Conceptualization, methodology, investigation, data curation, writing—review and editing, G.M. (Gaetano Messina) and G.M. (Giuseppe Modica). All authors have read and agreed to the published version of the manuscript.

Funding: This research received no external funding.

Acknowledgments: The authors are grateful to DR-One S.r.l. (http://www.dr-onesrl.com—Belmonte Calabro, Cosenza, Italy) for providing UAV thermal surveys. The authors would like to express their great appreciation to the three anonymous reviewers for their very constructive comments provided during the revision of the paper, and that contributes to improving it significantly.

Conflicts of Interest: The authors declare no conflicts of interest.

References

1. Lillesand, T.; Kiefer, R.W.; Chipman, J. *Remote Sensing and Image Interpretation*, 7th ed.; Wiley and sons: New York, NY, USA, 2015; ISBN 978-1-118-91947-7.
2. Pajares, G. Overview and Current Status of Remote Sensing Applications Based on Unmanned Aerial Vehicles (UAVs). *Photogramm. Eng. Remote Sens.* **2015**, *81*, 281–330. [CrossRef]
3. Chen, S.; Laefer, D.F.; Mangina, E. State of Technology Review of Civilian UAVs. *Recent Patents Eng.* **2016**, *10*, 160–174. [CrossRef]
4. Nex, F.; Remondino, F. UAV for 3D mapping applications: A review. *Appl. Geomat.* **2014**, *6*, 1–15. [CrossRef]
5. Shakhatreh, H.; Sawalmeh, A.; Al-Fuqaha, A.; Dou, Z.; Almaita, E.; Khalil, I.; Othman, N.S.; Khreishah, A.; Guizani, M. Unmanned Aerial Vehicles: A Survey on Civil Applications and Key Research Challenges. *arXiv* **2018**, arXiv:1805.00881, 1–58. [CrossRef]
6. Zhang, C.; Kovacs, J.M. The application of small unmanned aerial systems for precision agriculture: A review. *Precis. Agric.* **2012**, *13*, 693–712. [CrossRef]
7. Khanal, S.; Fulton, J.; Shearer, S. An overview of current and potential applications of thermal remote sensing in precision agriculture. *Comput. Electron. Agric.* **2017**, *139*, 22–32. [CrossRef]
8. Anderson, M.C.; Hain, C.; Otkin, J.; Zhan, X.; Mo, K.; Svoboda, M.; Wardlow, B.; Pimstein, A. An Intercomparison of Drought Indicators Based on Thermal Remote Sensing and NLDAS-2 Simulations with U.S. Drought Monitor Classifications. *J. Hydrometeorol.* **2013**, *14*, 1035–1056. [CrossRef]
9. Sagan, V.; Maimaitijiang, M.; Sidike, P.; Eblimit, K.; Peterson, K.; Hartling, S.; Esposito, F.; Khanal, K.; Newcomb, M.; Pauli, D.; et al. UAV-Based High Resolution Thermal Imaging for Vegetation Monitoring, and Plant Phenotyping Using ICI 8640 P, FLIR Vue Pro R 640, and thermoMap Cameras. *Remote Sens.* **2019**, *11*, 330. [CrossRef]
10. Ludovisi, R.; Tauro, F.; Salvati, R.; Khoury, S.; Mugnozza, G.S.; Harfouche, A. Uav-based thermal imaging for high-throughput field phenotyping of black poplar response to drought. *Front. Plant Sci.* **2017**, *8*, 1–18. [CrossRef]
11. Costa, J.M.; Grant, O.M.; Chaves, M.M. Thermography to explore plant-environment interactions. *J. Exp. Bot.* **2013**, *64*, 3937–3949. [CrossRef]
12. Gago, J.; Douthe, C.; Coopman, R.E.; Gallego, P.P.; Ribas-Carbo, M.; Flexas, J.; Escalona, J.; Medrano, H. UAVs challenge to assess water stress for sustainable agriculture. *Agric. Water Manag.* **2015**, *153*, 9–19. [CrossRef]
13. Radoglou-Grammatikis, P.; Sarigiannidis, P.; Lagkas, T.; Moscholios, I. A compilation of UAV applications for precision agriculture. *Comput. Netw.* **2020**. [CrossRef]

14. Calderón, R.; Navas-Cortés, J.A.; Lucena, C.; Zarco-Tejada, P.J. High-resolution airborne hyperspectral and thermal imagery for early detection of Verticillium wilt of olive using fluorescence, temperature and narrow-band spectral indices. *Remote Sens. Environ.* **2013**, *139*, 231–245. [CrossRef]

15. Baluja, J.; Diago, M.P.; Balda, P.; Zorer, R.; Meggio, F.; Morales, F.; Tardaguila, J. Assessment of vineyard water status variability by thermal and multispectral imagery using an unmanned aerial vehicle (UAV). *Irrig. Sci.* **2012**, *30*, 511–522. [CrossRef]

16. Pádua, L.; Marques, P.; Adão, T.; Guimarães, N.; Sousa, A.; Peres, E.; Sousa, J.J. Vineyard variability analysis through UAV-based vigour maps to assess climate change impacts. *Agronomy* **2019**, *9*, 581. [CrossRef]

17. Idso, S.B.; Jackson, R.D.; Pinter, P.J.; Reginato, R.J.; Hatfield, J.L. Normalizing the stress-degree-day parameter for environmental variability. *Agric. Meteorol.* **1981**, *24*, 45–55. [CrossRef]

18. Alderfasi, A.A.; Nielsen, D.C. Use of crop water stress index for monitoring water status and scheduling irrigation in wheat. *Agric. Water Manag.* **2001**, *47*, 69–75. [CrossRef]

19. Berni, J.A.J.; Zarco-Tejada, P.J.; Sepulcre-Cantó, G.; Fereres, E.; Villalobos, F. Mapping canopy conductance and CWSI in olive orchards using high resolution thermal remote sensing imagery. *Remote Sens. Environ.* **2009**, *113*, 2380–2388. [CrossRef]

20. Bellvert, J.; Zarco-Tejada, P.J.; Girona, J.; Fereres, E. Mapping crop water stress index in a "Pinot-noir" vineyard: Comparing ground measurements with thermal remote sensing imagery from an unmanned aerial vehicle. *Precis. Agric.* **2014**, *15*, 361–376. [CrossRef]

21. Quebrajo, L.; Perez-Ruiz, M.; Pérez-Urrestarazu, L.; Martínez, G.; Egea, G. Linking thermal imaging and soil remote sensing to enhance irrigation management of sugar beet. *Biosyst. Eng.* **2018**, *165*, 77–87. [CrossRef]

22. Romano, G.; Zia, S.; Spreer, W.; Sanchez, C.; Cairns, J.; Araus, J.L.; Müller, J. Use of thermography for high throughput phenotyping of tropical maize adaptation in water stress. *Comput. Electron. Agric.* **2011**, *79*, 67–74. [CrossRef]

23. Liu, T.; Li, R.; Zhong, X.; Jiang, M.; Jin, X.; Zhou, P.; Liu, S.; Sun, C.; Guo, W. Estimates of rice lodging using indices derived from UAV visible and thermal infrared images. *Agric. For. Meteorol.* **2018**, *252*, 144–154. [CrossRef]

24. Sullivan, D.G.; Fulton, J.P.; Shaw, J.N.; Bland, G. Evaluating the sensitivity of an unmanned thermal infrared aerial system to detect water stress in a cotton canopy. *Trans. ASABE* **2007**, *50*, 1963–1969. [CrossRef]

25. Gómez-Candón, D.; Virlet, N.; Labbé, S.; Jolivot, A.; Regnard, J.L. Field phenotyping of water stress at tree scale by UAV-sensed imagery: New insights for thermal acquisition and calibration. *Precis. Agric.* **2016**, *17*, 786–800. [CrossRef]

26. Prakash, A. Thermal Remote Sensing: Concepts, issues and applications. *Int. Arch. Photogramm. Remote Sens.* **2000**, *33*, 239–243.

27. Jensen, J.R. *Remote Sensing of the Environment: An Earth Resource Perspective*, 2nd ed.; Pearson: Harlow, UK, 2014; Volume 1, ISBN 9780131889507.

28. Vinet, L.; Zhedanov, A. A "missing" family of classical orthogonal polynomials. *Geogr. J.* **2010**, *146*, 448. [CrossRef]

29. Richter, R.; Schlapfer, D. Atmospheric and Topographic Correction: Model ATCOR3. *Aerospace* **2019**, *3*, 1–144.

30. Walker, J.; Halliday, D.; Resnick, R. *Fundamentals of Physics*, 10th ed.; Wiley: New York, NY, USA, 2015; ISBN 9781118230725.

31. Kuenzer, C.; Dech, S.; Zhang, J.; Jing, L.; Huadong, G. Thermal infrared remote sensing: Sensors, Methods, Applications. In *Remote Sensing and Digital Image Processing*; Springer: Heidelberg, Germany, 2013; Volume 17.

32. Kuenzer, C.; Zhang, J.; Jing, L.; Huadong, G.; Dech, S. Thermal infrared remote sensing of surface and underground coal fires. In *Remote Sensing and Digital Image Processing*; Springer: Heidelberg, Germany, 2013; Volume 17, pp. 429–451.

33. Sabin, F. *Remote Sensing: Principles and Interpretation, (Floyd F. Sabins)*; W.H.Freeman & Co: New York, NY, USA, 1997; ISBN 0716724421.

34. Schmugge, T.; French, A.; Ritchie, J.C.; Rango, A.; Pelgrum, H. Temperature and emissivity separation from multispectral thermal infrared observations. *Remote Sens. Environ.* **2002**, *79*, 189–198. [CrossRef]

35. Jacob, F.; Petitcolin, F.; Schmugge, T.; Vermote, É.; French, A.; Ogawa, K. Comparison of land surface emissivity and radiometric temperature derived from MODIS and ASTER sensors. *Remote Sens. Environ.* **2004**, *90*, 137–152. [CrossRef]

36. Campbell e Wynne. *Introduction to Remote Sensing*; The Guiford Press: New York, NY, USA, 2017; ISBN 9781609181765.

37. Salisbury, J.W.; D'Aria, D.M. Emissivity of terrestrial materials in the 8-14 µm atmospheric window. *Remote Sens. Environ.* **1992**, *42*, 83–106. [CrossRef]

38. Gates, D.M.; Keegan, H.J.; Schleter, J.C.; Weidner, V.R. Spectral Properties of Plants. *Appl. Opt.* **1965**, *4*, 11. [CrossRef]

39. Chen, C. Determining the leaf emissivity of three crops by infrared thermometry. *Sensors* **2015**, *15*, 11387–11401. [CrossRef] [PubMed]

40. López, A.; Molina-Aiz, F.D.; Valera, D.L.; Peña, A. Determining the emissivity of the leaves of nine horticultural crops by means of infrared thermography. *Sci. Hortic.* **2012**, *137*, 49–58. [CrossRef]

41. Kaplan, H. *Practical Applications of Infrared Thermal Sensing and Imaging Equipment*, 3rd ed.; SPIE: Bellingham, WA, USA, 2007; ISBN 9780819479020.

42. Gade, R.; Moeslund, T.B. Thermal cameras and applications: A survey. *Mach. Vis. Appl.* **2014**, *25*, 245–262. [CrossRef]

43. FLIR. *Tech Note: Cooled Versus Uncooled Cameras for Long Range Surveillance*; FLIR: Breda, The Netherlands, 2011.

44. Mesas-Carrascosa, F.J.; Pérez-Porras, F.; Meroño de Larriva, J.; Mena Frau, C.; Agüera-Vega, F.; Carvajal-Ramírez, F.; Martínez-Carricondo, P.; García-Ferrer, A. Drift Correction of Lightweight Microbolometer Thermal Sensors On-Board Unmanned Aerial Vehicles. *Remote Sens.* **2018**, *10*, 615. [CrossRef]

45. Jensen, A.M.; McKee, M.; Chen, Y. Procedures for processing thermal images using low-cost microbolometer cameras for small unmanned aerial systems. *Int. Geosci. Remote Sens. Symp.* **2014**, 2629–2632. [CrossRef]

46. Luhmann, T.; Piechel, J.; Roelfs, T. Geometric calibration of thermographic cameras. *Remote Sens. Digit. Image Process.* **2013**, *17*, 27–42. [CrossRef]

47. Sizov, F.F. IR region challenges: Photon or thermal detectors? Outlook and means. *Semicond. Phys. Quantum Electron.* **2015**, *15*, 193–199. [CrossRef]

48. Hyseni, G.; Caka, N.; Hyseni, K. Infrared thermal detectors parameters: Semiconductor bolometers versus pyroelectrics. *WSEAS Trans. Circuits Syst.* **2010**, *9*, 238–247.

49. Bhan, R.K.; Saxena, R.S.; Jalwania, C.R.; Lomash, S.K. Uncooled infrared microbolometer arrays and their characterisation techniques. *Def. Sci. J.* **2009**, *59*, 580–589. [CrossRef]

50. Fièque, B.; Tissot, J.L.; Trouilleau, C.; Crastes, A.; Legras, O. Uncooled microbolometer detector: Recent developments at Ulis. *Infrared Phys. Technol.* **2007**, *49*, 187–191. [CrossRef]

51. Bieszczad, G.; Kastek, M. Measurement of thermal behavior of detector array surface with the use of microscopic thermal camera. *Metrol. Meas. Syst.* **2011**, *18*, 679–690. [CrossRef]

52. FLIR. *Tech Note: Uncooled Detectors for Thermal Imaging Cameras*; FLIR: Breda, The Netherlands, 2015.

53. Budzier, H.; Gerlach, G. Calibration of uncooled thermal infrared cameras. *J. Sens. Sens. Syst.* **2015**, *4*, 187–197. [CrossRef]

54. Kelly, J.; Kljun, N.; Olsson, P.-O.; Mihai, L.; Liljeblad, B.; Weslien, P.; Klemedtsson, L.; Eklundh, L. Challenges and best practices for deriving temperature data from an uncalibrated UAV thermal infrared camera. *Remote Sens.* **2019**, *11*, 567. [CrossRef]

55. Manfreda, S.; McCabe, M.F.; Miller, P.E.; Lucas, R.; Madrigal, V.P.; Mallinis, G.; Dor, E.B.; Helman, D.; Estes, L.; Ciraolo, G.; et al. On the use of unmanned aerial systems for environmental monitoring. *Remote Sens.* **2018**, *10*, 641. [CrossRef]

56. Sheng, H.; Chao, H.; Coopmans, C.; Han, J.; McKee, M.; Chen, Y. Low-cost UAV-based thermal infrared remote sensing: Platform, calibration and applications. In Proceedings of the 2010 IEEE/ASME International Conference on Mechatronic and Embedded Systems and Applications, Qingdao, China, 15–17 July 2010; pp. 38–43.

57. Ribeiro-Gomes, K.; Hernández-López, D.; Ortega, J.; Ballesteros, R.; Poblete, T.; Moreno, M. Uncooled Thermal Camera Calibration and Optimization of the Photogrammetry Process for UAV Applications in Agriculture. *Sensors* **2017**, *17*, 2173. [CrossRef]

58. Gallo, M.A.; Willits, D.S.; Lubke, R.A.; Thiede, E.C. Low-cost uncooled IR sensor for battlefield surveillance. In Proceedings of the SPIE: International Symposium on Optic, Imaging, and Instrumentation, San Diego, CA, USA, 11–16 July 1993; Andresen, B.F., Shepherd, F.D., Eds.; Volume 2020, p. 351.

59. Maes, W.; Huete, A.; Steppe, K. Optimizing the Processing of UAV-Based Thermal Imagery. *Remote Sens.* **2017**, *9*, 476. [CrossRef]

60. Stark, B.; Smith, B.; Chen, Y. Survey of thermal infrared remote sensing for Unmanned Aerial Systems. In Proceedings of the 2014 International Conference on Unmanned Aircraft Systems (ICUAS), Orlando, FL, USA, 27–30 May 2014; pp. 1294–1299. [CrossRef]

61. Olbrycht, R.; Wiecek, B.; De Mey, G. Thermal drift compensation method for microbolometer thermal cameras. *Appl. Opt.* **2012**, *51*, 1788–1794. [CrossRef]

62. FLIR. *Tech Note: Radiometric Temperature Measurements Surface Characteristics and Atmospheric Compensation*; FLIR: Breda, The Netherlands, 2012.

63. Alchanatis, V.; Cohen, Y.; Cohen, S.; Moller, M.; Sprinstin, M.; Meron, M.; Tsipris, J.; Saranga, Y.; Sela, E. Evaluation of different approaches for estimating and mapping crop water status in cotton with thermal imaging. *Precis. Agric.* **2010**, *11*, 27–41. [CrossRef]

64. Sepulcre-Cantó, G.; Zarco-Tejada, P.J.; Jiménez-Muñoz, J.C.; Sobrino, J.A.; De Miguel, E.; Villalobos, F.J. Detection of water stress in an olive orchard with thermal remote sensing imagery. *Agric. For. Meteorol.* **2006**, *136*, 31–44. [CrossRef]

65. Berni, J.; Zarco-Tejada, P.J.; Suarez, L.; Fereres, E. Thermal and Narrowband Multispectral Remote Sensing for Vegetation Monitoring From an Unmanned Aerial Vehicle. *IEEE Trans. Geosci. Remote Sens.* **2009**, *47*, 722–738. [CrossRef]

66. Kelly, J.; Eklundh, L.; Kljun, N. Radiometric Calibration of a UAV Thermal Camera. Available online: https://pdfs.semanticscholar.org/3c00/560ae50c9c34187904dcb01af863a7c3088c.pdf (accessed on 1 March 2020).

67. Dupin, S.; Gobrecht, A.; Tisseyre, B. Tisseyre Airborne Thermography of Vines Canopy: Effect of the Atmosphere and Mixed Pixels on Observed Canopy Temperature. *8 Conf. Eur. Agric. Precis.* **2011**, *1*, 1–9.

68. Messina, G.; Praticò, S.; Siciliani, B.; Curcio, A.; Di Fazio, S.; Modica, G. Monitoring onion crops using UAV multispectral and thermal imagery. In *Conference AIIA Mid-Term 2019 Biosystems Engineering for Sustainable Agriculture, Forestry and Food Production, Matera, Italy, 12–13 September 2019*; Springer: Cham, Switzerland, 2019.

69. Boesch, R. Thermal remote sensing with UAV-based workflows. *Int. Arch. Photogramm. Remote Sens. Spat. Inf. Sci.* **2017**, *42*, 41–46. [CrossRef]

70. Park, S.; Ryu, D.; Fuentes, S.; Chung, H.; Hernández-Montes, E.; O'Connell, M. Adaptive estimation of crop water stress in nectarine and peach orchards using high-resolution imagery from an unmanned aerial vehicle (UAV). *Remote Sens.* **2017**, *9*, 828. [CrossRef]

71. Chio, S.H.; Lin, C.H. Preliminary study of UAS equipped with thermal camera for volcanic geothermal monitoring in Taiwan. *Sensors* **2017**, *17*, 1649. [CrossRef]

72. Maes, W.H.; Pashuysen, T.; Trabucco, A.; Veroustraete, F.; Muys, B. Does energy dissipation increase with ecosystem succession? Testing the ecosystem exergy theory combining theoretical simulations and thermal remote sensing observations. *Ecol. Modell.* **2011**, *222*, 3917–3941. [CrossRef]

73. Berk, A.; Anderson, G.P.; Acharya, P.K.; Chetwynd, J.H.; Bernstein, L.S.; Shettle, E.P.; Matthew, M.W.; Adler-Golden, S. MODTRAN4 User's manual. In *Hanscom AFB*; Air Force Res. Lab.: MA, USA, 1999; Available online: ftp://ftp.pmodwrc.ch/pub/Vorlesung%20K+S/MOD4_user_guide.pdf (accessed on 1 March 2020).

74. Hoffmann, H.; Nieto, H.; Jensen, R.; Guzinski, R.; Zarco-Tejada, P.; Friborg, T. Estimating evaporation with thermal UAV data and two-source energy balance models. *Hydrol. Earth Syst. Sci.* **2016**, *20*, 697–713. [CrossRef]

75. Pech, K.; Stelling, N.; Karrasch, P.; De, H.M. Generation of Multitemporal Thermal Orthophotos From UAV Data. *Int. Arch. Photogramm. Remote Sens.* **2013**, *1*, 4–6. [CrossRef]

76. Turner, D.; Lucieer, A.; Malenovský, Z.; King, D.H.; Robinson, S.A. Spatial co-registration of ultra-high resolution visible, multispectral and thermal images acquired with a micro-UAV over antarctic moss beds. *Remote Sens.* **2014**, *6*, 4003–4024. [CrossRef]

77. Lee, W.S.; Alchanatis, V.; Yang, C.; Hirafuji, M.; Moshou, D.; Li, C. Sensing technologies for precision specialty crop production. *Comput. Electron. Agric.* **2010**, *74*, 2–33. [CrossRef]

78. Gerhards, M.; Rock, G.; Schlerf, M.; Udelhoven, T. Water stress detection in potato plants using leaf temperature, emissivity, and reflectance. *Int. J. Appl. Earth Obs. Geoinf.* **2016**, *53*, 27–39. [CrossRef]

79. Gautam, D.; Pagay, V. A review of current and potential applications of remote sensing to study thewater status of horticultural crops. *Agronomy* **2020**, *10*, 140. [CrossRef]

80. Ballester, C.; Zarco-Tejada, P.J.; Nicolás, E.; Alarcón, J.J.; Fereres, E.; Intrigliolo, D.S.; Gonzalez-Dugo, V. Evaluating the performance of xanthophyll, chlorophyll and structure-sensitive spectral indices to detect water stress in five fruit tree species. *Precis. Agric.* **2017**, 1–16. [CrossRef]

81. Zarco-Tejada, P.J.; González-Dugo, V.; Williams, L.E.; Suárez, L.; Berni, J.A.J.; Goldhamer, D.; Fereres, E. A PRI-based water stress index combining structural and chlorophyll effects: Assessment using diurnal narrow-band airborne imagery and the CWSI thermal index. *Remote Sens. Environ.* **2013**, *138*, 38–50. [CrossRef]

82. Hsiao, T.C. Plants response to water stress. *Ann. Rev. Plant Physiol.* **1973**, *24*, 519–570. [CrossRef]

83. Gerhards, M.; Schlerf, M.; Mallick, K.; Udelhoven, T. Challenges and future perspectives of multi-/Hyperspectral thermal infrared remote sensing for crop water-stress detection: A review. *Remote Sens.* **2019**, *11*, 1240. [CrossRef]

84. Jackson, R.D.; Idso, S.B.; Reginato, R.J.; Pinter, J.P.J. Canopy temperature as a crop water stress indicator. *Water Resour. Res.* **1981**, *17*, 1133–1138. [CrossRef]

85. Lapidot, O.; Ignat, T.; Rud, R.; Rog, I.; Alchanatis, V.; Klein, T. Use of thermal imaging to detect evaporative cooling in coniferous and broadleaved tree species of the Mediterranean maquis. *Agric. For. Meteorol.* **2019**, *271*, 285–294. [CrossRef]

86. Gerhards, M.; Schlerf, M.; Rascher, U.; Udelhoven, T.; Juszczak, R.; Alberti, G.; Miglietta, F.; Inoue, Y. Analysis of airborne optical and thermal imagery for detection of water stress symptoms. *Remote Sens.* **2018**, *10*, 1139. [CrossRef]

87. Jones, H.G. Thermal imaging and infrared sensing in plant ecophysiology. *Adv. Plant Ecophysiol. Tech.* **2018**, 135–151. [CrossRef]

88. Cohen, Y.; Alchanatis, V.; Meron, M.; Saranga, Y.; Tsipris, J. Estimation of leaf water potential by thermal imagery and spatial analysis. *J. Exp. Bot.* **2005**, *56*, 1843–1852. [CrossRef] [PubMed]

89. Testi, L.; Goldhamer, D.A.; Iniesta, F.; Salinas, M. Crop water stress index is a sensitive water stress indicator in pistachio trees. *Irrig. Sci.* **2008**, *26*, 395–405. [CrossRef]

90. Rud, R.; Cohen, Y.; Alchanatis, V.; Levi, A.; Brikman, R.; Shenderey, C.; Heuer, B.; Markovitch, T.; Dar, Z.; Rosen, C.; et al. Crop water stress index derived from multi-year ground and aerial thermal images as an indicator of potato water status. *Precis. Agric.* **2014**, *15*, 273–289. [CrossRef]

91. Gonzalez-Dugo, V.; Zarco-Tejada, P.; Nicolás, E.; Nortes, P.A.; Alarcón, J.J.; Intrigliolo, D.S.; Fereres, E. Using high resolution UAV thermal imagery to assess the variability in the water status of five fruit tree species within a commercial orchard. *Precis. Agric.* **2013**, *14*, 660–678. [CrossRef]

92. Cohen, Y.; Alchanatis, V.; Saranga, Y.; Rosenberg, O.; Sela, E.; Bosak, A. Mapping water status based on aerial thermal imagery: Comparison of methodologies for upscaling from a single leaf to commercial fields. *Precis. Agric.* **2017**, *18*, 801–822. [CrossRef]

93. Jones, H.G. Use of infrared thermometry for estimation of stomatal conductance as a possible aid to irrigation scheduling. *Agric. For. Meteorol.* **1999**, *95*, 139–149. [CrossRef]

94. Maes, W.H.; Steppe, K. Estimating evapotranspiration and drought stress with ground-based thermal remote sensing in agriculture: A review. *J. Exp. Bot.* **2012**, *63*, 4671–4712. [CrossRef]

95. Jones, H.G.; Serraj, R.; Loveys, B.R.; Xiong, L.; Wheaton, A.; Price, A.H. Thermal infrared imaging of crop canopies for the remote diagnosis and quantification of plant responses to water stress in the field. *Funct. Plant Biol.* **2009**, *36*, 978–989. [CrossRef]

96. Maes, W.H.; Achten, W.M.J.; Reubens, B.; Muys, B. Monitoring stomatal conductance of Jatropha curcas seedlings under different levels of water shortage with infrared thermography. *Agric. For. Meteorol.* **2011**, *151*, 554–564. [CrossRef]

97. Maes, W.H.; Baert, A.; Huete, A.R.; Minchin, P.E.H.; Snelgar, W.P.; Steppe, K. A new wet reference target method for continuous infrared thermography of vegetations. *Agric. For. Meteorol.* **2016**, *226–227*, 119–131. [CrossRef]

98. Agam, N.; Cohen, Y.; Berni, J.A.J.; Alchanatis, V.; Kool, D.; Dag, A.; Yermiyahu, U.; Ben-Gal, A. An insight to the performance of crop water stress index for olive trees. *Agric. Water Manag.* **2013**, *118*, 79–86. [CrossRef]

99. Jones, H.; Sirault, X. Scaling of Thermal Images at Different Spatial Resolution: The Mixed Pixel Problem. *Agronomy* **2014**, *4*, 380–396. [CrossRef]

100. Hipps, L.E.; Asrar, G.; Kanemasu, E.T. A theoretically-based normalization of environmental effects on foliage temperature. *Agric. For. Meteorol.* **1985**, *35*, 113–122. [CrossRef]

101. Jones, H.G.; Vaughan, R.A. *Remote Sensing of Vegetation Principles, Techniques, and Applications*; Oxford University Press: Oxford, UK, 2010; ISBN 9780199207794.

102. Herwitz, S.R.; Johnson, L.F.; Dunagan, S.E.; Higgins, R.G.; Sullivan, D.V.; Zheng, J.; Lobitz, B.M.; Leung, J.G.; Gallmeyer, B.A.; Aoyagi, M.; et al. Imaging from an unmanned aerial vehicle: Agricultural surveillance and decision support. *Comput. Electron. Agric.* **2004**, *44*, 49–61. [CrossRef]

103. Sugiura, R.; Noguchi, N.; Ishii, K. Remote-sensing technology for vegetation monitoring using an unmanned helicopter. *Biosyst. Eng.* **2005**, *90*, 369–379. [CrossRef]

104. Bian, J.; Zhang, Z.; Chen, J.; Chen, H.; Cui, C.; Li, X.; Chen, S.; Fu, Q. Simplified Evaluation of Cotton Water Stress Using High Resolution Unmanned Aerial Vehicle Thermal Imagery. *Remote Sens.* **2019**, *11*, 267. [CrossRef]

105. Martínez, J.; Egea, G.; Agüera, J.; Pérez-Ruiz, M. A cost-effective canopy temperature measurement system for precision agriculture: A case study on sugar beet. *Precis. Agric.* **2017**, *18*, 95–110. [CrossRef]

106. Zhang, L.; Niu, Y.; Zhang, H.; Han, W.; Li, G.; Tang, J.; Peng, X. Maize Canopy Temperature Extracted From UAV Thermal and RGB Imagery and Its Application in Water Stress Monitoring. *Front. Plant Sci.* **2019**, *10*, 1–18. [CrossRef]

107. Crusiol, L.G.T.; Nanni, M.R.; Furlanetto, R.H.; Sibaldelli, R.N.R.; Cezar, E.; Mertz-Henning, L.M.; Nepomuceno, A.L.; Neumaier, N.; Farias, J.R.B. UAV-based thermal imaging in the assessment of water status of soybean plants. *Int. J. Remote Sens.* **2020**, *41*, 3243–3265. [CrossRef]

108. Raeva, P.L.; Šedina, J.; Dlesk, A. Monitoring of crop fields using multispectral and thermal imagery from UAV. *Eur. J. Remote Sens.* **2019**, *52*, 192–201. [CrossRef]

109. Sagan, V.; Maimaitiyiming, M.; Sidike, P.; Maimaitiyiming, M.; Erkbol, H.; Peterson, K.T.; Peterson, J.; Burken, J.; Fritschi, F. UAV/Satellite Multiscale Data Fusion for Crop Monitoring and Early Stress Detection. *Int. Arch. Photogramm. Remote Sens. Spat. Inf. Sci.* **2019**. [CrossRef]

110. Bellvert, J.; Marsal, J.; Girona, J.; Gonzalez-Dugo, V.; Fereres, E.; Ustin, S.L.; Zarco-Tejada, P.J. Airborne thermal imagery to detect the seasonal evolution of crop water status in peach, nectarine and Saturn peach orchards. *Remote Sens.* **2016**, *8*, 39. [CrossRef]

111. Ballester, C.; Jiménez-Bello, M.A.; Castel, J.R.; Intrigliolo, D.S. Usefulness of thermography for plant water stress detection in citrus and persimmon trees. *Agric. For. Meteorol.* **2013**, *168*, 120–129. [CrossRef]

112. Gonzalez-Dugo, V.; Zarco-Tejada, P.; Berni, J.A.J.; Suárez, L.; Goldhamer, D.; Fereres, E. Almond tree canopy temperature reveals intra-crown variability that is water stress-dependent. *Agric. For. Meteorol.* **2012**, *154–155*, 156. [CrossRef]

113. Gonzalez-Dugo, V.; Zarco-Tejada, P.J.; Fereres, E. Applicability and limitations of using the crop water stress index as an indicator of water deficits in citrus orchards. *Agric. For. Meteorol.* **2014**, *198–199*, 94–104. [CrossRef]

114. Zarco-Tejada, P.J.; González-Dugo, V.; Berni, J.A.J. Fluorescence, temperature and narrow-band indices acquired from a UAV platform for water stress detection using a micro-hyperspectral imager and a thermal camera. *Remote Sens. Environ.* **2012**, *117*, 322–337. [CrossRef]

115. Solano, F.; Di Fazio, S.; Modica, G. Int J Appl Earth Obs Geoinformation A methodology based on GEOBIA and WorldView-3 imagery to derive vegetation indices at tree crown detail in olive orchards. *Int J. Appl. Earth Obs. Geoinf.* **2019**, *83*, 101912. [CrossRef]

116. Poblete-Echeverría, C.; Sepulveda-Reyes, D.; Ortega-Farias, S.; Zuñiga, M.; Fuentes, S. Plant water stress detection based on aerial and terrestrial infrared thermography: A study case from vineyard and olive orchard. *Acta Hortic.* **2016**, *1112*, 141–146. [CrossRef]

117. Egea, G.; Padilla-Díaz, C.M.; Martinez-Guanter, J.; Fernández, J.E.; Pérez-Ruiz, M. Assessing a crop water stress index derived from aerial thermal imaging and infrared thermometry in super-high density olive orchards. *Agric. Water Manag.* **2017**, *187*, 210–221. [CrossRef]

118. Ortega-Farías, S.; Ortega-Salazar, S.; Poblete, T.; Kilic, A.; Allen, R.; Poblete-Echeverría, C.; Ahumada-Orellana, L.; Zuñiga, M.; Sepúlveda, D. Estimation of Energy Balance Components over a Drip-Irrigated Olive Orchard Using Thermal and Multispectral Cameras Placed on a Helicopter-Based Unmanned Aerial Vehicle (UAV). *Remote Sens.* **2016**, *8*, 638. [CrossRef]

119. Santesteban, L.G.; Di Gennaro, S.F.; Herrero-Langreo, A.; Miranda, C.; Royo, J.B.; Matese, A. High-resolution UAV-based thermal imaging to estimate the instantaneous and seasonal variability of plant water status within a vineyard. *Agric. Water Manag.* **2017**, *183*, 49–59. [CrossRef]

120. Matese, A.; Baraldi, R.; Berton, A.; Cesaraccio, C.; Di Gennaro, S.F.; Duce, P.; Facini, O.; Mameli, M.G.; Piga, A.; Zaldei, A. Estimation of Water Stress in grapevines using proximal and remote sensing methods. *Remote Sens.* **2018**, *10*, 114. [CrossRef]

121. Pádua, L.; Adão, T.; Sousa, A.; Peres, E.; Sousa, J.J. Individual Grapevine Analysis in a Multi-Temporal Context Using UAV-Based Multi-Sensor Imagery. *Remote Sens.* **2020**, *12*, 139. [CrossRef]

122. Feng, A.; Zhou, J.; Vories, E.D.; Sudduth, K.A.; Zhang, M. Yield estimation in cotton using UAV-based multi-sensor imagery. *Biosyst. Eng.* **2020**, *193*, 101–114. [CrossRef]

123. Maimaitijiang, M.; Sagan, V.; Sidike, P.; Hartling, S.; Esposito, F.; Fritschi, F.B. Soybean yield prediction from UAV using multimodal data fusion and deep learning. *Remote Sens. Environ.* **2020**, *237*. [CrossRef]

124. Sangha, H.S.; Sharda, A.; Koch, L.; Prabhakar, P.; Wang, G. Impact of camera focal length and sUAS flying altitude on spatial crop canopy temperature evaluation. *Comput. Electron. Agric.* **2020**, *172*. [CrossRef]

125. Tucci, G.; Parisi, E.I.; Castelli, G.; Errico, A.; Corongiu, M.; Sona, G.; Viviani, E.; Bresci, E.; Preti, F. Multi-sensor UAV application for thermal analysis on a dry-stone terraced vineyard in rural Tuscany landscape. *ISPRS Int. J. Geo-Inf.* **2019**, *8*, 87. [CrossRef]

126. Parisi, E.I.; Suma, M.; Güleç Korumaz, A.; Rosina, E.; Tucci, G. Aerial platforms (uav) surveys in the vis and tir range. Applications on archaeology and agriculture. *ISPRS Ann. Photogramm. Remote Sens. Spat. Inf. Sci.* **2019**, *42*, 945–952. [CrossRef]

127. Filippo, S.; Gennaro, D.; Matese, A.; Gioli, B.; Toscano, P.; Zaldei, A.; Palliotti, A.; Genesio, L. Multisensor approach to assess vineyard thermal dynamics combining high- resolution unmanned aerial vehicle (UAV) remote sensing and wireless sensor network (WSN) proximal sensing. *Sci. Hortic.* **2017**, *221*, 83–87. [CrossRef]

128. Modica, G.; Praticò, S.; Di Fazio, S. Abandonment of traditional terraced landscape: A change detection approach (a case study in Costa Viola, Calabria, Italy). *Land Degrad. Dev.* **2017**, *28*, 2608–2622. [CrossRef]

129. Yang, G.; Liu, J.; Zhao, C.; Li, Z.; Huang, Y.; Yu, H.; Xu, B.; Yang, X.; Zhu, D.; Zhang, X.; et al. Unmanned Aerial Vehicle Remote Sensing for Field-Based Crop Phenotyping: Current Status and Perspectives. *Front. Plant Sci.* **2017**, *8*. [CrossRef] [PubMed]

130. Neely, L.; Rana, A.; Bagavathiannan, M.V.; Henrickson, J.; Putman, E.B.; Popescu, S.; Burks, T.; Cope, D.; Ibrahim, A. Unmanned Aerial Vehicles for High- Throughput Phenotyping and Agronomic. *PLoS ONE* **2016**, 1–26. [CrossRef]

131. Sankaran, S.; Khot, L.R.; Espinoza, C.Z.; Jarolmasjed, S.; Sathuvalli, V.R.; Vandemark, G.J.; Miklas, P.N.; Carter, A.H.; Pumphrey, M.O.; Knowles, N.R.; et al. Low-altitude, high-resolution aerial imaging systems for row and field crop phenotyping: A review. *Eur. J. Agron.* **2015**, *70*, 112–123. [CrossRef]

132. Natarajan, S.; Basnayake, J.; Wei, X.; Lakshmanan, P. High-throughput phenotyping of indirect traits for early-stage selection in sugarcane breeding. *Remote Sens.* **2019**, *11*, 2952. [CrossRef]

133. Gracia-Romero, A.; Kefauver, S.C.; Fernandez-Gallego, J.A.; Vergara-Díaz, O.; Nieto-Taladriz, M.T.; Araus, J.L. UAV and ground image-based phenotyping: A proof of concept with durum wheat. *Remote Sens.* **2019**, *11*, 1244. [CrossRef]

134. Perich, G.; Hund, A.; Anderegg, J.; Roth, L.; Boer, M.P.; Walter, A.; Liebisch, F.; Aasen, H. Assessment of Multi-Image Unmanned Aerial Vehicle Based High-Throughput Field Phenotyping of Canopy Temperature. *Front. Plant Sci.* **2020**, *11*, 1–17. [CrossRef]

135. Berliner, P.; Oosterhuis, D.M.; Green, G.C. Evaluation of the infrared thermometer as a crop stress detector. *Agric. For. Meteorol.* **1984**, *31*, 219–230. [CrossRef]

136. Blum, A.; Shpiler, L.; Golan, G.; Mayer, J. Yield stability and canopy temperature of wheat genotypes under drought-stress. *Field Crops Res.* **1989**, *22*, 289–296. [CrossRef]

137. Amani, I.; Fischer, R.A.; Reynolds, M.P. Canopy temperature depression association with yield of irrigated spring wheat cultivars in a hot climate. *J. Agron. Crop Sci.* **1996**, *176*, 119–129. [CrossRef]

138. Lopes, M.S.; Reynolds, M.P. Partitioning of assimilates to deeper roots is associated with cooler canopies and increased yield under drought in wheat. *Funct. Plant Biol.* **2010**, *37*, 147–156. [CrossRef]

139. Maimaitijiang, M.; Ghulam, A.; Sidike, P.; Hartling, S.; Maimaitiyiming, M.; Peterson, K.; Shavers, E.; Fishman, J.; Peterson, J.; Kadam, S.; et al. Unmanned Aerial System (UAS)-based phenotyping of soybean using multi-sensor data fusion and extreme learning machine. *ISPRS J. Photogramm. Remote Sens.* **2017**, *134*, 43–58. [CrossRef]

140. Pavelis, G.A. Economic Survey of Farm Drainage. In *Farm Drainage in the United States: History, Status, and Prospects*; Georg, A., Ed.; Doc. RESUME CE 050 265 Pavelis, Miscellaneous Publication Number 1455; U.S. Government Printing Office: Washington, DC, USA, 1987.

141. Fausey, N.R. Drainage management for humid regions. *Int. Agric. Eng. J.* **2005**, *14*, 209–214.

142. Lal, R.; Taylor, G.S. Drainage and Nutrient Effects in a Field Lysimeter Study: II. Mineral Uptake by Corn. *Soil Sci. Soc. Am. J.* **1970**, *34*, 245–248. [CrossRef]

143. Freeland, R.; Allred, B.; Eash, N.; Martinez, L.; Wishart, D.B. Agricultural drainage tile surveying using an unmanned aircraft vehicle paired with Real-Time Kinematic positioning—A case study. *Comput. Electron. Agric.* **2019**, *165*, 104946. [CrossRef]

144. Cannell, R.Q.; Gales, K.; Snaydon, R.W.; Suhail, B.A. Effects of short-term waterlogging on the growth and yield of peas (Pisum sativum). *Ann. Appl. Biol.* **1979**, *93*, 327–335. [CrossRef]

145. Du, B.; Arnold, J.G.; Saleh, A.; Jaynes, D.B. Development and application of SWAT to landscapes with tiles and potholes. *Trans. Am. Soc. Agric. Eng.* **2005**, *48*, 1121–1133. [CrossRef]

146. Allred, B.; Martinez, L.; Fessehazion, M.K.; Rouse, G.; Williamson, T.N.; Wishart, D.B.; Koganti, T.; Freeland, R.; Eash, N.; Batschelet, A.; et al. Overall results and key findings on the use of UAV visible-color, multispectral, and thermal infrared imagery to map agricultural drainage pipes. *Agric. Water Manag.* **2020**, *232*, 106036. [CrossRef]

147. Mira, M.; Valor, E.; Boluda, R.; Caselles, V.; Coll, C. Influence of soil water content on the thermal infrared emissivity of bare soils: Implication for land surface temperature determination. *J. Geophys. Res. Earth Surf.* **2007**, *112*, 1–11. [CrossRef]

148. Allred, B.; Eash, N.; Freeland, R.; Martinez, L.; Wishart, D.B. Effective and efficient agricultural drainage pipe mapping with UAS thermal infrared imagery: A case study. *Agric. Water Manag.* **2018**, *197*, 132–137. [CrossRef]

149. Williamson, T.N.; Dobrowolski, E.G.; Meyer, S.M.; Frey, J.W.; Allred, B.J. Delineation of tile-drain networks using thermal and multispectral imagery—Implications for water quantity and quality differences from paired edge-of-field sites. *J. Soil Water Conserv.* **2019**, *74*, 1–11. [CrossRef]

150. Woo, D.K.; Song, H.; Kumar, P. Mapping subsurface tile drainage systems with thermal images. *Agric. Water Manag.* **2019**, *218*, 94–101. [CrossRef]

151. Ezenne, G.I.; Jupp, L.; Mantel, S.K.; Tanner, J.L. Current and potential capabilities of UAS for crop water productivity in precision agriculture. *Agric. Water Manag.* **2019**, *218*, 158–164. [CrossRef]

152. Maes, W.H.; Steppe, K. Perspectives for Remote Sensing with Unmanned Aerial Vehicles in Precision Agriculture. *Trends Plant Sci.* **2019**, *24*, 152–164. [CrossRef] [PubMed]

153. Zhang, J.; Huang, Y.; Pu, R.; Gonzalez-Moreno, P.; Yuan, L.; Wu, K.; Huang, W. Monitoring plant diseases and pests through remote sensing technology: A review. *Comput. Electron. Agric.* **2019**, *165*. [CrossRef]

154. Khaliq, A.; Comba, L.; Biglia, A.; Ricauda Aimonino, D.; Chiaberge, M.; Gay, P. Comparison of satellite and UAV-based multispectral imagery for vineyard variability assessment. *Remote Sens.* **2019**, *11*, 436. [CrossRef]

155. McCabe, M.F.; Houborg, R.; Lucieer, A. High-resolution sensing for precision agriculture: From Earth-observing satellites to unmanned aerial vehicles. *Remote Sens. Agric. Ecosyst. Hydrol.* **2016**, *9998*, 999811. [CrossRef]

156. Houborg, R.; McCabe, M.F. High-Resolution NDVI from planet's constellation of earth observing nano-satellites: A new data source for precision agriculture. *Remote Sens.* **2016**, *8*, 768. [CrossRef]

157. Selva, D.; Krejci, D. A survey and assessment of the capabilities of Cubesats for Earth observation. *Acta Astronaut.* **2012**, *74*, 50–68. [CrossRef]

158. Tsouros, D.C.; Bibi, S.; Sarigiannidis, P.G. A review on UAV-based applications for precision agriculture. *Information* **2019**, *10*, 349. [CrossRef]

159. Borgogno Mondino, E.; Gajetti, M. Preliminary considerations about costs and potential market of remote sensing from UAV in the Italian viticulture context. *Eur. J. Remote Sens.* **2017**, *50*, 310–319. [CrossRef]

 *remote sensing*

Article

Biomass and Crop Height Estimation of Different Crops Using UAV-Based Lidar

Jelle ten Harkel, Harm Bartholomeus and Lammert Kooistra *

Wageningen University & Research, Laboratory of Geo-Information Science and Remote Sensing, Droevendaalsesteeg 3, 6708 PB Wageningen, The Netherlands; jelle.tenharkel@wur.nl (J.t.H.); harm.bartholomeus@wur.nl (H.B.)
* Correspondence: lammert.kooistra@wur.nl

Received: 27 September 2019; Accepted: 17 December 2019; Published: 18 December 2019

Abstract: Phenotyping of crops is important due to increasing pressure on food production. Therefore, an accurate estimation of biomass during the growing season can be important to optimize the yield. The potential of data acquisition by UAV-LiDAR to estimate fresh biomass and crop height was investigated for three different crops (potato, sugar beet, and winter wheat) grown in Wageningen (The Netherlands) from June to August 2018. Biomass was estimated using the 3DPI algorithm, while crop height was estimated using the mean height of a variable number of highest points for each m^2. The 3DPI algorithm proved to estimate biomass well for sugar beet ($R^2 = 0.68$, RMSE = 17.47 g/m^2) and winter wheat ($R^2 = 0.82$, RMSE = 13.94 g/m^2). Also, the height estimates worked well for sugar beet ($R^2 = 0.70$, RMSE = 7.4 cm) and wheat ($R^2 = 0.78$, RMSE = 3.4 cm). However, for potato both plant height ($R^2 = 0.50$, RMSE = 12 cm) and biomass estimation ($R^2 = 0.24$, RMSE = 22.09 g/m^2), it proved to be less reliable due to the complex canopy structure and the ridges on which potatoes are grown. In general, for accurate biomass and crop height estimates using those algorithms, the flight conditions (altitude, speed, location of flight lines) should be comparable to the settings for which the models are calibrated since changing conditions do influence the estimated biomass and crop height strongly.

Keywords: UAV-based LiDAR; biomass; crop height; field phenotyping

---

1. Introduction

Phenotyping of crops is important to estimate biomass and the potential yield of new varieties of agricultural crops. Due to the increasing need to increase food production and improve the associated quality, it is important to optimize the yield, for which accurate estimation of biomass during the growing season is needed. In this context, phenotyping focuses on the characterization of morphological as well as physiological crop traits. Morphological parameters, such as plant height, stem diameter, leaf area or leaf area index (LAI), leaf angle, stalk length, and in-plant space [1], can be determined with LiDAR (light detection and ranging). Research on phenotyping using LiDAR often focusses on one specific crop, for example, wheat [2,3] or cotton [4].

Phenotyping of individual plants can be done in very high detail with LiDAR, analyzing complex phenotypical properties, such as leaf area, leaf width, and leaf angle [5,6]. For this, plants were put on a slowly turning platform while a fixed LiDAR instrument was used. A drawback of this setup is the low throughput of the scanning system while there is also a need to evaluate phenotypes under field conditions.

For high-throughput phenotyping, traits such as plant-height, LAI, and leaf cover fraction are determined directly in the field, using a LiDAR-based system mounted on a vehicle or RGB cameras mounted on a UAV (unmanned aerial vehicle) [3,4,7]. Tractor based LiDAR systems data have shown good correlation with in situ field measurements of plant height. Sun et al. [4] published an R^2 of 0.98

for cotton plants, [2] published an R^2 of 0.99 for wheat, and [7] showed an R^2 of 0.90 for wheat. These studies show the capability of LiDAR to measure basic phenotypes such as plant height. However, LiDAR systems mounted on tractors can be unsuitable for labour-intensive crops such as rice or in orchards [8], for example, due to compaction of the soil [9]. A UAV equipped with a LiDAR system can overcome those limitations.

Earlier research on the relation between plant height and biomass was based on varying approaches for plant height measurements. Madec et al. [7] found an R^2 of 0.88 for the correlation between plant height and field-measured biomass, using a tractor based LiDAR system. Bendig et al. [10] used a structure from a motion (SfM) technique on UAV acquired imagery to derive plant height and found an R^2 of 0.81 between field-measured height and SfM derived height.

In the last few years, LiDAR systems have been miniaturized, resulting in lower weights and reduced dimensions, and as a result, can be operated from UAVs. This development opens the way towards high throughput derived, more complex products like biomass and yield, thus, improving the speed and frequency at which these plant traits can be acquired in the field in an undisturbed way.

LiDAR-based biomass estimates of agricultural crops can be derived in different ways. Based on the Lambert-Beer LAI model of [11], the authors of [2] developed a biomass prediction model called the 3-Dimensional Profile Index (3DPI). Where the LIDAR 3D point cloud is divided into layers and for each layer, the fraction of points divided by the total amount of points is calculated. These layers are then summed, and the 3DPI values can be related to biomass with a linear function. As follow up [2] also proposed a voxel-based method (3DVI) to estimate the biomass of wheat. The 3DVI method divides the LIDAR 3D point cloud in voxels of equal size and calculates the ratio between the number of voxels containing points and the number of subdivisions in the horizontal plane. They showed that 3DVI could estimate wheat biomass accurately with an R^2 of 0.91, and for 3DPI, an R^2 of 0.93 was achieved. Jimenez et al. [2] used a tractor based LiDAR system. An alternative approach using airborne LiDAR was proposed by [12], who used Pearson's correlation analysis and structural equation modelling (SEM) to estimate plant height and LAI, which proved to be the best predictors of the biomass of maize (R^2 of 0.87).

The goal of this study was to investigate the potential of UAV-LiDAR for estimation of crop height and fresh weight biomass for three different agricultural crops. For this, the RIEGL RiCOPTER with a VUX-SYS LiDAR system was flown over fields with sugar beet, wheat, and potatoes, on a number of moments during the growing season. First, it was investigated how accurate this system can estimate crop height and biomass. Next, we answered the question if it is possible to create models that are generally applicable for different crops. Finally, we did experiments with different UAV flight patterns, altitudes, and speeds to investigate if this influences the LiDAR-derived plant height and biomass estimation. Optimization of the flight parameters was beyond the scope of this paper.

2. Materials and Methods

2.1. Study Area

Three fields with different crops, each covering approximately 2 ha, were selected at the experimental farm of Wageningen University, located just north of Wageningen in the Netherlands (Figure 1). The three types of arable crops in this study are sugar beet, winter wheat, and potatoes (Table 1), which were planted on a Placic Podzol [13]. The variation in elevation in the fields is very small with a height difference of less than 10 cm going from east to west in the area. Orientation of the main crop rows for all three crops is north–south. The plant density of winter wheat was 260 plants/m^2. Planting distance for potato and sugar beet was 30 cm and 23 cm, respectively.

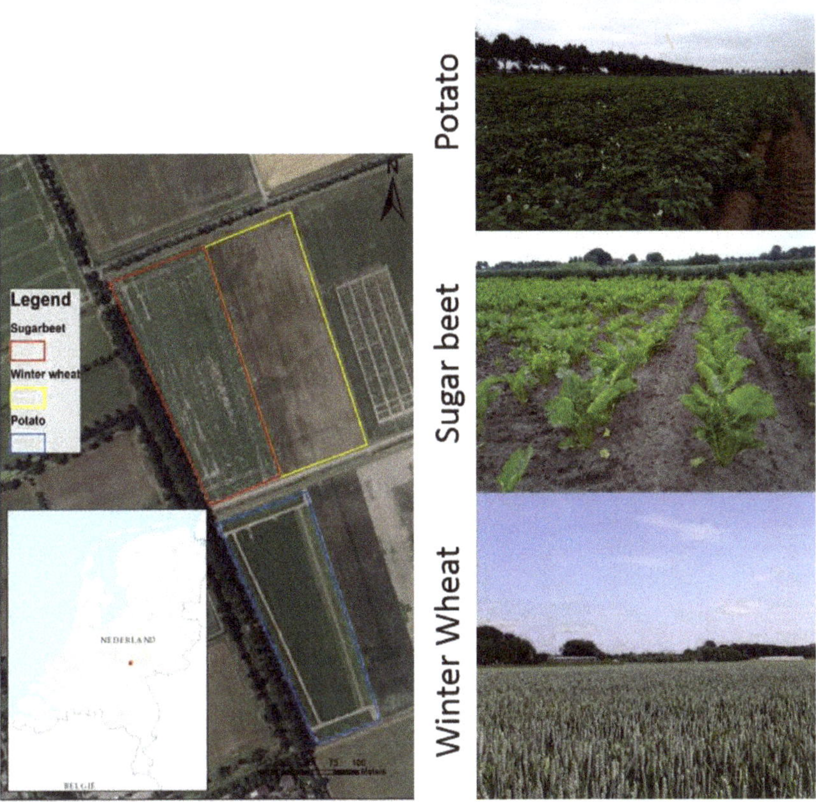

Figure 1. Left: Overview of fields within the study area: indicated are the three different fields and crops that are studied. The image was taken on 4 June 2018. The large inset shows the location of Wageningen in the Netherlands. Right: Impression of the three different crops used in this research. Potato and sugar beet images were taken on 5 June 2018. The winter wheat image was taken on 25 June 2018.

Table 1. General information about the three crops planted in the study area.

Crop Type	Sugar Beet	Potato	Winter Wheat
Fieldname	Dijkgraaf	Com02	Braam
Planting month	April 2018	April 2018	October 2017
Harvesting month	October 2018	September 2018	August 2018
Size (hectare)	1.99	2.0	1.78

2.2. Data Acquisition

UAV-LiDAR data and field measurements of crop height and biomass were collected simultaneously on four dates (Table 2). For the 1st, 2nd, and 4th flight date, field data were collected on the same day, directly starting after the flight and finished within 3 to 4 hours. The 3rd flight day was not planned initially, so no field measurements were done on that day, but this flight was conducted only 2 days after the 2nd flight. Therefore, the field data collected at the day of the 2nd flight were used for both those flights.

Table 2. Flying dates and corresponding growth stages for the crops in selected fields.

Day	Date of Flight	Sugar Beet	Potato	Winter Wheat
1	07-06-2018	Vegetative	Vegetative/Tuber initiation	Heading
2	25-06-2018	Vegetative	Tuber bulking	Ripening
3	27-06-2018	Vegetative	Tuber bulking	Ripening
4	11-07-2018	Vegetative	Tuber bulking	Ripening

2.2.1. Field Data Collection

Plant height was measured using a ruler at 14–21 locations in each field. Locations were chosen in such a way that differences within the field were well represented. On each location, an average height was calculated based on three height measurements of randomly selected plants taken within 15 cm of each other. Biomass was determined through destructive sampling at different locations within the field, differing from the plant height sampling locations [14]. For potato and sugar beet, five biomass samples per field were taken and for winter wheat three. Biomass was harvested after the flights. Biomass samples were taken for the first, second, and fourth flight day only, which resulted in 15 samples for both potato and sugar beet and 9 samples for winter wheat. For potato, biomass was harvested for one meter along the potato ridge and converted to biomass per 1 m^2. The same procedure was used for harvesting the sugar beet, where one meter in the planting direction was harvested and converted to biomass per 1 m^2. For wheat, an area of 1 m^2 area was harvested. Biomass was measured as fresh biomass weight directly after harvesting. Table 3 shows the mean and standard deviation of every field campaign. The location of every crop trait measurement was measured with a Topcon HIPER V RTK-GNSS system (TOPCON, Japan), which has a sub centimeter accuracy.

Table 3. Mean and standard deviation of field measurements separated per crop and sampling date.

Crop	Sampling Date	Sample Size	Plant Height Mean (m)	Standard Deviation (m)	Sample Size	Biomass Fresh Weight Mean (g/m^2)	Standard Deviation (g/m^2)
	07 June 2018	21	0.599	0.073	5	3364.93	461.65
Potato	25 June 2018	20	0.690	0.229	5	3546.40	846.19
	11 July 2018	20	0.547	0.151	5	3749.07	889.75
	07 June 2018	16	0.275	0.098	5	1237.55	650.85
Sugar beet	25 June 2018	20	0.501	0.099	5	3834.40	1374.40
	11 July 2018	15	0.445	0.105	5	3656.00	1191.38
	07 June 2018	17	0.483	0.063	3	3477.37	751.71
Winter wheat	25 June 2018	14	0.489	0.076	3	3371.67	940.45
	11 July 2018	20	0.449	0.076	3	1583.00	473.11

2.2.2. UAV-LiDAR Data

LiDAR data were collected using the RIEGL RiCOPTER UAV system (Riegl, Austria) with the VUX-SYS laser scanner mounted underneath. The authors of [15] provided a description of the details of the system. The VUX LiDAR unit has an accuracy of 1 cm and precision of 0.5 cm. The absolute positioning error on the ground also depends on the IMU accuracy, which was <5 cm in the horizontal and <10 cm in vertical direction. Analysis of previous datasets acquired with the same setup resulted in positioning errors of <5 cm. The scanner pulse repetition rate was set at 550 kHz, the scanner angle from 30–330 degrees, and the scanner speeds were synchronized with the UAV forward speed to create a regular point spacing. On the predefined dates, UAV-LiDAR data was acquired by flying at 40 m above ground level (m.a.g.l.) with a programmed speed of 6 m/s. The actual flight altitude and speed can deviate slightly from those preset values and were calculated from the flight recordings afterwards (Table 4: mean flying height and speed).

Further, to assess the influence of the flying height and speed on the accuracy of acquired phenological parameters, additional flights were done within a few days after the second flight day, with heights varying from 20 to 90 m above ground level, and with programmed speeds ranging from

3 m/s to 8 m/s. An overview of the flights and the actual flight data is shown in Table 4. The code given in Table 4 is later used to identify the data acquired from each flight.

Table 4. Overview of flight dates, including flight codes and corresponding general flight details The flight codes are 1) indicating the flight number; 2) the flight altitude (LA = Low Altitude, MA = Medium Altitude, HA = High Altitude; 3) and flight speed (LS = Low Speed, MS = Medium Speed, HS = High Speed). Those codes will be used throughout the paper.

Flight Date	Flight Number	Related Field Data	Code	Mean Flying Height (m.a.g.l.)	Mean Flying Speed (m/s)	Number of Flight Lines	Average Point Density (points/m^2) Wheat	Potato	Sugar Beet
07-6-2018	1	07-6-2018	Day1-MA-MS	41.84	5.85	11	997	833	933
25-6-2018	1	25-6-2018	Day2-MA-MS	45.32	5.88	9	960	777	664
25-6-2018	2	25-6-2018	Day2-LA-MS	24.16	4.17	7	1516	659	805
25-6-2018	3	25-6-2018	Day2-LA-LS	20.05	2.73	5	2755	1402	1262
27-6-2018	1	25-6-2018	Day3-HA-HS	92.68	7.39	11	575	446	432
27-6-2018	2	25-6-2018	Day3-LA-LS	22.71	2.93	9	2257	2097	804
11-7-2018	1	11-7-2018	Day4-MA-MS	42.14	5.78	11	1011	873	1003

If during the analysis it appeared that field measurements of plant height or biomass ended up in the same pixel as a driving path as seen from the Lidar 3D point cloud, the point was moved 1 pixel away from the driving path. This was the case for one point in the winter wheat dataset. Further, the field measurement locations were selected before the flights were made. Points that appeared to be in areas with a much lower LiDAR point density were excluded from the height and biomass analysis. Two winter wheat points during the last flight date had to be excluded for this reason.

2.3. Data Pre-Processing

Riegl RiCOPTER data was pre-processed to co-registered point cloud datasets (Figure 2) using the standard Riegl processing chain as outlined by Brede et al. [15]. This includes trajectory processing, for which a virtual GNSS base station was used. The movements of the UAV platform were corrected using the POSPac software (Applanix, Canada), which combines the GNSS correction and the movement data recorded by the IMU, with a reported precision of ~1–2 cm for all flights. To produce a geo-referenced point cloud, the flight path, and raw point cloud data were combined in RiPROCESS (Riegl, Austria). Further refinement of the point cloud, using automatically detected surfaces, was done by RiPRECISION (Riegl, Austria). The final co-registered and geo-referenced point cloud datasets were stored in LAS format.

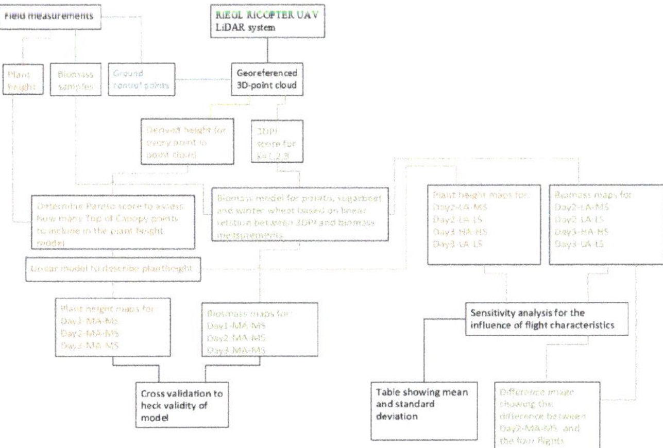

Figure 2. Overview of methodology for height and biomass retrieval from UAV-LiDAR observations for crops.

The point clouds were clipped for the boundaries of the three fields (Figure 1). The ground points were selected using LasTools (Rapidlasso, USA) with the lasground function. From the points classified as ground, a triangulated irregular network (TIN) was created. The height of each non-ground classified point was calculated as the height above this TIN, using the "replace z" option in lasground. The same procedure was applied for all the three fields.

2.4. Data Analysis

2.4.1. LiDAR Based Plant Height

Due to the difference in the structure of the crop canopies, the highest point in the point cloud is not always the best representation for the field-measured crop height. Therefore, for each grid cell of 1 m^2 the average plant height was derived by including a certain number of top-of canopy points. To determine the amount of top-of-canopy points needed to get the best plant height estimation, a theory was used that is more commonly used in economics to decide on the best allocation of products: the Pareto efficiency score [16]. This method compares different alternatives and based on preset criteria, calculates a score of optimal performance. In this case, we used this method here to decide based on two criteria, namely the highest R^2 and lowest root mean squared error (RMSE), which are calculated for each alternative of measured plant height and a certain number of top-of canopy points. The method then only selects the combination for which no better combination exists. This can result in multiple optimal Pareto efficiency scores. Next, from these Pareto efficient combinations, the combination with the highest R^2 is selected. This was done once for the whole growing season for every crop. The corresponding number of top-of canopy points of this combination was used to calculate the plant height of each raster cell within the field. A linear model relating field plant height measurements to LiDAR-derived plant height was built using the 3D LiDAR point cloud by taking the optimal number of top-of-canopy points into account.

2.4.2. Crop Biomass

To estimate biomass from the LiDAR point cloud, the 3DPI method as developed by the authors of [2] and described by Equation (1) was used and calculated for a cell size of 1 m^2:

$$3DPI = \sum_{i=0}^{i=n}\left(\frac{p_i}{p_t}e^{k\frac{p_{cs}}{p_t}}\right) \tag{1}$$

where i is a given 10 mm vertical layer with 0 being the ground layer and n the uppermost layer respectively; k is a tuning parameter changing from -1 to 5 by steps of 0.05; p_i is the number of LiDAR points for a given layer of 50 mm; p_t is the total number of LiDAR points for all layers; and p_{cs} is the cumulative sum of LiDAR points intercepted above a given layer of 50 mm, as described in [2].

To derive the biomass prediction model, a linear regression was performed between the dimensionless 3DPI indicator and the biomass measured in the field based on the 3DPI indicator. First, the optimal value for the parameter k was determined. For each individual crop and flight, the 3DPI score was calculated by varying the k-parameter from -1 to 5 by increments of 0.05. The best k-parameter was chosen based on the correlation between the field measurements and the 3DPI score. The data for each crop was combined into a crop-specific seasonal prediction model for the flights Day1-MA-MS, Day2-MA-MS, and DAY4-MA-MS. Secondly, all the data of these flights and crops were combined into a general non-crop-specific seasonal prediction model.

Next, the flight data from DAY2-LA-MS, DAY2-LA-LS, DAY3-HA-HS, and DAY3-LA-LS were used to determine the effects of different flight specification and were compared to Day2-MA-MS. The 3DPI for these four flights was calculated using the same field measurements of plant height and biomass as for Day2-MA-MS. This enabled the analysis of different flight settings because only the 3DPI indicator differed between the flights so a direct comparison can be made. To visualize the differences between these four flights, maps of predicted biomass were derived and difference maps of each flight with flight Day2-MA-MS were calculated.

2.4.3. Model Validation

To assess the performance of the models (Figure 2) for determining biomass, plant height, and the general models, three statistical indicators were used; the coefficient of determination (R^2), the mean absolute error (MAE), and the RMSE. We included a cross validated MAE and RMSE, using a k-fold cross validation. For each k-fold, the out-sample RMSE and MAE were calculated, after which they were averaged over the n-number of k-folds used in the cross validation. These averaged out-sample RMSE and MAE were compared to the in-sample RMSE and MAE. This comparison shows the predictive power of the model (out-sample errors).

For the plant height analysis, a 10-fold cross validation was performed, using 61, 51, and 51 samples for potato, sugar beet, and wheat, respectively (Table 3). For the biomass estimation, a repeated 5×3-fold cross validation was performed. The reason for this repeated cross validation for biomass is the limited amount of biomass samples available for the statistical analysis. In this case, we used 15, 15, and 9 samples for potato, sugar beet, and wheat, respectively (Table 3).

For the biomass error analysis, the MAE and RMSE were then normalized by dividing by Xmax-Xmin, where Xmax is the highest measured biomass sample in the dataset and Xmin the lowest measured biomass sample in the dataset, resulting in a normalized mean absolute error (NMAE) and normalized RMSE (NRMSE).

3. Results

3.1. Crop Specific Models

3.1.1. Plant Height

Based on the analysis of the Pareto efficiency scores, the LiDAR plant height for each grid cell of 1 m^2 is based on the top 10 points for potato, on the top 90 points for sugar beet, and for winter wheat on the top 30 points. The average height of these points serves as the LiDAR-based plant height estimation.

The correlation between field-measured height and LiDAR-based plant height is highest for sugar beet and winter wheat (Figure 3). Plant height was most accurately derived for winter wheat with an RMSE of 3.4 cm. Determining plant height for potato proved to be the most difficult, showing relatively large residuals.

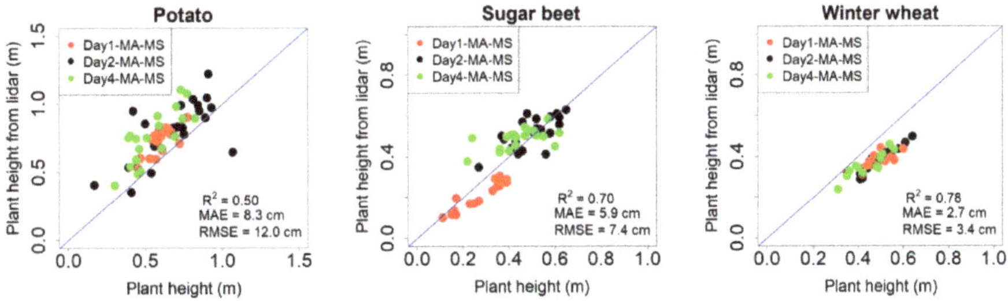

Figure 3. Scatter plots showing measured plant height vs derived plant height, where potato is based on the 10 highest points within a pixel, sugar beet on top 90 highest points, and winter wheat on the top 30 highest points. The blue line in the Figure is the 1:1 line.

Figure 3 shows that there is a general over prediction of plant height for potato while winter wheat shows an underestimation. As winter wheat has a relatively open and erectophile structure, the number of returns in the top of the canopy could be lower, resulting in an underestimation of height. Although sugar beet falls on the 1:1 line, the prediction of plant height has a higher spread than winter wheat, which is shown by the larger MAE and RMSE. The over prediction of potato is most likely the result of the complex canopy, which is explained in more detail in the discussion.

3.1.2. Biomass

Choosing k = 1 for potato and winter wheat and k = 3 for sugar beet results in the best fit for the biomass models. Values lower than k = 1 proved to be very unstable in the performance for the 3DPI algorithm, showing large fluctuation in R^2 values. Values higher than k = 3 became saturated and showed only minor increases in R^2. For potato and winter wheat the model errors, NMAE and NRMSE are the smallest for k = 1 (Figure 4), while for sugar beets the errors are smallest for k = 3.

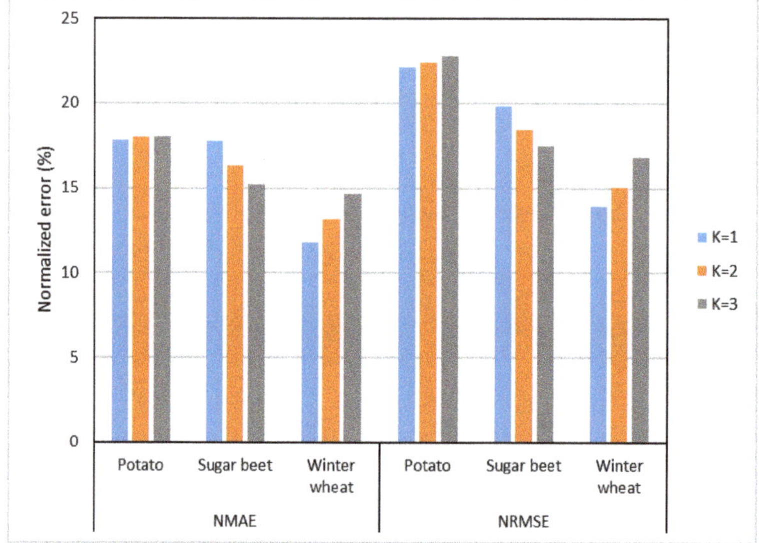

Figure 4. Error analysis for biomass derivation based on the 3-Dimensional Profile Index (3DPI) indicator, deviating K between 1, 2, or 3. Showing biomass estimations for potato, sugar beet, and winter wheat, including the dimensionless normalized in-sample.

Potato shows a relatively large difference between the NMAE and NRMSE compared to those of sugar beet and winter wheat. This indicates that the low R^2 is most likely the result of large residuals (Figure 4). Furthermore, the spread in points for potato is less equally distributed between large and small residuals compared to the spread of points for sugar beet and winter wheat (Figure 5). This explains the larger NRMSE and NMAE of potato.

The prediction model for potato does not provide reliable estimates of biomass, especially for values higher than 3500 g/m^2 (Figure 5) in which biomass is underestimated. Furthermore, the scatterplot does not show a strong relation. For sugar beet, there is a good correspondence between the fresh biomass in the field and the LiDAR predicted biomass (Figure 5). Also, the crop development during the growing season is captured well: during flight 1, the sugar beets were still small, and the field was only partially covered, while during flights 2 and 4 the sugar beets had grown, which is also derived from the UAV-LiDAR observations.

The biomass of winter wheat was determined most accurately. For flights 1 and 2, the points are practically on the 1:1 line (Figure 5). During the 4th flight, the wheat had ripened, which lowered the amount of fresh biomass.

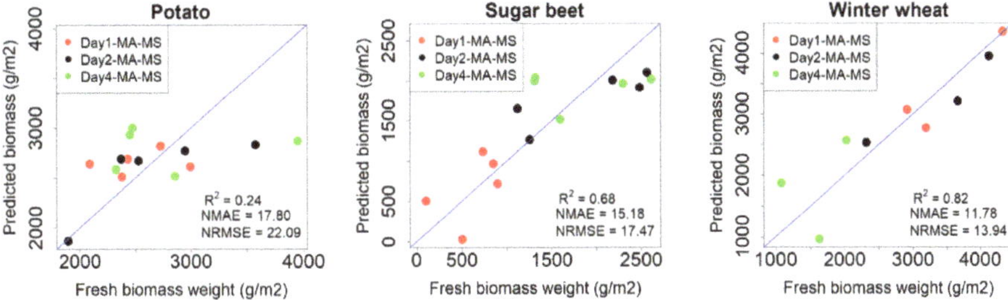

Figure 5. Scatter plots showing the relation between measured biomass and predicted for potato (k = 1), sugar beet (k = 3) and winter wheat (k = 1). The blue line in the Figure is the 1:1 line.

3.2. General Prediction Models

3.2.1. Crop Height

When combining the crops into a general prediction model for plant height, the RMSE is 10.1 cm (Figure 6). This error is much larger compared to the crop-specific models. For example, this accuracy was 3.4 cm for winter wheat in the single crop model (Figure 3).

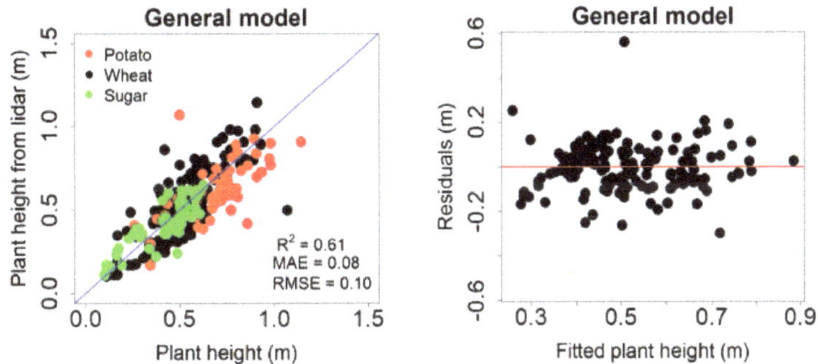

Figure 6. Scatterplot showing measured plant height versus LiDAR-derived plant height and the residual plot for the general plant height. Showing the model residuals versus fitted plant height. The colours in the left Figure show each separate crop. The blue line indicates the 1:1 line.

The Pareto method is used to determine the optimized number of points to be used to calculate the height per pixel, which showed that using 100 points per pixel yielded the best fit for the general model, resulting in an R^2 of 0.61 for the general model. This is lower than the R^2 for sugar beet and winter wheat (0.70 and 0.78, respectively), but higher than that of potato (0.49). The residual plot shows a random distribution of points with one outlier (Figure 6). This point comes from the potato dataset, but no valid reason was found to exclude the point.

3.2.2. Biomass

The general model for biomass performed worse for biomass prediction of winter wheat; the normalized error increased to 17.07% for the general model compared to 13.9% for the crop-specific model (Figure 7) based on tuning-parameter k = 3. The general models were fitted using different k-factors, where the R^2 was 0.47 for k = 1, 0.49 for k = 2 and 0.50 for k = 3. The generic model saturates for higher biomass predictions, especially for winter wheat (Figure 7). This indicates that a new model should be built for the higher biomass estimates. However, the general model can be used to predict biomass for potatoes (NRMSE = 22.1%) and sugar beet (NRMSE = 17.47%), in which NRMSE values are larger than those of the general model (NRMSE = 17.07%).

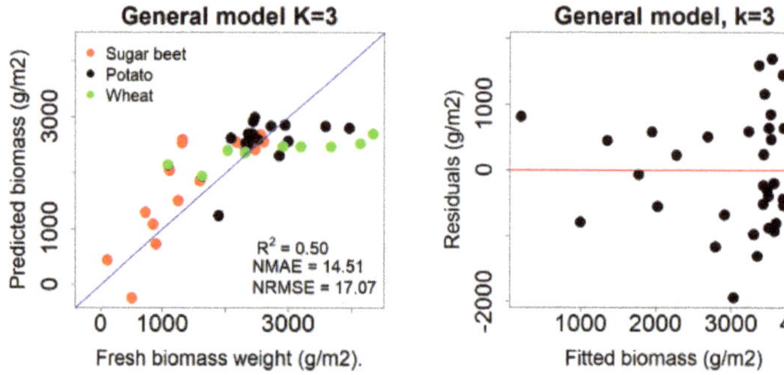

Figure 7. Scatter plot showing the relation between measured biomass and fitted biomass, for k = 3. The colours indicate the original crop data used for the model. The blue line in the Figure is the 1:1 line. On the right, the corresponding residual plot is shown. Showing the fitted plant height plotted against the model residuals.

The cross validated NRMSE shows that biomass can be predicted on a new dataset with an NRMSE_cv of 17.61%. The NRMSE_cv is 0.54 larger than the NRMSE for k = 3. This indicates that the model could be used for fields, where no biomass samples were taken for calibration.

3.3. Influence of Flight Characteristics on Plant Height

Changes in flight characteristics (altitude and speed) result in differences in point cloud density (Figure 8), where a cross section (10 × 0.2 m) of the sugar beet point clouds of four flights with different flight characteristics are shown. The flights were done on two days, with one day in between, but we assumed that the crops did not change between those two days.

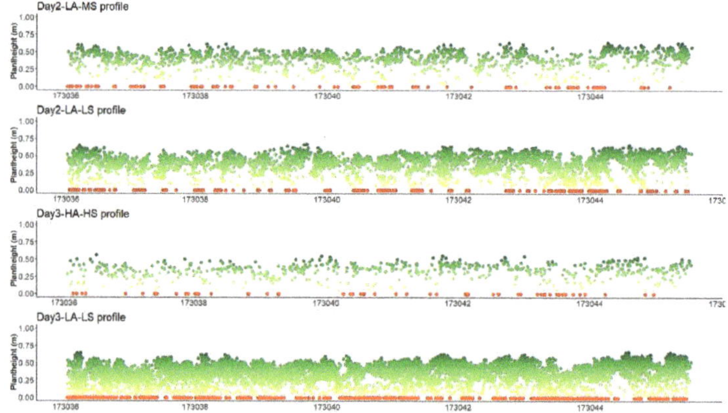

Figure 8. Profile plot for sugar beet showing a profile of 0.20 x 10 m. From top to bottom, it shows the flights DAY2-LA-MS, DAY2-LA-LS, DAY3-HA-HS, and DAY3-LA-LS. The green colours indicate plant material, where darker colours indicate a higher elevation of the point. The red points indicate the points classified as ground points.

Measured plant height is consistent for the different flight characteristics, except for the high and fast flight (DAY3-HA-HS: 90 m, 8 m/s). As can be seen in Figure 8, this results in a much lower point density and an underestimation of the crop height, shown by the lighter green colours in Figure 9. However, the spatial patterns of crop height between the four sets of flight characteristics are comparable. The locations with low and high crop height show a comparable location and distribution.

For DAY3-LA-LS height map, the north-western corner of the field was not covered well by the pre-programmed flight path, which resulted in a lower point density and, thus, an underestimation of crop height. This indicates that the flight characteristics do not influence the spatial patterns of estimated crop height, but it is important to keep them consistent for measuring temporal changes in crop height.

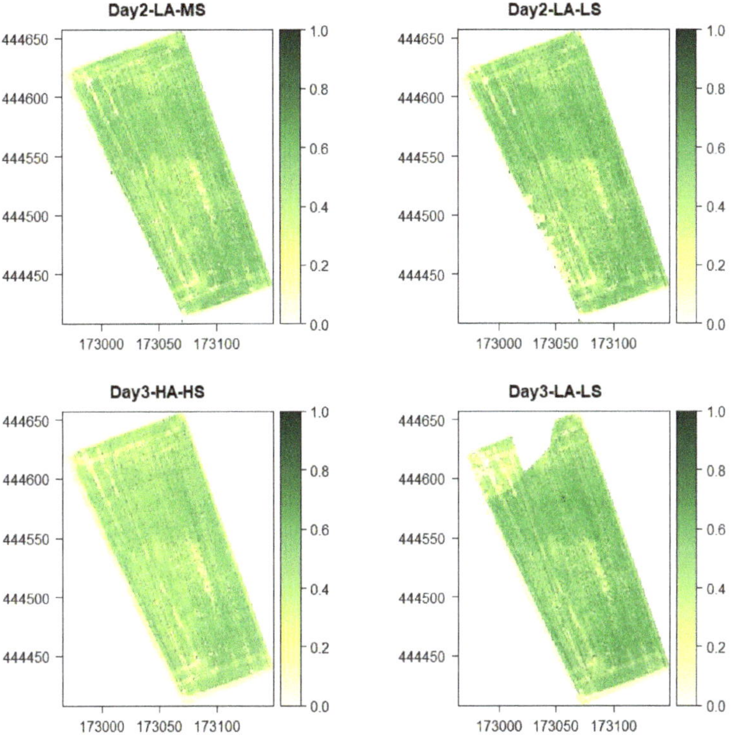

Figure 9. Maps indicating the estimated plant height in meters for sugar beet for DAY2-LA-MS, DAY2-LA-LS, DAY3-HA-HS, and DAY3-LA-LS. Darker greens indicate a higher plant height.

3.4. Influence of Flight Characteristics on Biomass

Biomass maps based on the 3DPI scores show varying patterns of biomass, dependent on the flight characteristics (Figure 10). The absolute differences, compared to flight Day2-MA-MS (40 m, 6 m/s), are more clearly visible in Figure 10. Flying higher and faster (DAY3-HA-HS), results in an underestimation of the amount of biomass. For flight DAY3-HA-HS, the average biomass estimation was 1622.1 g/m², where for Day2-MA-MS, the average biomass estimation was 2479.7 g/m² (Table 5). For the other flight characteristics, the overall estimate of biomass does not deviate too much from the standard flight settings, but there are locations where the estimated biomass for the adapted flights is lower compared to the standard flight. Further investigations learned that those areas are located exactly between two flight lines. It is important to indicate that the exact position of the flight lines was not consistent for the different flights. For each of the flights with the standard settings, the same pre-programmed flight paths were used, but for the flights with adapted altitude and speed, new flight plans were made, resulting in different flight patterns. This results in a different point density and point distribution along the dataset, which appears to influence the estimated biomass more than the altitude and speed.

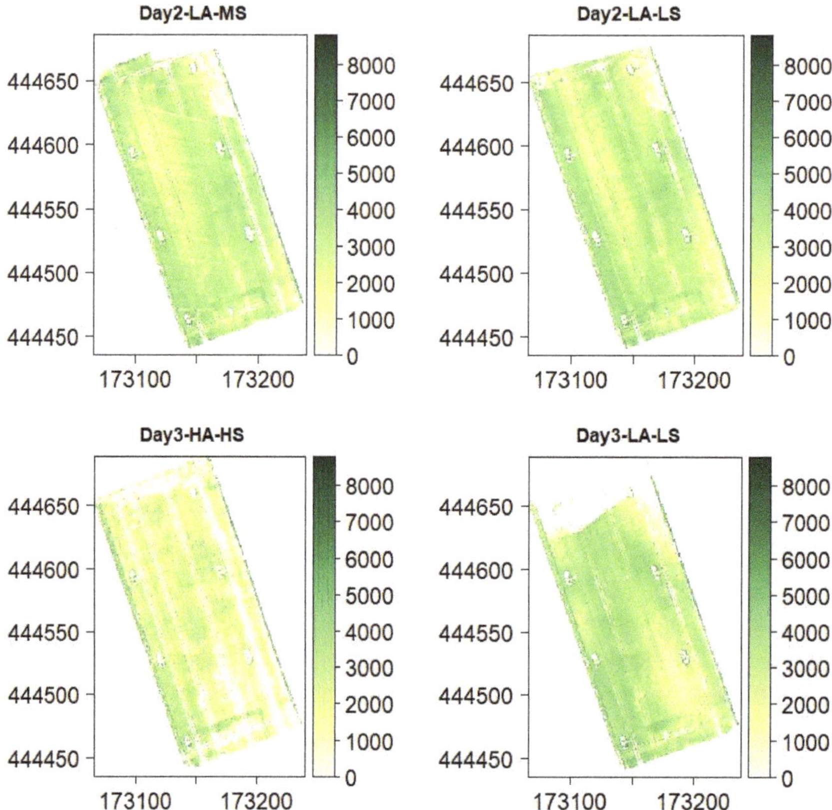

Figure 10. Maps indicating for flight DAY2-LA-MS, DAY2-LA-LS, DAY3-HA-HS, and DAY3-LA-LS the estimated biomass in grams for winter wheat. Darker greens mean a larger biomass prediction.

Table 5. Overview of mean and standard deviation for the flights discussed in Figures 8–10. The data for plant heights are from sugar beet, and the data for the biomass data are from the winter wheat maps. For Day1-MA-MS, Day2-MA-MS and Day4-MA-MS, there is no data (nd) in the difference cell because these were not compared to Day2-MA-MS.

Flight	Plant Height (m)				Biomass (g/m^2)			
	Data		Difference		Data		Difference	
	Mean	SD	Mean	SD	Mean	SD	Mean	SD
Day1-MA-MS	0.186	0.096	nd	nd	2181.38	1182.68	nd	nd
Day2-MA-MS	0.443	0.153	nd	nd	2479.68	1133.56	nd	nd
Day2-LA-MS	0.445	0.149	0.002	0.050	2603.81	1164.33	96.80	1014.95
Day2-LA-LS	0.466	0.160	0.022	0.085	2720.69	1157.64	202.70	929.70
Day3-HA-HS	0.403	0.134	0.040	0.056	1622.14	1132.21	−857.79	871.38
Day3-LA-LS	0.472	0.164	0.026	0.069	2435.03	1421.60	−44.97	1018.49
Day4-MA-MS	0.445	0.136	nd	nd	2164.12	1261.95	nd	nd

The biomass prediction for flight Day3-HA-HS shows an under prediction compared to the other flights (Table 5), where for plant height the difference in the mean between the flights are smaller. The sugar beet plant height estimation shows almost no variation between the flight characteristics describing the field data from 29 June 2018. The biomass estimations for winter wheat does appear to be sensitive to a difference in flight characteristics. There are large differences in the mean prediction between the different flight specifications, where for faster and higher flights (Figure 11 and Table 5: DAY3-HA-HS), an under prediction of biomass occurs compared to flight Day2-MA-MS.

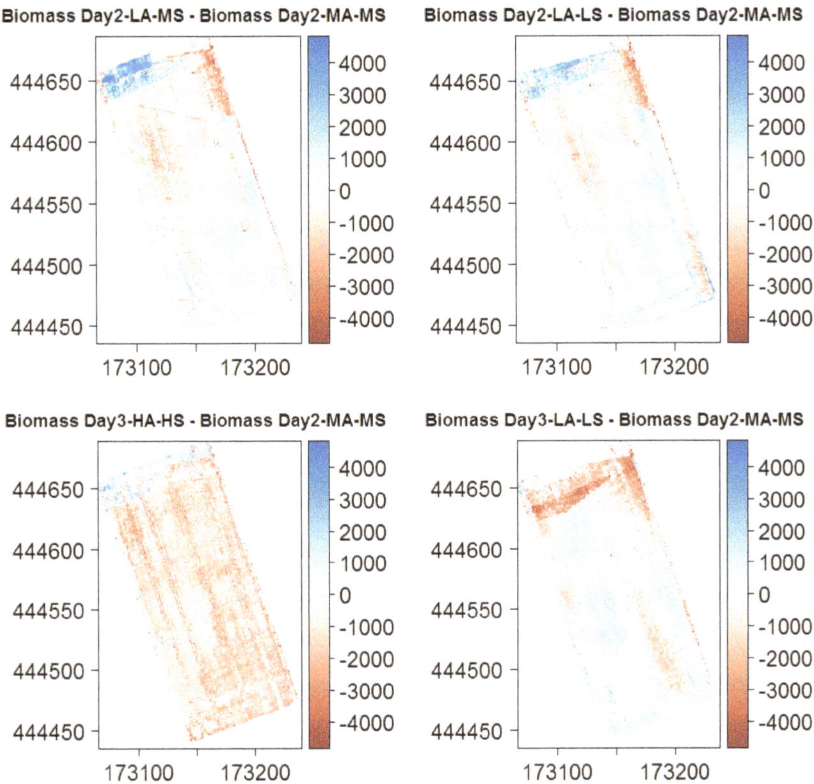

Figure 11. Maps indicating pixel differences between the biomass prediction for winter wheat of Flight Day2-MA-MS and, respectively, DAY2-LA-MS, DAY2-LA-LS, DAY3-HA-HS, and DAY3-LA-LS. Positive values (blue) indicate a higher prediction for, respectively, DAY2-LA-MS, DAY2-LA-LS, DAY3-HA-HS, and DAY3-LA-LS compared to Day2-MA-MS.

4. Discussion

4.1. Plant Height

Depending on the type of crop, the accuracy for plant height retrieval varied between 3.4 cm for winter wheat, 7.4 cm for sugar beet, and 12 cm for potato (Figure 3). Previous research efforts on determining plant height via LiDAR have focused on winter wheat, where Jimenez-Berni et al. [2] found an accuracy of 1.7 cm for a ground-based LiDAR system. Madec et al. [7] compared plant height derived from a LiDAR scanner on a ground vehicle with estimates from SfM analysis on RGB-UAV images. The latter method provided an accuracy of 3.47 cm for winter wheat plant height, proving that the methodology presented within this paper proves just as accurate. Homan et al. [17] showed that using a terrestrial laser scanner (TLS), the RMSE was 2.7 cm for wheat, which is comparable to the achieved accuracy by the RiEGL RiCOPTER UAV LiDAR system. This, and previous research, shows that using a UAV-based LiDAR system can be used to accurately assess plant height for winter wheat fields.

However, the accuracy drops for crops with a more complicated canopy structure, such as potato and sugar beet, where it proves to be harder to provide a plant height estimate with high accuracy.

This starts with the procedure for field measurements. Winter wheat is an easy-to-measure crop in the field due to its erectophile structure and a homogeneous pattern in height. Potato and sugar beet and their planophile leaf angle distribution prove to be harder to measure within the field, which is further complicated for potatoes since it is grown on ridges. Therefore, it is visually harder to determine the highest point of a specific location. We tried to account for this by averaging field measurements over the three highest points in close proximity, but we cannot ignore the fact that there are uncertainties in the field measurements (Table 3).

In particular, potato proved to be a difficult crop because it grows on ridges, showing the need for a good digital terrain model (DTM) of these ridges. A flight should be performed just after planting the potato to get a good DTM, which can be used to normalize the 3D LIDAR point cloud of the flights further during the growing season. This research started when the potato plants were already developed, resulting in a closed canopy, making it impossible to create a good DTM. This could be one of the reasons for the worse performance of potato height estimation.

As a result, the standard deviations for potato are generally larger than for the other two crops (Table 3). Especially on 25 June when the potato plants were fully grown, it proved more difficult to get accurate measurements.

4.2. Biomass

The 3DPI algorithm proved to estimate the biomass of winter wheat (NRMSE = 13.9% and R^2 = 0.82) and sugar beet (NRMSE = 17.47% and R^2 = 0.68) well, showing slightly higher accuracies compared to the paper of [2], where an NRMSE of 19.82% was achieved for winter wheat. It should be noticed that the early growth stages are not included in this study, which partially explains the differences in NRMSE. For winter wheat, similar results were obtained in [18], where a linear relation was also examined between LiDAR-derived plant height and biomass, reporting an R^2 of 0.88.

Possible improvements in the amount of biomass for winter wheat could be made using the dry weight of the plant material. Especially during the ripening phase, the water content of winter wheat decreases [19], resulting in a mismatch between the 3DPI algorithm and the measured fresh weight biomass. The 3DPI method is developed on plant structure and does not account for the changing water content within the winter wheat. Also, research by [20] showed that LiDAR laser scans proved to be useful in determining leaf water content. This shows that there is an influence on the return signal, which is influenced by the water content that has not been accounted for now. This effect depends on the architecture, size, and density of the crop and, as such, is crop dependent.

The 3DPI algorithm proved to be unsuitable for estimating the biomass of potato. The 3DPI algorithm was developed for winter wheat and not tested on other crops so far. Probably the main reason is the very dense canopy, resulting in very few points in the lower parts of the canopy, which resulted in a small number of returns lower in the canopy. These points underneath the canopy are needed to create a complete height profile relative to the 30 cm high ridges where potatoes are grown on, resulting in a better representation of the above ground biomass. The reason the algorithm works better for sugar beet, which also shows a relatively dense canopy, is that sugar beet has larger inter-row spacing which leads to more hits from larger scan angles.

The biomass prediction could possibly be improved for areas with a low points density by changing the p_i layer size in the 3DPI algorithm. In this study, a p_i layer size of 50 mm was chosen for two reasons. Firstly, because our point density was lower than of the experiment done by [2], choosing a p_i of 10 mm would result in layers with no points in it. This could possibly influence the 3DPI algorithm negatively. Thus, increasing the p_i layer size resulted in an increase in points per p_i layer, but how this influences the final 3DPI indicator should be further researched. Secondly, choosing a larger p_i resulted in a faster computation time, which was useful for the large files we had.

Possible further research could be done using machine learning, where the 3DPI may be one of many predictors to estimate biomass. Using, for example, a PLS regression as mentioned in [21], or using machine learning approaches like [22], where a deep convolutional network was used to predict biomass using RGB imagery. Or the machine learning approach of [23], where they used a range of spectral indices as predictors to estimate above ground biomass with an NRMSE of 24.95%. Combining these different methodologies with a predictor such as the 3DPI indicator could result in a better prediction of biomass.

4.3. General Models

The accuracy of the prediction of plant height using the general prediction model is 10.1 cm, which is lower than those for the crop-specific prediction models of sugar beet (7.4 cm) and winter wheat (3.4 cm).

The general model for biomass prediction displays a strong influence of sugar beet on the total prediction model (Figure 7). The general model has the best predictive power at k = 3, showing this influence. Sugar beet had the highest predictive power for k = 3, with the other two crops at k = 1 (Figure 5). The reason for this over-representation of sugar beet is due to larger sample size of 15 samples compared to nine for winter wheat and the better performance of the sugar beet biomass model compared to that of potato. Combining these two explains why at k = 3, the model has the highest predictive power. When combining data from different crops, extra attention should be paid to include an equal sample size of the multiple crops. Therefore, using a crop-specific model to estimate crop height from UAV-LiDAR data proves to increase the prediction accuracy. However, depending on the accuracy requirements, a general model could provide plant height estimations with an accuracy of 10.1 cm and a biomass estimation with an accuracy of 17.07%. The results of this research show that crop-specific biomass models have a higher retrieval accuracy; however, further research is required to evaluate if general models can be relevant for mixed cropping systems.

Again, the trade-off is visible between a general biomass model, compared to the specialized models for each individual crop. Furthermore, the model now has to be fitted through three kinds of crops, which are completely different from each other. It was also indicated that different K-values were needed to get a good fit of the specialized biomass models. A certain combination of 3DPI indicator and measured biomass for potato does not necessarily correspond to the same combination of the 3DPI indicator and measured biomass for winter wheat. Therefore, limiting this general applicability of the model, the 3DPI appears to be crop sensitive. Possible other factors, such a height statistics derived from the LiDAR 3D point cloud as analyzed in [21] or the spectral vegetation indices of [23], could be helpful to make the 3DPI method less crop sensitive, which is something for further research.

4.4. Influence of Flight Altitude and Speed on Biomass and Plant Height Estimation

Plant height estimation from 3D points can be accurately achieved when the data is acquired at high speed and relatively high altitude. The LiDAR-derived plant height from DAY3-HA-HS, which was acquired at 92.68 m.a.g.l. and a speed of 7.39 m/s, still shows the same patterns as with DAY2-LA-MS and DAY2-LA-LS, but the general plant height is lower (Figure 9). The LiDAR-derived plant height underestimates the plant height compared to DAY2-LA-MS on average with 4.03 cm. This underestimation is still smaller than the sugar beet plant height model error of 7.2 cm, showing that acquiring plant height could be done with these settings. Therefore, speeding up data collection and covering larger areas is feasible.

Applying the generic model for sugar beet on the 3D points clouds from the flights DAY2-LA-MS, DAY2-LA-LS, DAY3-HA-HS, and DAY3-LA-LS showed that between DAY2-LA-MS, DAY2-LA-LS and DAY3-LA-LS, there is no real difference in LiDAR-derived plant height. This indicates that there is no real benefit in slowing down to 2.53 m/s or decreasing flying height. The different flight specifications do result in different point densities and point distributions. However, this results mostly in more points underneath the canopy, which has only a minor influence on the top of canopy points. To increase accuracy, a DTM could be created before plants start to grow, decreasing the dependence on getting enough returns underneath the canopy.

For biomass estimation, the under the canopy points are important, where for DAY3-HA-HS, a large under estimation is made averaging 858.8 g/m2 compared to the biomass prediction for DAY2-LA-MS. There are not enough points to represent the full height of the plant. The non-normalised accuracy of the winter wheat biomass model was 415.8 g/m2. Showing the inaccuracy of data acquired at 92.68 m.a.g.l. and 7.39 m/s compared to that of DAY2-LA-MS, acquired at 24.16 m.a.g.l. and 4.17 m/s.

Changing flight characteristics shows some areas with patches where the biomass for wheat is underestimated (Figure 10). These patches appear to result from areas that are not covered with any flight lines and almost no perpendicular flight lines. It appears to be crop-specific as these patches do not appear for sugar beet and potato. A possible reason why it affects winter wheat more is that laser pulses from the LiDAR UAV penetrate more easily due to its erectophile leaf structure compared to the planophile leaf structure of potato and sugar beet. Sofonia et al. [24] showed that for a LiDAR-based application, a cross-flight pattern worked best. A better cross pattern and closer flight lines could, therefore, possibly improve biomass estimation in general because laser pulses are acquired from multiple directions increasing the chance of hitting parts of the plant underneath the canopy and remove these unwanted patches. For an accurate biomass prediction, a homogeneous point density is needed.

To allow a good estimation of biomass and crop height over the growing season, it is necessary to keep the flight patterns consistent for all flights. Our results show that biomass and crop height estimation errors increase if the point density and point distribution vary from the circumstances used to calibrate the models. To determine the optimal flight pattern, altitude, and speed, more intensive experiments should be done.

4.5. Outlook

Our results show the possibilities for UAV LiDAR to estimate plant phenotypes, such as biomass and plant height, accurately and with high throughput. This fills the gap between slower but highly accurate tractor-based LiDAR systems and high-throughput but less-detailed manned airborne systems as indicated by [25]. Therefore, providing a solution to situations where a quick analysis of the field is required, tractor-based solutions are not suitable. Furthermore, this study shows that crop trait monitoring can be done throughout the season, using the same model trained on data from the whole season. Moreover, this research showed that under uncontrolled conditions, relevant biomass and plant height estimations can be made, which is marked as a bottleneck in the review paper of [26]. When using UAV-LiDAR for high-throughput estimation of plant height and biomass [27], time should be invested in creating a detailed and structured flight plan. A flight plan would ideally consist of a cross-flight pattern and it will cover the whole field, where flights lines are created alongside the boundaries.

This research furthermore showed that there are limitations to the biomass estimation for certain crops and that models developed for a specific crop cannot directly be used for other crops, and generic models should be used with care. For a fully operational approach, an effort should be made towards combining LiDAR with hyperspectral data, as mentioned by [22,23,26], so models can be trained for a range of crops. This will increase both the accuracy and general applicability of high-throughput biomass estimation models. Also, alternative methods for canopy height determination from UAV-based 3D point cloud datasets have recently been published [28].

5. Conclusions

Retrieval of plant height and biomass using UAV-LiDAR proved to be possible for sugar beet and winter wheat. While for potato both plant height and biomass estimation proved to be hard due to the complex canopy structure and the ridges on which potatoes are grown. For plant height, the data acquisition can be performed relatively fast and at high altitudes increasing opportunities for high-throughput approaches. However, for accurate biomass estimates, the flight conditions (altitude, speed, location of flight lines) should be kept constant. The higher acquisition speed, compared to, for example, tractor-based LiDAR systems, means that UAV-LiDAR can be used to assess large areas and can provide data quickly. Creating a reliable general model to predict biomass for different crops results in a lower accuracy, especially when crops with a dense canopy like potato are included. To increase the predictive performance of both LiDAR-derived plant height and biomass, a clear DTM should be created before germination of the plants. This accurate DTM could then be used to accurately perform the height normalization of the 3D LiDAR point cloud.

Author Contributions: Conceptualization, H.B., L.K., J.t.H.; methodology, H.B., L.K. and J.t.H.; software, H.B., J.t.H.; validation, H.B., L.K., J.t.H.; formal analysis, J.t.H.; investigation, H.B., J.t.H.; resources, L.K.; writing—original draft preparation, J.t.H.; writing—review and editing, H.B., L.K., J.t.H.; visualization, J.t.H.; supervision, H.B., L.K.; project administration, L.K.; funding acquisition, L.K. All authors have read and agreed to the published version of the manuscript.

Funding: The access to the RiCOPTER was made possible by the Shared Research Facilities of Wageningen University & Research. This work was supported by the SPECTORS project (143081), which is funded by the European cooperation program INTERREG Deutschland-Nederland.

Acknowledgments: We would like to acknowledge all the people that assisted with this research, where without them, the field campaign would have taken much longer. Therefore, special thanks to the following persons without whom this research could not be performed. Berry Onderstal for assisting with the biomass sampling. Marcello Novani, Kim Calders, Tessa Rozemuller, Tim Jak, Sophie Stuhler and Hannah Stuhler for assisting in the field campaigns. And Matthijs ten Harkel for correction and editing of the paper manuscript.

Conflicts of Interest: The authors declare no conflict of interest.

References

1. Qiu, R.; Wei, S.; Zhang, M.; Li, H.; Sun, H.; Liu, G.; Li, M. Sensors for measuring plant phenotyping: A review. *Int. J. Agric. Biol. Eng.* **2018**, *11*, 1–17. [CrossRef]

2. Jimenez-Berni, J.A.; Deery, D.M.; Rozas-Larraondo, P.; Condon, A.G.; Rebetzke, G.J.; James, R.A.; Bovill, W.D.; Furbank, R.T.; Sirault, X.R.R. High Throughput Determination of Plant Height, Ground Cover, and Above-Ground Biomass in Wheat with LiDAR. *Front. Plant Sci.* **2018**, *9*, 237. [CrossRef] [PubMed]

3. Liu, S.; Baret, F.; Abichou, M.; Boudon, F.; Thomas, S.; Zhao, K.; Fournier, C.; Andrieu, B.; Irfan, K.; Hemmerlé, M.; et al. Estimating wheat green area index from ground-based LiDAR measurement using a 3D canopy structure model. *Agric. For. Meteorol.* **2017**, *247*, 12–20. [CrossRef]

4. Sun, S.; Li, C.; Paterson, A.H. In-field high-throughput phenotyping of cotton plant height using LiDAR. *Remote Sens.* **2017**, *9*, 377. [CrossRef]

5. Hui, F.; Zhu, J.; Hu, P.; Meng, L.; Zhu, B.; Guo, Y.; Li, B.; Ma, Y. Image-based dynamic quantification and high-accuracy 3D evaluation of canopy structure of plant populations. *Ann. Bot.* **2018**, *121*, 1079–1088. [CrossRef]

6. Thapa, S.; Zhu, F.; Walia, H.; Yu, H.; Ge, Y. A Novel LiDAR-Based Instrument for High-Throughput, 3D Measurement of Morphological Traits in Maize and Sorghum. *Sensors* **2018**, *18*, 1187. [CrossRef]

7. Madec, S.; Baret, F.; de Solan, B.; Thomas, S.; Dutartre, D.; Jezequel, S.; Hemmerlé, M.; Colombeau, G.; Comar, A. High-Throughput Phenotyping of Plant Height: Comparing Unmanned Aerial Vehicles and Ground LiDAR Estimates. *Front. Plant Sci.* **2017**, *8*, 2002. [CrossRef]

8. Gnyp, M.; Panitzki, M.; Reusch, S.; Jasper, J.; Bolten, A.; Bareth, G. Comparison between tractor-based and UAV-based spectrometer measurements in winter wheat. In Proceedings of the 13th International Conference on Precision Agriculture, St. Louis, MS, USA, 31 July–3 August 2016.

9. Naderi-Boldaji, M.; Kazemzadeh, A.; Hemmat, A.; Rostami, S.; Keller, T. Changes in soil stress during repeated wheeling: A comparison of measured and simulated values. *Soil Res.* **2018**, *56*, 204–214. [CrossRef]

10. Bendig, J.; Bolten, A.; Bennertz, S.; Broscheit, J.; Eichfuss, S.; Bareth, G. Estimating Biomass of Barley Using Crop Surface Models (CSMs) Derived from UAV-Based RGB Imaging. *Remote Sens.* **2014**, *6*, 10395–10412. [CrossRef]

11. Richardson, J.J.; Moskal, L.M.; Kim, S.-H. Modeling approaches to estimate effective leaf area index from aerial discrete-return LIDAR. *Agric. For. Meteorol.* **2009**, *149*, 1152–1160. [CrossRef]

12. Li, W.; Niu, Z.; Huang, N.; Wang, C.; Gao, S.; Wu, C. Airborne LiDAR technique for estimating biomass components of maize: A case study in Zhangye City, Northwest China. *Ecol. Indic.* **2015**, *57*, 486–496. [CrossRef]

13. Krasilnikov, P.; Ibánez, J.; Arnold, R.; Shoba, S. *A Handbook of Soil Terminology, Correlation and Classification*; Routledge: Abingdon, UK, 2009.

14. Tackenberg, O. A New Method for Non-destructive Measurement of Biomass, Growth Rates, Vertical Biomass Distribution and Dry Matter Content Based on Digital Image Analysis. *Ann. Bot.* **2007**, *99*, 777–783. [CrossRef] [PubMed]

15. Brede, B.; Lau, A.; Bartholomeus, H.M.; Kooistra, L. Comparing RIEGL RiCOPTER UAV LiDAR derived canopy height and DBH with terrestrial LiDAR. *Sensors* **2017**, *17*, 2371. [CrossRef] [PubMed]

16. Mock, W.B.T. Pareto Optimality. In *Encyclopedia of Global Justice*; Chatterjee, D.K., Ed.; Springer: Dordrecht, The Netherlands, 2011; pp. 808–809. [CrossRef]

17. Holman, F.H.; Riche, A.B.; Michalski, A.; Castle, M.; Wooster, M.J.; Hawkesford, M.J. High Throughput Field Phenotyping of Wheat Plant Height and Growth Rate in Field Plot Trials Using UAV Based Remote Sensing. *Remote Sens.* **2016**, *8*, 1031. [CrossRef]

18. Eitel, J.U.H.; Magney, T.S.; Vierling, L.A.; Brown, T.T.; Huggins, D.R. LiDAR based biomass and crop nitrogen estimates for rapid, non-destructive assessment of wheat nitrogen status. *Field Crop. Res.* **2014**, *159*, 21–32. [CrossRef]

19. Harcha, C.I.; Calderini, D.F. Dry matter and water dynamics of wheat grains in response to source reduction at different phases of grain filling. *Chil. J. Agric. Res.* **2014**, *74*, 380–390. [CrossRef]

20. Zhu, X.; Wang, T.; Skidmore, A.K.; Darvishzadeh, R.; Niemann, K.O.; Liu, J. Canopy leaf water content estimated using terrestrial LiDAR. *Agric. For. Meteorol.* **2017**, *232*, 152–162. [CrossRef]

21. Luo, S.; Chen, J.M.; Wang, C.; Xi, X.; Zeng, H.; Peng, D.; Li, D. Effects of LiDAR point density, sampling size and height threshold on estimation accuracy of crop biophysical parameters. *Opt. Express* **2016**, *24*, 11578–11593. [CrossRef]

22. Ma, J.; Li, Y.; Chen, Y.; Du, K.; Zheng, F.; Zhang, L.; Sun, Z. Estimating above ground biomass of winter wheat at early growth stages using digital images and deep convolutional neural network. *Eur. J. Agron.* **2019**, *103*, 117–129. [CrossRef]

23. Han, L.; Yang, G.; Dai, H.; Xu, B.; Yang, H.; Feng, H.; Li, Z.; Yang, X. Modeling maize above-ground biomass based on machine learning approaches using UAV remote-sensing data. *Plant Methods* **2019**, *15*, 10. [CrossRef]

24. Sofonia, J.J.; Phinn, S.; Roelfsema, C.; Kendoul, F.; Rist, Y. Modelling the effects of fundamental UAV flight parameters on LiDAR point clouds to facilitate objectives-based planning. *ISPRS J. Photogramm. Remote Sens.* **2019**, *149*, 105–118. [CrossRef]

25. Yang, G.; Liu, J.; Zhao, C.; Li, Z.; Huang, Y.; Yu, H.; Xu, B.; Yang, X.; Zhu, D.; Zhang, X.; et al. Unmanned aerial vehicle remote sensing for field-based crop phenotyping: Current status and perspectives. *Front. Plant Sci.* **2017**, *8*, 1111. [CrossRef] [PubMed]

26. Roitsch, T.; Cabrera-Bosquet, L.; Fournier, A.; Ghamkhar, K.; Jiménez-Berni, J.; Pinto, F.; Ober, E.S. Review: New sensors and data-driven approaches—A path to next generation phenomics. *Plant Sci.* **2019**, *282*, 2–10. [CrossRef] [PubMed]

27. Ravi, R.; Lin, Y.; Shamseldin, T.; Elbahnasawy, M.; Crawford, M.; Habib, A. Implementation of UAV-Based Lidar for High Throughput Phenotyping. In Proceedings of the IGARSS 2018—2018 IEEE International Geoscience and Remote Sensing Symposium, Valencia, Spain, 22–27 July 2018; pp. 8761–8764. [CrossRef]

28. Song, Y.; Wang, J. Winter Wheat Canopy Height Extraction from UAV-Based Point Cloud Data with a Moving Cuboid Filter. *Remote Sens.* **2019**, *11*, 1239. [CrossRef]

Article

In-Field Detection of Yellow Rust in Wheat on the Ground Canopy and UAV Scale

David Bohnenkamp [1],*, Jan Behmann [1] and Anne-Katrin Mahlein [2]

[1] INRES-Plant Diseases and Plant Protection, University of Bonn, Nussallee 9, D-53115 Bonn, Germany; jbehmann@uni-bonn.de

[2] Institute of Sugar Beet Research (IfZ), Holtenser Landstraße 77, D-37079 Göttingen, Germany; mahlein@ifz-goettingen.de

* Correspondence: david.bohnenkamp@uni-bonn.de or davidb@uni-bonn.de

Received: 26 September 2019; Accepted: 23 October 2019; Published: 25 October 2019

Abstract: The application of hyperspectral imaging technology for plant disease detection in the field is still challenging. Existing equipment and analysis algorithms are adapted to highly controlled environmental conditions in the laboratory. However, only real time information from the field scale is able to guide plant protection measures and to optimize the use of resources. At the field scale, many parameters such as the optimal measurement distance, informative feature sets, and suitable algorithms have not been investigated. In this study, the hyperspectral detection and quantification of yellow rust in wheat was evaluated using two measurement platforms: a ground-based vehicle and an unmanned aerial vehicle (UAV). Different disease development stages and disease severities were provided in a plot-based field experiment. Measurements were performed weekly during the vegetation period. Data analysis was performed by three prediction algorithms with a focus on the selection of optimal feature sets. In this context, the across-scale application of optimized feature sets, an approach of information transfer between scales, was also evaluated. Relevant aspects for an on-line disease assessment in the field integrating affordable sensor technology, sensor spatial resolution, compact analysis models, and fast evaluation have been outlined and reflected upon. For the first time, a hyperspectral imaging observation experiment of a plant disease was comparatively performed at two scales, ground canopy and UAV.

Keywords: feature selection; spectral angle mapper; support vector machine; support vector regression; hyperspectral imaging; UAV; cross-scale; yellow rust; spatial resolution; winter wheat

1. Introduction

Today's demands of agricultural cropping systems are high. Agroecosystems have to be highly productive, while the undesirable impact on the environment has to be as low as possible. Resource-conserving methods with a minimum of chemical input are in favor. One vision able to approximate this goal is the use site-specific cropping measures. Site-specific management has the potential to lead to a higher or constant productivity with a constant or reduced input of resources [1]. One group of for site-specific applications are plant protection measures [2].

The spatial occurrence of plant diseases in the field, especially in the early season, is often heterogeneous, while in most cases, plant protection compounds are applied homogeneously onto the crop. This spatial heterogeneity of disease occurrence might lead to diverse fungicide demands that are often not considered. Many diseases first occur in patches before they start spreading in the field. One approach for a site-specific application of plant protection measures might be the application of fungicides to patches of disease occurrence [3–5]. This could prevent or stop disease spreading without applying a fungicide to the whole field [1].

Spectral sensors might be tools able to contribute to site-specific disease management [6,7]. Spectral sensors measure the light reflected from the crop canopy [1]. During pathogen attack and disease development on the crop leaf, diseases establish a spectral fingerprint in the reflected leaf signature [8–10]. These shifts of the signature can be detected using spectral sensors, particularly in the electromagnetic spectrum from 400–2500 nm [11]. Spectral sensors can be divided into hyperspectral and multispectral sensors, depending on their spectral resolution. The number and width of measured wavebands mainly characterize the spectral resolution [11].

Non-imaging hyperspectral sensors average the spectral information over a certain area, while imaging sensors contain the spectral information for each pixel [7]. Hyperspectral imaging sensors (HSIs) provide spectral information in a spatial resolution. Multispectral sensors typically cover the RGB range with an additional NIR band. These sensors are less cost-intensive and the generated data are less complex, but do not cover the broad spectral range like a hyperspectral sensor.

Spectral sensors have been applied on different scales [12]. For field approaches, a hyperspectral camera can be mounted to a ground based vehicle or to a UAV [1,3,11,13]. Depending on the interrogation and measuring setup, each scale can have advantages and disadvantages. On the ground scale, it is possible to detect small features of a few mm through high resolution on close range, while the throughput on the UAV scale is much higher, with still higher resolution compared to satellite imagery [5,14–16]. For field applications of spectral sensors, depending on the scale, the resolution or the measurement time can become a limiting factor. Most field applications for disease detection focused on the calculation of vegetation indices (VIs) [17–19] using multispectral sensors. VIs are developed by accounting certain band ratios to highlight one factor and reduce the impact of another factor [20]. Depending on the wavelength, these indices can be indicators for crop vitality, general crop stress, pigment content, or a specific plant disease [18,21]. Few works have demonstrated an approach for disease detection using imaging hyperspectral sensors under field conditions [10]. This might be because spectral measurements under field conditions are challenging and the complexity of hyperspectral data is higher than multispectral data [1]. The features of multispectral sensors might result in lower image acquisition durations and lower susceptibility to environmental factors during measurements. The image quality of field data in general is influenced by various factors. Beside suitable weather conditions, the field crop species and the disease symptom type are of high relevance for successful measurements. The leaf architecture and disease occurrence on the plant mainly determine the detectability of the disease. Disease presence on lower plant and leaf levels results in a decreased reflected signal. Disturbing weather conditions such as wind and rain can easily obscure spectral images obtained in the field. One elusive and eminently important factor is the illumination. Changing illumination conditions over time, caused by clouds or solar altitude, can lead to uninterpretable data, because spectra of different images cannot be compared with one another anymore [1,3,22]. The detection of diseases on different leaf levels is also challenging because of inhomogeneous illumination conditions through the leaf altitude in the crop stand and upper leaves that cast shadow. These leaves might also be in different developmental stages, and a senescent leaf has to be differentiated from a healthy green or a diseased leaf. The leaf angle to the camera influences the spectral signal. Not least, the image quality is essentially determined by the spatial resolution of the sensor system used. Small symptoms of a disease can only be visualized when the spatial resolution in combination with the measurement distance is appropriate for the desired data quality.

So far, various field measurements on the ground canopy scale of cereal crops have focused on the detection of biotrophic diseases such as yellow rust [3,10,13,23–25], brown rust [18,26], and powdery mildew of wheat [19]. This might be due to the fact that biotrophic diseases are more likely to appear on the upper leaf layers because of wind distribution and a preference for fresh and healthy leaf tissue [27]. Necrotrophic diseases are most severe on lower leaf levels, and therefore more difficult to detect with remote sensors. A detection and quantification of septoria tritici blotch with a hyperspectral radiometer has been demonstrated in the field [28].

The analysis and interpretation of sensor data is crucial for future implementation. Algorithms from machine and deep learning, in combination with suitable sensors and measurement platforms are promising techniques. These methods are particularly able to cope with the number of wavebands provided in hyperspectral data, and can be used for the detection of plant diseases [7,29–32].

This work presents a new approach for field trial studies using innovative and machine learning for a pixel wise detection of crop diseases. A winter wheat trial was conducted in the vegetation period of 2018. The crop was infected with *Puccinia striiformis*, the causal agent of yellow rust (YR). Weekly hyperspectral measurements were performed on the ground-canopy and the UAV scale to monitor the spectral dynamic of crop stands during the vegetation period. Measurements were performed using a mobile field platform with a distance of 50 cm to the crop canopy and with a UAV drone at 20 m height over the plots to work on and compare different scales. Hyperspectral images were captured using a line scanner attached to a linear stage in a measurement booth and a frame-based hyperspectral camera for UAV applications. Field data were preprocessed and normalized, and then analyzed using the supervised classification methods spectral angle mapper (SAM) and support vector machine (SVM) to detect yellow rust of wheat. Additionally, a feature selection was performed on the hyperspectral data to verify the potential for a waveband reduction from hyperspectral to multispectral data for disease detection.

2. Materials and Methods

2.1. Field Trial Layout

In the vegetation period 2017/2018, a field trial with winter wheat was conducted at trial station Campus Klein-Altendorf 50°37′31.00″N, 6°59′20.54″E (Rheinbach, Germany). In 2016/2017, a first field trial was performed to specify the measuring setup and routine (data are not shown). The cultivars JB Asano (Limagrain GmbH, Edemissen, Germany) and Bussard (KWS SAAT SE, Einbeck, Germany) were sown on 26.10.2017 with 320 kernels/m². Field emergence was on 14th November 2017, while harvest took place on 24.07.2018. JB Asano was chosen because of its susceptibility to YR. The field trial was designed in 10 treatments per cultivar with two repetitions, resulting in 40 plots (plot size: 3 × 7 m). The plot design was randomized within each cultivar. With a change of the cultivar after each plot, the direct proximity of plots with the same cultivar was avoided. This was designed to arrange the field trial into two long rows of plots with cultivars alternating one after another in 20 plots per repetition (Figure 1). The treatments within one cultivar were randomized. Two fertilizer intensities (160 kg N/ha and 30 kg N/ha) were applied per cultivar. For cultivar Asano, two treatments were used for additional inoculation with YR. The whole field trial was aligned from northwest to southeast.

2.1.1. Inoculations

Additional inoculations of *Puccinia striiformis* were performed in April and May (25 April 2019; 12 May 2019). To establish high disease infections, the inoculations were repeatedly performed by applying a spore suspension to the plants immediately after rainfall incidences. Inoculations were timed to forecasted infection risks after the xarvio field manager (BASF Digital Farming GmbH, Münster, Germany). The spore suspension was applied with a garden pump sprayer and contained 8 × 10⁴ spores/mL. Two liters of spore suspension were homogeneously applied over one plot.

2.1.2. Visual Disease Ratings

Visual disease ratings were performed weekly from calendar week 17–23. One plot was rated two times at the same locations where hyperspectral measurements took place. Disease incidence was assessed by the eye of a human rater. The diseased leaf area (%) was rated on 15 leaves per leaf level. Additionally, the growing stage and the visible leaf area of each leaf level in the hyperspectral image (from top view) was ascertained.

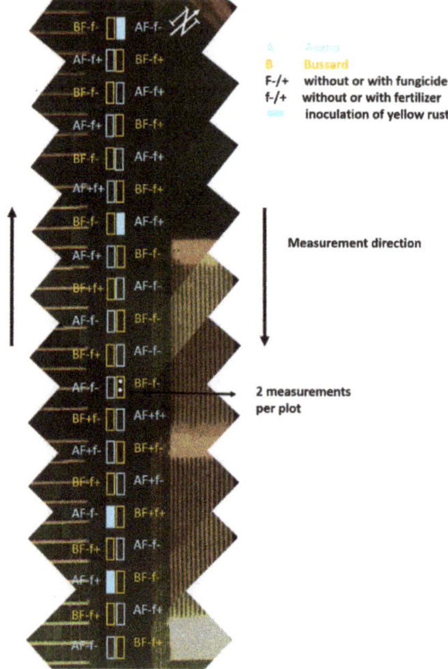

Figure 1. Field trial layout (RGB stitch from unmanned aerial vehicle (UAV) images) at the research station Campus Klein-Altendorf (Rheinbach, Germany) of winter wheat varieties Bussard and Asano in 2018, with treatment declaration and measurement strategy. The trial was designed in two long rows of plots to reduce the number of turns and keep continuous measurements.

2.1.3. Crop Stand and Disease Development

The vegetation in 2018 started late in March and was denoted by a drought that especially affected the length of the growing season and led to an early harvest in July (Figure 2). *Septoria tritici* blotch, tan spot, and powdery mildew were insignificant throughout the growing season. YR could establish a significant infection on cultivar Asano, and measurements were not aggravated due to mixed infections. The first YR symptoms were found in mid of April and were based on natural infection incidences. Until the beginning of May (BBCH 31), a serious increase of YR was rated. Warm days and cold nights seemed to favor infection incidences through dew formation. Until the middle of May (BBCH 37-39), YR was the dominating disease.

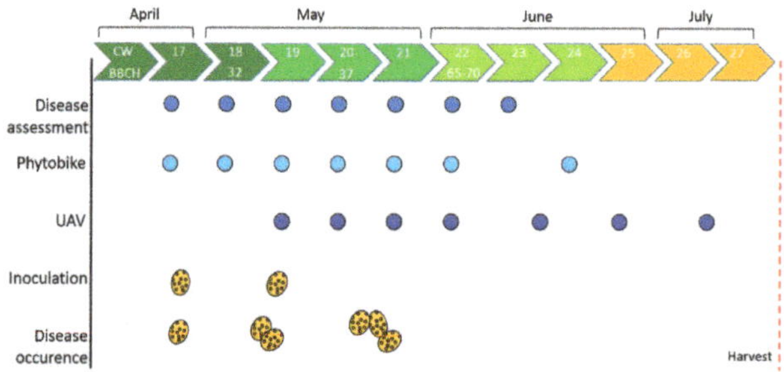

Figure 2. Schematic overview of the course of the 2018 vegetation period. For each calendar week, the actions are presented and disease occurrence of yellow rust (YR) is shown. Weekly assessments and measurements were performed during the vegetation period (CW = calendar week; BBCH = growing stage).

2.2. Measurement Platforms

2.2.1. Field Platform Phytobike

The measurement platform, based on a square steel construction with four wheels and provided by Forschungszentrum Jülich (Jülich, Germany), covered a 3 m wide experimental plot (Figure 3). Sensors for reflectance characteristics, localization systems, power supply, and control of the sensors via a control laptop were mounted to the steel frame. All sensors were variable in height by a moveable aluminum profile construction. In this way, the sensor platform could be adapted to the growth stages of the plants and a constant distance between sensors and crop canopy was enabled. With a weight of around 150 kg, the construction approached the limit of the platforms without steering, and could still be moved by the physical strength of two people.

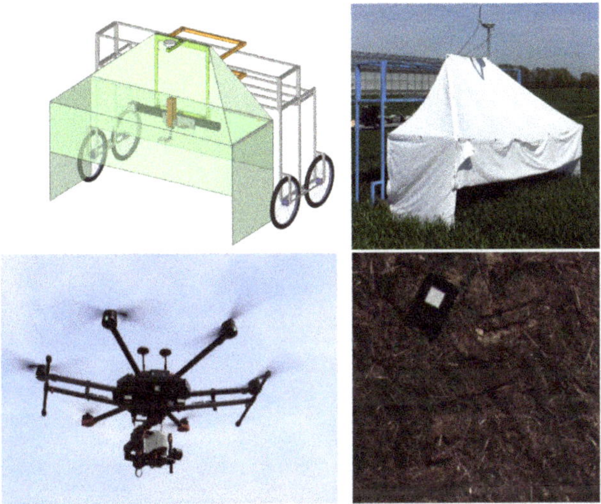

Figure 3. Construction plan of the Phytobike (**top left**) and the final appearance in the field including the cotton diffuser (**top right**). The UAV system used, consisting of a DJI Matrice 600 and a Rikola hyperspectral camera (**bottom left**). Normalization was performed using a 50% grey reference panel (**bottom right**).

As a hyperspectral sensor, the Specim V10E line camera (Specim Oy, Oulu, Finland) was used. The motion required for the Specim V10E camera was realized by a linear stage (Velmex, Bloomfield, USA). Measurements were triggered via the control computer, allowing a flexible reaction to changing light situations by an adapted integration time. The Specim V10E camera measured the electromagnetic spectrum in a range from 400 to 1000 nm with a spectral resolution of 2.73 nm. Sunlight was used as a natural light source. A canvas measuring cabin was constructed to avoid shadows cast by the sensors and equipment of the Phytobike.

2.2.2. UAV Measurements

The UAV allowed overview images of whole experiments or at least of parts of the experiment to be collected. Recent technologies have enabled hyperspectral imaging at UAV scale. We combined a UAV DJI Matrice 600 (Da-Jiang Innovations Science and Technology Co., Shenzen, China) with a Rikola hyperspectral camera (Senop Oy, Oulu, Finland) (Figure 3). The Rikola camera measured the reflected light in a range from 500 to 900 nm. The measured wavebands were selectable and spectral resolution was set to 7 nm using 55 wavebands. With flight times of around 20 min, the whole experiment was captured within one battery capacity. For plot observations, a 20 m flight height was selected and the UAV hovered over each plot center for a duration of 10 s. Sunlight was used as a natural light source.

2.3. Data Preprocessing

This study focused on the information about relevant wavebands as the central outcome. We used a data flow to assess the ability to transfer this information between observation scales (Figure 4). We built two data sets for yellow rust prediction—a classification data set on field scale and a regression data set on UAV scale. Multiple prediction models and feature selection results were derived. In the final step, models were optimized using the selected features, and feature selection information was also exchanged. The resulting four classification models with selected features, two on the field scale (features selected on the field scale and on the UAV scale) and two on the UAV scale (features selected on the field scale and on the UAV scale), were evaluated. This allowed the values of feature selection and, more specifically, the values of feature information obtained at a different observational scales to be evaluated.

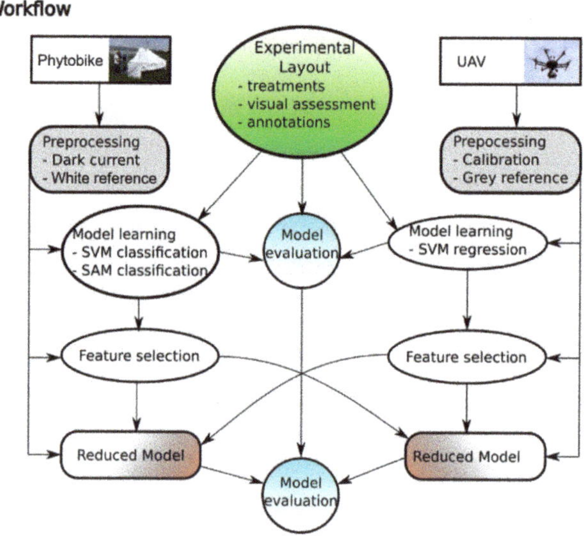

Figure 4. Underlying workflow of data assessment to compare different prediction algorithms and evaluate the value of information about selected features at different scales. SVM = support vector machine, SAM = spectral angle mapper.

2.3.1. Spectral Preprocessing

The derivation of the physical surface property reflectance from observed intensity values is an essential part of hyperspectral image processing. The normalization procedure has to be adapted to the measurement platform and is based on a spectral reference panel with a known homogeneous reflectance in the observed wavelengths. At all scales, the following equation was applied to calculate the reflectance R from the observation Im, reference Im_{ref}, and the corresponding dark currents DC_{Im} and DC_{ref}. In the field, the additional DC_{ref} was omitted for practical reasons.

$$R = \frac{\Im - DC_{\Im}}{\Im_{ref} - DC_{ref}},\tag{1}$$

On the ground canopy scale, a 50% spectral reference panel was measured within each image of the line scanner. A separate dark current was observed for the Specim V10E camera before every image.

For practical reasons, UAV flight sequences were started with the acquisition of a single dark current, and one image of the reference panel immediately before and after flying the frame-based Rikola camera over the wheat plots. Image quality always suffers from motion of the object to be measured or motion of the sensor. To avoid this, images were taken in conditions as calm as possible on the ground canopy scale. The Rikola camera was hovered for at least 10 images over the reference panel to ensure that image quality was sufficient for data normalization. The use of cross-sensor normalization, e.g., by using a separate spectrometer that continuously logs the incoming light intensity, was tested but was not successful due to a deviating response characteristic between the different sensors.

To remove high frequency noise at the spectral border regions of the Specim V10e, the bands 1–20 (400–450 nm) and 181–211 (910–1000 nm) were excluded from further analysis, resulting in 161 used spectral bands. In addition, a Savitzky–Golay filter using 15 centered points and a polynomial of degree 3 smoothed the data of the Specim V10e.

2.3.2. Data Normalization

Plant geometry can present severe distortions due to varying leaf angles, leaf distances to the camera, and specular reflections on particular parts of the leaves. To compare the reflectance characteristics, omitting the additive and multiplicative factors, the standard normal variate (SNV) has been developed [33]. It is able to remove scaling factors due to varying distance or leaf angle, as well as additional factors like specular reflection, e.g., on leaf tips. The normalization was performed on both the ground canopy and field data. The SNV representation was calculated per spectrum S and focuses the shape of the spectral curve:

$$SNV = \frac{S - mean(S)}{std(S)}, \tag{2}$$

2.4. Prediction Algorithms

Multiple algorithms can perform predicting a class or continuous value based on features of a sample. In general, they use a vector representation as input. In this study, the classifiers spectral angle mapper (SAM) and support vector machine (SVM), as well as the regression algorithm—support vector regression (SVR)—were applied to the ground canopy data (taken with the phytobike). To train and evaluate the models, four images of one measuring day were annotated to be used as training data and four images were annotated to be used as test data. The number of annotated pixels differed in the different images due to natural heterogeneity in the crop stand. Pixel numbers were at least several thousand for each class, up to several hundred thousand pixels for all classes in one image. Based on the huge number of annotated pixels, models were trained on a subsampled data set, to make them trainable and to rebalance the classes. With the exception of the water class, all classes were trained with 1000 samples per class after subsampling of training data. The SAM was used because it has been described in the literature to work resiliently under inhomogeneous light conditions [34]. The development of the classification model was easy and fast. The SVM was used because, in theory, it is trained on the whole data set and considers the spectrum of each pixel as training data. Vegetation indices (VIs) were used because various published works have focused on VIs as tool for disease detection. VIs can be seen as established representatives for optical measurements of plant parameters. The models were trained using three data representations: full spectra, SNV normalization, and 20 spectral VIs. The results were compared to a SAM that represented the base line accuracy. The comparison was performed on the YR test data from 23 May 2018. The evaluation of different feature representations showed a small advantage of SNV normalizations, whereas it was treated as standard representation in the following. As performance measures, we applied the overall accuracy using six classes for the model, combining the background and the old leaves/straw class. Furthermore, we evaluated the F1 score (Table 1) for the class disease, providing a homogenized combination of precision and recall. The F1 score declares the number of pixels of one class that are correctly classified into this class after the formula 2 × (precision × recall) ÷ (precision + recall)). The two performance measures corresponded in tendency; however, the F1 score decreased faster as the large number of background pixels stabilized the overall accuracy.

Table 1. Comparison of evaluation parameters obtained on test data for different data representations and prediction algorithms on the ground scale for the support vector machine (SVM) and the spectral angle mapper (SAM).

	SVM Raw	SVM SNV	SVM Indices	SAM
Accuracy	91.9%	92.9%	90.2%	81.4%
F1 disease score	83.2%	84.0%	76.4%	48.3%

2.4.1. Spectral Angle Mapper

The SAM is a prediction algorithm developed for the efficient classification of high- dimensional spectral data. The assignment uses the angle between the spectra to classify and reference spectra, treating a spectrum as high-dimensional vector [34]. The spectrum to be classified was assigned to the reference spectrum/class with the smallest angular distance. In addition, a threshold prohibited the assignment of spectra with a large angular distance.

2.4.2. Support Vector Algorithms

The SVM andSVR are established machine learning methods that have been proven to deal well in situations with many features but a very limited number of samples [35]. This is a common situation in hyperspectral data analysis, and following it is a suitable approach for hyperspectral remote sensing as well as close range imaging. A critical point for the application of SVM and SVR is the selection of the hyper parameters Cost C, kernel parameter γ (SVM) or C, and complexity control v (v-SVR). They were selected by grid search combined with a cross validation. Grid points were $10^{-5}\ldots10^{10}$ for C, $10^{-8}\ldots10^{2}$ for v and $0.05\ldots0.50$ for n. The optimization algorithm was the sequential minimal optimization SMO, and LIBSVM 3.18 with Matlab was used for as implementation [36].

2.5. Vegetation Indices

On hyperspectral images of ground canopy data, 20 VIs were tested to visualize crop heterogeneity and to detect yellow rust in the field. The composition was used because it has previously been successfully tested as an indicator for crop vitality [37]. The composition of VIs was chosen according to Behmann et al. [37]. Due to the limitation of available bands of the Rikola camera, only 16 VIs were used for UAV data and the ARVI, mRESR, mRENDVI, and SIPI were excluded.

2.6. Model Evaluation

To compare the performance of the different models on the respective data sets, different measures were applied depending on the model type. All measures were calculated on a test set that was not used in training. For classification models that determined the discrete classes y as a function f(X) of the data X, the accuracy was defined as the percentage of correctly classified data points. The F1 score, in contrast, was based on the precision and recall of each class. In regression tasks with continuous target variables, the coefficient of determination R^2, correlation, and root mean square error (RMSE) were applied. Due to the limited number of data, we applied leave-one-out cross validation to generate the test predictions. This procedure learns a model on the whole data set except for one sample. This was repeated for all samples in the data set.

2.7. Feature Selection

There are multiple approaches for feature selection, feature subset selection, and feature weighting. Filter approaches like Relief are very fast and provide a weight for each feature. In contrast, wrapper approaches have the big advantage of dealing well with high levels of redundancy and selecting the best subset with minimal size [38]. A major drawback is the high computational load. Feature selection at all scales (on ground-canopy and UAV images) was performed using a wrapper approach comprising a SVM or SVR, respectively. A sequential forward feature selection (Statistics Toolbox, Matlab2013a) was used, and the called criterion function minimizing the prediction error was implemented based on LIBSVM 3.18. For the SVM, the accuracy was maximized and for the SVR, the RMSE was minimized. Due to the limited number of samples in the UAV data set, a leave-one-out cross-validation was performed to generate the test predictions to calculate the criterion.

2.8. Spatial Resolution as a Key Parameter for Disease Detection

Besides the relevant wavelengths, the required spatial resolution or ground sampling distance (GSD) is highly important for the definition of a sensor capable of detecting different wheat diseases in the field. Based on the test and training data sets, simulations were performed where the test data were extracted from subsampled spectra by a factor of 2, 10, 20, and 100. A knn and an aggressive subsampling approach were compared to visualize the effects of different annotation strategies on the F1 score for the detection of YR.

3. Results and Discussion

3.1. Supervised Classification of Hyperspectral Pixels at the Ground Canopy Scale

One approach used to analyze hyperspectral data on the field scale is the pixel-wise classification into usual pixel (background, straw, healthy leaf tissue) and in disease specific symptoms. In the field experiment 2018, YR had a significant disease severity and classifiers for this disease were derived.

The use of 16 VIs reached a reasonable but not competitive performance. However, it has to be noted that we compared 161 features to 16, meaning a significant reduction in dimensionality. The results can be integrated in a later discussion of the various feature sets obtained by feature selection.

Based on these results, SVM SNV (Table 1) was selected as most appropriate approach. A visual comparison of the SAM results and the SVM SNV result is shown in Figure 5. Significant differences were apparent. The SVM detected many more senescent leaves, e.g., all leaves from the lower leaf levels, whereas the SAM assigned these to the background or the healthy leaves. The SVM was more sensitive for ear detection, which caused major problems in the SAM image, where they were partly assigned to YR. Overall, both approaches were sensitive to YR, but the SVM was much more accurate in the very bright image parts as well as the darker background parts, while the class YR was overrepresented in the SAM classification. The visually most significant aspect was the large number of blue pixels in the visualization of the SVM result. YR disease was present at all leaf levels and led to early senescence in lower leaf levels.

The classification models were validated via two approaches: (1) pixel-wise classification of the hold-out test data set consisting of manually annotated pixels of new images of separate plots, and (2) prediction of pixel classes of all images obtained on a respective day and comparing the total % disease class from all plant pixels to the visual assessment done by the expert (Figure 6). Approach 1 resulted in a confusion matrix allowing the calculation of multiple performance measures such as the overall accuracy, the sensitivity, and the recall, whereas Approach 2 provided the R^2 value, correlation coefficient, and a regression plot. Table 1 shows the overall accuracy and the F1 score for the different classification methods.

Presumably, due to light reflections and transmission or a deviating weighting of the different canopy levels, the SVM prediction overestimated the ratio of diseased pixels. To compensate for this, a linear regression model was applied. However, deviations between the predicted disease severity and the visual assessment could have various reasons, e.g., the section of the plot observed by the sensor did not represent the true status that was evaluated by the visual assessment. The viewing angle produced a variable composition of different leafs and leaf layers in the field of view of the human and the sensors. Furthermore, the visibility of the lower leaf levels was low for imaging system from the top, and more accurate if the human rater could go deep into the crop stand for individual leaf disease rating. Further points are that the visual assessment produced a single value averaging the affected leaf area. From repeated disease assessments with multiple experts, different deviations have been observed depending on the literature [39,40]. The method of disease detection is subjective to the individuals performing the assessment. Another prime factor for deviations and classification inaccuracies is the biological heterogeneity. This has to be considered as highly dynamic within one field, one plot, and one location, and even on different leaf layers and single leaves. The biological heterogeneity can be affected by many factors, e.g., the leaf color and status, stem elongation (distance of leaf layers), the density of the canopy, and other biological growth processes.

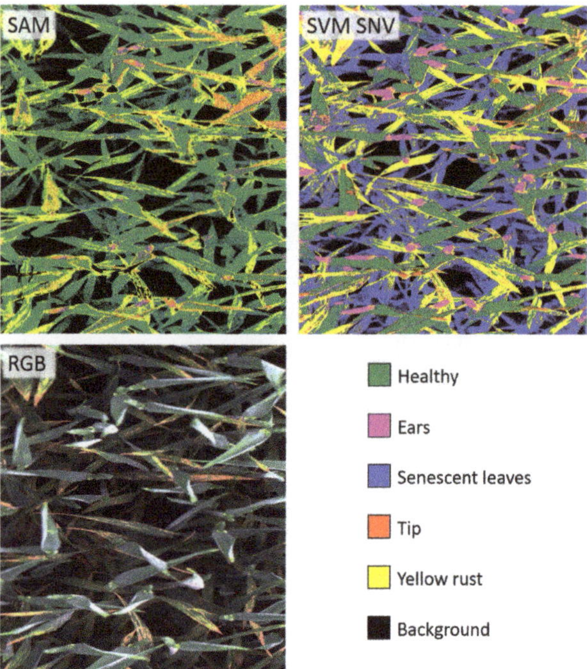

Figure 5. Visual comparison of the representative spectral angle mapper (SAM) classification (**top left**) and the support vector machine standard normal variate (SVM SNV) classification (**top right**) with the original RGB visualization of the corresponding ground-based image of one representative measurement location of a plot inoculated with YR (**bottom left**). The image is captured with the hyperspectral camera Specim V10. The classes (**bottom right**) were generated from manual annotation of train and test data.

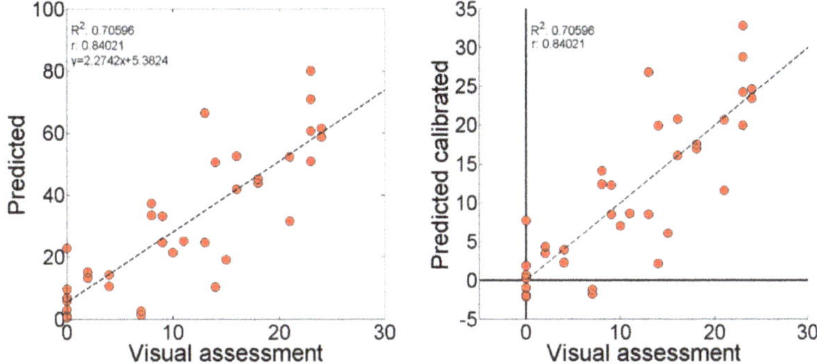

Figure 6. Scatter plot of the relation between visual assessment and predicted disease ratios for yellow rust on 23 May 2018 before (**left**) and after the application of a linear calibration model (**right**). The calibration model had the purpose of compensating for scale differences in the prediction values.

3.2. Evaluation of Hyperspectral UAV Observations Using a Filter-System Hyperspectral Camera

To characterize the reflectance characteristics of the field plots, the spectra of the central 4×2 m of each plot were averaged. Intra-plot variations were neglected. Multiple traits were predicted with reasonable accuracy based on SVM and SVR analysis of the 55 recorded bands from 500–900 nm. Table 2 shows the obtained performance parameters based on a SNV representation and the integration of all 55 bands.

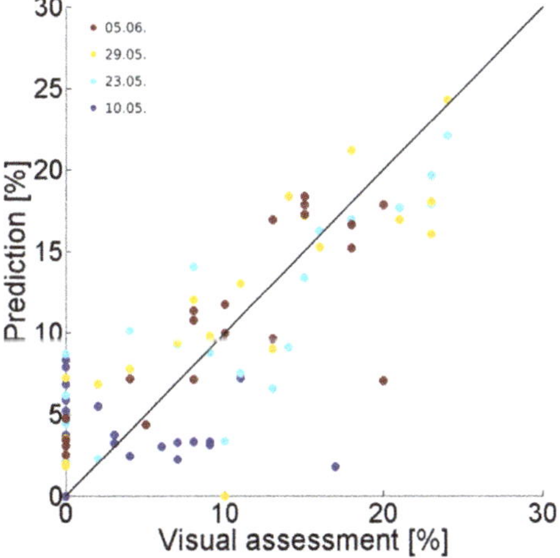

Figure 7. Prediction results of a YR infestation obtained by applying a leave-one-out procedure to the support vector regression (SVR) on the UAV scale. Each point represents a plot observation, and its color, the observation date (DD-MM).

Table 2. Performance values for different SVM and SVR models predicting the treatments of the wheat field experiment based on UAV observations.

Trait/Treatment	Performance
Fertilizer level	Accuracy = 82.3%
Fungicide	Accuracy = 91.5%
Fungicide + fertilizer level (four classes)	Accuracy = 71.4%
Disease detection (severity > 0)	Accuracy = 90.0%
Disease severity estimation	Correlation = 70.6%

Using the SVR approach, a prediction of the disease severity in percent was possible with reasonable accuracy (Figure 7). The interpretation of this result has to take into account that a value from the visual assessment might not represent the average plot value because diseases may occur at zoned locations in the plot, and assessment locations may or may not be in these spots.

3.3. Selection of Relevant Features at Different Scales

One of the main motivations for the application of hyperspectral imaging technology is the potential to find the most relevant wavelength for a specific task, and to subsequently design a specific sensor. Reference [41] showed that specific wavelengths might be useful to identify certain leaf diseases in sugar beet. In wheat, VIs have been described that are capable of detecting brown rust [18]. This shows that a selection of specific wavelengths can be specific for one disease. We applied the introduced technique to the data sets on the ground-canopy and UAV scale and derived important wavelength for the detection of disease symptoms as well as the prediction of disease severity.

3.3.1. Ground Scale

Feature selection on the field scale was performed for the detection of YR. The models were trained on a homogenized sample of training data and validated by a five-fold cross-validation. The final accuracy was determined by the hold-out test set. To reduce the computational complexity, the feature was regularly subsampled by a factor of 5. The resulting 33 bands were ranked and an optimal band number was selected (Figure 8).

For YR, an optimal number of 16 features reached 91% accuracy. However, to allow a comparison with the UAV scale selection, we selected the best 10 features, providing an accuracy of 88%. The waveband of 780 nm in the NIR was the most important for YR detection. The next two bands were also in in the NIR, followed by a band in the blue/green spectral region. Less important was the NIR wavebands > 800 nm and the red part of the spectrum. Various works have shown that VIs using wavelengths out of these spectral regions can be successfully used to detect rust diseases of wheat [17,18,25], or even for necrotrophic diseases of other crop plants such as groundnuts [42]. In the literature, it has been described that pigments and water influence the absorbance and reflectance of light with plant interactions [43–45]. The measured reflectance signal is always a mixed signal and the result of complex biochemical interactions [43,46,47]. The visible region is mainly influenced by the light absorption of leaf pigments [48]. Healthy wheat canopies appear dark green because of high amounts of chlorophyll in the leaves [10]. With YR infection in the leaf tissue, a degradation of chlorophyll happens, while the urediniospores of rust fungi are pigmented through the formation of carotenoids [49]. This could explain the importance of certain absorption or reflection bands of pigments for YR detection in the visible range. The effect of chlorophyll degradation and the formation of chlorosis, and a resulting detectability for the disease has also been described for *Septoria tritici* blotch [28]. The NIR region is strongly influenced by the leaf and cell structures, the architecture of the canopy, and water absorption bands [43,50]. High YR incidence leads to an early senescence of leaves in the upper, but particularly in the lower leaf levels. This changes the appearance of the crop architecture, reduces the vitality of leaves and water content, and could explain the importance of specific wavebands for YR detection.

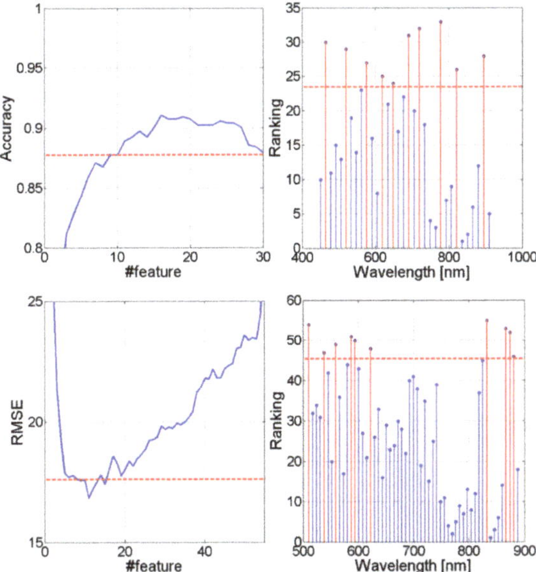

Figure 8. Results of the feature selection for the relevant wavebands for the classification of YR in the field on the ground (**top**) and UAV (**bottom**) scales. The accuracy reached for the different numbers of features (**left**) and the ranking of the inclusion within the feature subset (**right**) is displayed. RMSE = root mean square error.

3.3.2. UAV Scale

For the feature selection on the UAV scale, the detection and quantification of YR infections was investigated. Using the UAV and the filter-system Rikola hyperspectral camera, the mean spectrum of the central part of each plot was measured at multiple days. The first four dates were used, as a suitable disease estimation was not possible later due to the beginning of senescence.

The optimal number of features was 11 features, reaching an RMSE of 17.9 (i.e., to the visual assessment at the ground of around 70%) (Figure 8). Here, the most important bands were 830 nm and 510 nm, followed by NIR bands. Without significance were the red region 630–700 nm and the beginning of the NIR at 700–800 nm. The selection of the spectral border band would be a sign of fitting to noise if the Specim V10E line scanner had been used, but here, the Rikola camera was used without an increased noise at the spectral border regions.

Feature selection results for further traits are shown in Table 3. Important bands were also found in the green and NIR regions, which might have been triggered by the same biochemical reactions as on the ground scale. However, for the fertilizer, fungicide, and the combined treatment the spectral region 600–750 nm had a higher relevance.

Table 3. The six most important bands for selected plot traits at the UAV scale and ranking of the wavebands (in nm) for the importance of feature selection, beginning with the highest. Selected traits: fertilizer (Fert), fungicide (Fung), fungicide + fertilizer (Fert+Fung), yellow rust (YR) detection, yellow rust regression.

Ranking	Fert	Fung	Fert + Fung	YR Detection	YR Regression
1	767	727	734	797	832
2	725	804	887	881	510
3	648	762	545	601	867
4	557	648	559	706	874
5	627	594	517	874	587
6	704	767	748	594	594

3.3.3. Cross-Scale Interpretation

The cross-scale interpretation revealed significant inconsistencies but also some parallels. The inconsistencies were related to sensor characteristics, as the same sensor had not always been applied. Furthermore, additional factors at the different scales (leaf geometry, mixed pixels with background) were included at the higher scales that may have relied on further bands to be regarded properly by the prediction model.

The number of required features varied at the different scales. In a separate experiment (data not shown) with fixed leaves in the laboratory, a perfect differentiation was possible using two bands. Geometry was also not relevant, as the leaves were fixed in a horizontal position. The highest number required on the field scale was 18 on average, as the complex geometry and complex scattering effects in the canopy affected the recorded signal. At the UAV scale, the geometry was the same, but due to the physical smoothing by blur and high pixel size, the signal was simplified again. There, an optimum was reached at 11, omitting the spectral region 620–820 nm.

The red region had a low relevance for the classification of YR on the field and UAV scales. This might have been due to the fact that urediniospores *P. striiformis* appear more yellow than red (due to carotenoid composition) and do not show strong reflection in the red region. The NIR region had an increased relevance on the UAV scale. Presumably this was related to simple separability based on pigments on the lower scale, whereas in the field, the leaf geometry distorted this signal and the NIR region was required to compensate for this effect.

The differences and parallels of the different feature sets motivated the cross-scale application of feature sets. It was assumed that information about optimal feature sets could also be an advantage at a different scale. Therefore, the feature sets for the assessment of YR were exchanged between the ground scale with the Specim V10, and the UAV scale with the Rikola hyperspectral camera. To allow a comparison of the different feature sets, the number of included features was fixed to 10, based on the previous feature selection runs (Figure 8). Evaluation at the ground and UAV scales was performed following the same principle as for the feature selection. Table 4 shows the performance of multiple feature sets. The highest accuracy was reached by the full data set, followed by the 16 VIs. The feature sets with 10 features reached a slightly lower, but in direct comparison, very similar accuracy. The results indicate that the complex situation in the wheat canopy required more than 10 features. The good performance of the equidistant feature set can be explained by the resemblance to the 10 selected features that were nearly equidistantly distributed over the spectral range. Both feature sets applied wavebands out of the same spectral regions. Furthermore, the performance of the field-selected feature set points to the heterogeneity of reflectance characteristics even within the same treatment group. The test and training data were extracted from separate image sets. Following, this feature set was optimized on the training data, but had no advantages compared to the equidistant feature set on the test data.

Table 4. Performance of the different feature sets for the YR detection based on ground observations.

Ground Class.	All	UAV Select	Field Select	Equidistant	VI
# feature	210	10	10	10	16
Acc.	92.9%	87.4%	88.9%	89.2%	90.2%
F1 disease	0.84	0.694	0.751	0.732	0.764

The comparison of different feature sets showed the potential positive results of feature selection to a higher degree. The highest accuracy was obtained by the feature set optimized at the UAV scale, whereas the feature set from the ground scale obtained an even lower performance than the equidistant feature set (Table 5). For the UAV data set, a separation of test and training data was not possible due to the much smaller data base. Here, a leave-one-out cross validation was applied to obtain R^2 and correlation coefficients. The obtained feature set may have been more adapted to the evaluation procedure at the UAV scale compared to the ground scale. The UAV evaluation shows that it was possible to slightly increase the accuracy by feature selection compared to the full data set, and also that uninformed subsampling did not lead to optimal results.

Table 5. Performance of the different feature sets for the YR regression based on UAV observations.

UAV Regression	All	UAV Select	Field Select	Equidistant
# feature	55	10	10	10
R^2	0.63	0.69	0.57	0.61
Corr.	79.4%	83.0%	75.5%	78.1%

However, the data characteristics at the ground canopy and the UAV scale were so disparate that an advantage of feature set transfer is doubtful. The transferred feature set had a lower performance even compared to the uninformed equidistant sampling. There were multiple factors contributing to the deviating data characteristics expressed by different demands to the feature sets. One of the main points was the use of different sensors with different measurement principles, each adapted to its measurement scale. The noise characteristics of the ground camera showed an increased noise level at the spectral border regions and a noise optimum in the red range. The UAV camera showed a homogenous measurement quality for the whole range, despite some artifact bands around 630 nm, where optical refractions seem to occur at a beam splitter. The suitability of a spectral region can be significantly reduced by such sensor characteristics, but if the effect occurs only at one sensor, the optimal feature set changes. Further points regard the implicit spatial smoothing if a larger area is captured by a single pixel. At the ground scale, the feature set will directly point to the reflectance characteristics of the spores, whereas at the UAV scale, the reduced vitality and even morphological changes have to be taken into account. In contrast, the close-range observations at the ground scale were dominated by the leaf geometry, and more specifically by leaf angle and position within the crop stand. Therefore, the analysis model had to integrate these factors to enable predictions as robust as possible against the plant geometry. At the UAV scale, most of the pixels provided a mixed signal of multiple leaves and, in addition, the analysis was performed on the mean spectra of each plot. Most of the geometric effects averaged out as the characteristics of hundreds of leaves were averaged.

In general, there is no single waveband for individual diseases, but broad regions (blue, green, red, NIR I (700–800), NIR II 800–1000) with varying relevance for the different diseases. This is tightly coupled with the sensor characteristics. The Rikola camera was not able to measure the blue and NIR (900–1000 nm), but provided stable noise conditions over the whole measurement region. The Specim V10E camera had a larger measurement region (400–1000 nm), but the spectral border regions had a much higher noise level.

3.3.4. Spatial Resolution as Key Parameter for Disease Detection

The un-sampled data had a GSD of approximately 0.4 mm (for Specim as well as Rikola). The UAV observations (20 m flight height) had a GSD of approximately 8 mm (Figure 9).

This approach did not regard the adaptation of the model to the new classification scale. By retraining the prediction, the accuracy may be improved, as the smoothing here also affected the data characteristics. However, even then, the disease-specific information will vanish at a certain level. We omitted this evaluation as the performance measures of the retrained models were not comparable anymore as the number of training data declined drastically, e.g., to around 100 samples for YR at higher subsampling scales.

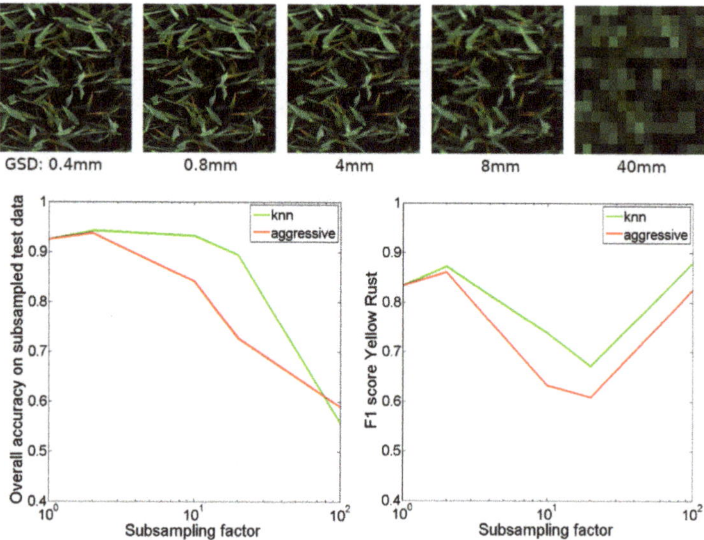

Figure 9. Visualization of spatial subsampling effects for the four investigated subsampling levels (images) and the effect of scale on accuracy and F1 score for the two different approaches knn (nearest neighbor) and an aggressive for subsampling the annotation.

The investigations allowed the definition of a minimal sampling distance at which the mixed information no longer allowed the prediction of plant diseases. Without retraining the model, the accuracy decreased at subsampling factors of 10 and 20. A low subsampling of 2 seems to have had no negative effects. Presumably, the included smoothing removed border cases and outliers which are hard to classify correctly. At higher subsampling levels, more and more mixed pixels occurred where the aggressive label subsampling tended to extend the image regions assigned to a class. Subsequently, the effect was more severe here. The accuracy of more than 50% at the final state was related to the dominant background, which provided a significant majority of test data at the high subsampling levels. It was not related to the ability to predict the presence of YR. This was demonstrated by the F1 score, a measure to quantify the performance of a multi-class prediction model on class level.

In this performance measure, the quality also decreased at subsampling factor of 10. Surprisingly, the F1 score increased to nearly optimal numbers at a subsampling factor of 100. Discussing this fact, it has to be noted that at the highest subsampling factor, only 119 YR samples were included in the data set, which are all correctly classified. Maybe this was related to the accuracy on the UAV scale, where the majority of geometric effects were averaged out. This point remains to be evaluated in further investigations, but it seems that subsampling by a factor of 100 removed the leaf structure completely, whereas at a lower subsampling factor, the leaf structure was still apparent but more and more effects of mixture become present. Without any leaf structure, the classification problem is simplified to the presence of YR within this part or even within the image. At low level disease severities, this will cause major problems, but here, the test data has been selected to show clear disease symptoms.

3.4. Optimal Sensor System for Plant Disease Detection

Two sensor systems were evaluated, both showing strengths and weaknesses. In direct comparison, the flying sensor system had strong advantages in usability, throughput, and commercial viability. The ground sensing system was much more sensitive, as a single symptom with a diameter of a few mm could be recorded. Such a spatial resolution could only be obtained by the flying system at a flight altitude of around 1 m above the canopy. However, with the used system this was not possible, as the downstream from the rotors would have strongly moved the canopy. Alternatively, an optical zoom could be applied, presumably reducing the light flux as well as the throughput of the overall system. This could be compensated for by an increased spatial resolution of the sensing array.

Summarizing, based on the experiments conducted during the presented study, we propose a focus on a UAV flying at low height in combination with a frame-based spectral camera sensing in around 15 equally distributed bands. A tunable band configuration would be an alternative that could use bands optimized for every single disease scenario, e.g., crop species, crop developmental stage, assumed disease setting, assumed symptom maturity. The spatial resolution should be set at around 1 mm GSD, a value that allows the detection small symptoms but neglects the high-frequency noise caused by the complex surface structure of plants [29,51].

4. Conclusions

This study investigated the detection of plant diseases using hyperspectral cameras at ground and UAV scales. In this context, the appropriate data analysis was decisively able to reach suitable results. Supervised classification has the advantage of separating disease-related signals from a huge amount of natural biological, geometrical, and sensor-related variability within a hyperspectral image of a crop canopy in the field. We proved that hyperspectral imaging in combination with supervised classification and regression showed good accordance to visual assessment at the ground. This allows questions to be addressed regarding the transfer of information between different scales and sensors. We showed that a feature selection was able to increase the prediction accuracy if it was performed on the analyzed data set. In contrast, scale or sensor transfer of selected feature sets was not successful, and was even less predictive than an uninformed regularly sampled feature set. This highlighted the importance of a precise specification of a prediction task by representative data samples. Deviations in data characteristics can significantly impair the performance of a data analysis pipeline or a tailored sensor in real-life applications.

This study sets a basis for ongoing research. New, upcoming sensors fulfilling the demands defined in this study might also cope with the current disadvantages. Consequently, there is a high probability that the defined flying sensor system with high resolution spectral camera, computing capabilities, and self-localization will be realized. Adapted legal conditions would allow an integrated system of field managing software, remote sensing based predictions, and current observations from the field using an automatized UAV.

Author Contributions: Conceptualization, D.B., J.B. and A.-K.M.; methodology, D.B. and J.B.; software, D.B. and J.B.; validation, D.B. and J.B.; formal analysis, D.B. and J.B.; investigation, D.B., J.B. and A.-K.M.; resources, D.B., J.B. and A.-K.M.; data curation, D.B., J.B. and A.-K.M.; writing—original draft preparation, D.B. and J.B.; writing—review and editing, D.B., J.B. and A.-K.M.; visualization, D.B., J.B. and A.-K.M.; supervision, J.B. and A.-K.M.; project administration, A.-K.M.; funding acquisition, A.-K.M.

Funding: This work was funded by BASF Digital Farming.

Acknowledgments: The authors would like to thank Onno Muller (Research Center Jülich, Germany) for providing the basic Phytobike frame, Thorsten Kraska for general support at Campus Klein-Altendorf and Winfried Bungert (Campus Klein-Altendorf, Germany) for implementation of cultivation measures during the vegetation period.

Remote Sens. **2019**, *11*, 2495

Conflicts of Interest: The authors declare no conflict of interest.

References

1. West, J.; Bravo, C.; Oberti, R.; Lemaire, D.; Moshou, D.; McCartney, A. The potential of optical canopy measurement for targeted control of field crop diseases. *Ann. Rev. Phytopathol.* **2003**, *41*, 593–614. [CrossRef] [PubMed]
2. Hillnhütter, C.; Mahlein, A.K.; Sikora, R.A.; Oerke, E.C. Remote sensing to detect plant stress induced by *Heterodera schachtii* and *Rhizoctonia solani* in sugar beet fields. *Field Crops Res.* **2011**, *122*, 70–77. [CrossRef]
3. Bravo, C.; Moshou, D.; West, J.; McCartney, A.; Ramon, H. Early disease detection in wheat fields using spectral reflectance. *Biosyst. Eng.* **2003**, *84*, 137–145. [CrossRef]
4. Mewes, T.; Franke, J.; Menz, G. Spectral requirements on airborne hyperspectral remote sensing data for wheat disease detection. *Precis. Agric.* **2011**, *12*, 795–812. [CrossRef]
5. Mirik, M.; Jones, D.C.; Price, J.A.; Workneh, F.; Ansley, R.J.; Rush, C.M. Satellite remote sensing of wheat infected by wheat streak mosaic virus. *Plant Dis.* **2011**, *95*, 4–12. [CrossRef]
6. Gebbers, R.; Adamchuk, V.I. Precision agriculture and food security. *Science* **2010**, *327*, 828–831. [CrossRef]
7. Mahlein, A.K.; Kuska, M.T.; Behmann, J.; Polder, G.; Walter, A. Hyperspectral sensors and imaging technologies in phytopathology: State of the art. *Ann. Rev. Phytopathol.* **2018**, *56*, 535–558. [CrossRef]
8. Wahabzada, M.; Mahlein, A.K.; Bauckhage, C.; Steiner, U.; Oerke, E.C.; Kersting, K. Plant phenotyping using probabilistic topic models: Uncovering the hyperspectral language of plants. *Sci. Rep.* **2016**, *6*, 22482. [CrossRef]
9. Whetton, R.; Hassall, K.; Waine, T.W.; Mouazen, A. Hyperspectral measurements of yellow rust and fusarium head blight in cereal crops: Part 1: Laboratory study. *Biosyst. Eng.* **2017**, *166*, 101–115. [CrossRef]
10. Whetton, R.; Waine, T.; Mouazen, A. Hyperspectral measurements of yellow rust and fusarium head blight in cereal crops: Part 2: On-line field measurement. *Biosyst. Eng.* **2018**, *167*, 144–158. [CrossRef]
11. Mahlein, A.K. Plant disease detection by imaging sensors—Parallels and scientific demands for precision agriculture and plant phenotyping. *Plant Dis.* **2016**, *100*, 241–251. [CrossRef] [PubMed]
12. Thomas, S.; Kuska, M.T.; Bohnenkamp, D.; Brugger, A.; Alisaac, E.; Wahabzada, M.; Behmann, J.; Mahlein, A.K. Benefits of hyperspectral imaging for plant disease detection and plant protection: A technical perspective. *J. Plant Dis. Prot.* **2018**, *125*, 5–20. [CrossRef]
13. Bravo, C.; Moshou, D.; Oberti, R.; West, J.; McCartney, A.; Bodria, L.; Ramon, H. Foliar disease detection in the field using optical sensor fusion. *Agric. Eng. Int.* **2004**, *6*.
14. Franke, J.; Menz, G. Multi-temporal wheat disease detection by multi-spectral remote sensing. *Precis. Agric.* **2007**, *8*, 161–172. [CrossRef]
15. Huang, W.; Lamb, D.W.; Niu, Z.; Zhang, Y.; Liu, L.; Wang, J. Identification of yellow rust in wheat using in-situ spectral reflectance measurements and airborne hyperspectral imaging. *Precis. Agric.* **2007**, *8*, 187–197. [CrossRef]
16. Lan, Y.B.; Chen, S.D.; Fritz, B.K. Current status and future trends of precision agricultural aviation technologies. *Int. J. Agric. Biol. Eng.* **2017**, *10*, 1–17.
17. Devadas, R.; Lamb, D.W.; Simpfendorfer, S.; Backhouse, D. Evaluating ten spectral vegetation indices for indentifying rust infection in individual wheat leaves. *Precis. Agric.* **2009**, *10*, 459–470. [CrossRef]
18. Ashourloo, D.; Mobasheri, M.R.; Huete, A. Developing two spectral indices for detection of wheat leaf rust (*Puccinia triticina*). *Remote Sens.* **2014**, *6*, 4723–4740. [CrossRef]
19. Cao, X.; Luo, Y.; Zhou, Y.; Fan, J.; Xu, X.; West, J.S.; Duan, X.; Cheng, D. Detection of powdery mildew in two winter wheat plant densities and prediction of grain yield using canopy hyperspectral reflectance. *PLoS ONE* **2015**, *10*. [CrossRef]
20. Blackburn, G.A. Hyperspectral remote sensing of plant pigments. *J. Exp. Bot.* **2007**, *58*, 855–867. [CrossRef]
21. Gitelson, A.A.; Keydan, G.P.; Merzlyak, M.N. Three-band model for noninvasive estimation of chlorophyll, carotenoids, and anthocyanin contents in higher plant leaves. *Geophys. Res. Lett.* **2006**, *33*, L11402. [CrossRef]
22. Roosjen, P.P.J.; Suomalainen, J.M.; Bartholomeus, H.M.; Clevers, J.G.P.W. Hyperspectral reflectance anisotropy measurements using a pushbroom spectrometer on an unmanned aerial vehicle—Results for barley, winter wheat and potato. *Remote Sens.* **2016**, *8*, 909. [CrossRef]

23. Moshou, D.; Bravo, C.; West, J.; Wahlen, T.; McCartney, A.; Ramon, H. Automatic detection of 'yellow rust' in wheat using reflectance measurements and neural networks. *Comput. Electron. Agric.* **2004**, *44*, 173–188. [CrossRef]

24. Zhang, J.; Pu, R.; Huang, W.J.; Yuan, L.; Luo, J.; Wang, J. Using in-situ hyperspectral data for detecting and discriminating yellow rust disease from nutrient stresses. *Field Crop Res.* **2012**, *134*, 165–174. [CrossRef]

25. Zheng, Q.; Huang, W.; Cui, X.; Dong, Y.; Shi, Y.; Ma, H.; Liu, L. Identification of wheat yellow rust using optimal three-band spectral indices in different growth stages. *Sensors* **2019**, *19*, 35. [CrossRef]

26. Azadbakht, M.; Ashourloo, D.; Aghighi, H.; Radiom, S.; Alimohammadi, A. Wheat leaf rust detection at canopy scale under different LAI levels using machine learning techniques. *Comput. Electron. Agric.* **2019**, *156*, 119–128. [CrossRef]

27. Chen, W.; Wellings, C.; Chen, X.; Kang, Z.; Liu, T. Wheat stripe (yellow) rust caused by *Puccinia striiformis* f. sp. *tritici*. *Mol. Plant. Pathol.* **2014**, *15*, 433–446. [CrossRef]

28. Yu, K.; Anderegg, J.; Mikaberidze, A.; Petteri, K.; Mascher, F.; McDonald, B.A.; Walter, A.; Hund, A. Hyperspectral canopy sensing of wheat *Septoria tritici* blotch disease. *Front. Plant Sci.* **2018**, *9*, 1195. [CrossRef]

29. Behmann, J.; Mahlein, A.K.; Rumpf, T.; Romer, C.; Plümer, L. A review of advanced machine learning methods for the detection of biotic stress in precision crop protection. *Precis. Agric.* **2015**, *16*, 239–260. [CrossRef]

30. Camps-Valls, G.; Tuia, D.; Bruzzone, L.; Benediktsson, J.A. Advances in hyperspectral image classification: Earth monitoring with statistical learning methods. *IEEE Signal Process. Mag.* **2014**, *31*, 45–54. [CrossRef]

31. Lowe, A.; Harrison, N.; French, A.P. Hyperspectral image analysis techniques for the detection and classification of the early onset of plant disease and stress. *Plant Methods* **2017**, *13*, 80. [CrossRef] [PubMed]

32. Singh, A.; Ganapathysubramanian, B.; Singh, A.K.; Sarkar, S. Machine learning for high-throughput stress phenotyping in plants. *Trends Plant Sci.* **2016**, *21*, 110–124. [CrossRef] [PubMed]

33. Barnes, R.J.; Dhanoa, M.S.; Lister, S.J. Standard normal variate transformation and de-trending of near-infrared diffuse reflectance spectra. *Appl. Spectrosc.* **1989**, *43*, 772–777. [CrossRef]

34. Kruse, F.A.; Lefkoff, A.B.; Boardman, J.W.; Heidebrecht, K.B.; Shapiro, A.T.; Barloon, P.J.; Goetz, A.F.H. The spectral image processing system (SIPS)—Interactive visualization and analysis of imaging spectrometer data. *Remote Sens. Environ.* **1993**, *44*, 145–163. [CrossRef]

35. Cortes, C.; Vapnik, V. Support-Vector networks. *Mach. Learn.* **1995**, *20*, 273–297. [CrossRef]

36. Chang, C.C.; Lin, C.J. LIBSVM: A library for support vector machines. *ACM Trans. Intell. Syst. Technol.* **2011**, *2*, 27. [CrossRef]

37. Behmann, J.; Steinrücken, J.; Plümer, L. Detection of early plant stress responses in hyperspectral images. *ISPRS* **2014**, *93*, 98–111. [CrossRef]

38. Guyon, I.; Elisseeff, A. An introduction to variable and feature selection. *J. Mach. Learn. Res.* **2003**, *3*, 1157–1182.

39. Parker, S.R.; Shaw, M.W.; Royle, D.J. The reliability of visual estimates of disease severity on cereal leaves. *Plant Pathol.* **1995**, *44*, 856–864. [CrossRef]

40. Nutter, F.W.; Gleason, M.L.; Jenco, J.H.; Christians, N.C. Assessing the accuracy, intra-rater repeatability, and inter-rater reliability of disease assessment systems. *Phytopathology* **1993**, *83*, 806–812. [CrossRef]

41. Mahlein, A.K.; Rumpf, T.; Welke, P.; Dehne, H.W.; Plümer, L.; Steiner, U.; Oerke, E.C. Development of spectral indices for detectiong and identifying plant diseases. *Remote Sens. Environ.* **2013**, *128*, 21–30. [CrossRef]

42. Chen, T.; Zhang, J.; Chen, Y.; Wan, S.; Zhang, L. Detection of peanut leaf spots disease using canopy hyperspectral reflectance. *Comput. Electron. Agric.* **2019**, *156*, 677–683. [CrossRef]

43. Gates, D.M.; Keegan, H.J.; Schelter, J.C.; Weidner, V.R. Spectral properties of plants. *Appl. Opt.* **1965**, *4*, 11–20. [CrossRef]

44. Curran, P.J. Remote sensing of foliar chemistry. *Remote Sens. Environ.* **1989**, *30*, 271–278. [CrossRef]

45. Heim, R.H.J.; Jurgens, N.; Große-Stoltenberg, A.; Oldeland, J. ; The effect of epidermal structures on leaf spectral signatures of ice plants (*Aizoaceae*). *Remote Sens.* **2015**, *7*, 16901–16914. [CrossRef]

46. Carter, G.A.; Knapp, A.K. Leaf optical properties in higher plants: Linking spectral characteristics to stress and chlorophyll concentrations. *Am. J. Bot.* **2001**, *88*, 677–684. [CrossRef]

47. Pandey, P.; Ge, Y.; Stoerger, V.; Schnable, J.C. High throughput in vivo analysis of plant leaf chemical properties using hyperspectral imaging. *Front. Plant Sci.* **2017**, *8*, 1348. [CrossRef]

48. Gay, A.; Thomas, H.; James, C.; Taylor, J.; Rowland, J.; Ougham, H. Nondestructive analysis of senescence in mesophyll cells by spectral resolution of protein synthesis-dependent pigment metabolism. *New Phytol.* **2008**, *179*, 663–674. [CrossRef]

49. Bohnenkamp, D.; Kuska, M.T.; Mahlein, A.K.; Behmann, J. Hyperspectral signal decomposition and symptom detection of wheat rust disease at the leaf scale using pure fungal spore spectra as reference. *Plant Pathol.* **2019**, *68*, 1188–1195. [CrossRef]

50. Elvidge, C.D. Visible and near infrared reflectance characteristics of dry plant materials. *Int. J. Remote Sens.* **1990**, *11*, 1775–1795. [CrossRef]

51. Kuska, M.; Wahabzada, M.; Leucker, M.; Dehne, H.W.; Kersting, K.; Oerke, E.C.; Steiner, U.; Mahlein, A.K. Hyperspectral phenotyping on the microscopic scale: Towards automated characterization of plant-pathogen interactions. *Plant Methods* **2015**, *11*, 28. [CrossRef] [PubMed]

Article

Experimental Evaluation and Consistency Comparison of UAV Multispectral Minisensors

Han Lu [1,2,3], **Tianxing Fan** [1,2,3], **Prakash Ghimire** [1,2,4] **and Lei Deng** [1,2,3,*]

[1]   College of Resource Environment and Tourism, Capital Normal University, Beijing 100048, China; 2180902125@cnu.edu.cn (H.L.); 2190902194@cnu.edu.cn (T.F.); 4183628002@cnu.edu.cn (P.G.)

[2]   College of Geospatial Information Science and Technology, Capital Normal University, Beijing 100048, China

[3]   Key Laboratory of 3D Information Acquisition and Application, Capital Normal University, Beijing 100048, China

[4]   Department of Survey, Ministry of Land Management, Cooperatives and Poverty Alleviation, Government of Nepal, Kathmandu 44600, Nepal

*   Correspondence: denglei@cnu.edu.cn

Received: 26 June 2020; Accepted: 5 August 2020; Published: 7 August 2020

Abstract: In recent years, the use of unmanned aerial vehicles (UAVs) has received increasing attention in remote sensing, vegetation monitoring, vegetation index (VI) mapping, precision agriculture, etc. It has many advantages, such as high spatial resolution, instant information acquisition, convenient operation, high maneuverability, freedom from cloud interference, and low cost. Nowadays, different types of UAV-based multispectral minisensors are used to obtain either surface reflectance or digital number (DN) values. Both the reflectance and DN values can be used to calculate VIs. The consistency and accuracy of spectral data and VIs obtained from these sensors have important application value. In this research, we analyzed the earth observation capabilities of the Parrot Sequoia (Sequoia) and DJI Phantom 4 Multispectral (P4M) sensors using different combinations of correlation coefficients and accuracy assessments. The research method was mainly focused on three aspects: (1) consistency of spectral values, (2) consistency of VI products, and (3) accuracy of normalized difference vegetation index (NDVI). UAV images in different resolutions were collected using these sensors, and ground points with reflectance values were recorded using an Analytical Spectral Devices handheld spectroradiometer (ASD). The average spectral values and VIs of those sensors were compared using different regions of interest (ROIs). Similarly, the NDVI products of those sensors were compared with ground point NDVI (ASD-NDVI). The results show that Sequoia and P4M are highly correlated in the green, red, red edge, and near-infrared bands (correlation coefficient (R^2) > 0.90). The results also show that Sequoia and P4M are highly correlated in different VIs; among them, NDVI has the highest correlation (R^2 > 0.98). In comparison with ground point NDVI (ASD-NDVI), the NDVI products obtained by both of these sensors have good accuracy (Sequoia: root-mean-square error (RMSE) < 0.07; P4M: RMSE < 0.09). This shows that the performance of different sensors can be evaluated from the consistency of spectral values, consistency of VI products, and accuracy of VIs. It is also shown that different UAV multispectral minisensors can have similar performances even though they have different spectral response functions. The findings of this study could be a good framework for analyzing the interoperability of different sensors for vegetation change analysis.

Keywords: reflectance; digital number (DN); vegetation index (VI); Parrot Sequoia (Sequoia); DJI Phantom 4 Multispectral (P4M)

---

1. Introduction

An unmanned aerial vehicle (UAV) is an unmanned aircraft operated by radio remote control equipment and self-provided program control device [1]. The combination of a UAV and a remote sensing sensor can constitute an ultralow-altitude remote sensing monitoring system. UAV remote sensing has many advantages, such as high image spatial resolution, instant information acquisition, convenient operation, high maneuverability, freedom from cloud interference, and low cost [2–4]. With the rapid development of UAV technology, UAV remote sensing has been widely used in agriculture, forestry, resource surveys, and vegetation monitoring [5–10]. Vegetation indices (VIs), as simple and effective measures of the surface vegetation condition, are widely used in vegetation monitoring via remote sensing [11–13]. Because of the unique response characteristics of vegetation in the near-infrared band, most vegetation indices (such as the normalized vegetation index [14] and the soil-adjusted vegetation index) are currently based on a combination of visible light and near-infrared bands [15].

At present, there are a variety of UAV-based multispectral minisensors on the market that can be used for vegetation monitoring [16–23] and can be selected according to the different needs of users. To make the VI products obtained from different sensors at different times comparable, the digital number (DN) of the collected image data is usually converted into reflectance, and then the reflectance is used to calculate the vegetation index [24]. For example, the Parrot Sequoia (Sequoia) multispectral sensor can help users to obtain the reflectance value, and then some conversions can be performed on the reflectance to calculate the VI [25]. Different sensors may use different conversion methods, but the conversion process may have a certain effect on the reflectance value which will further affect the calculated VI value. Unlike the Sequoia, the DJI Phantom 4 Multispectral (P4M) provides users with DN values, and then some conversions can be performed on the DN to calculate the VI. Although these two sensors provide users with different types of data for calculating VI values, the calculation method is the same; i.e., the VI value is the result of the spectral value (Sequoia: reflectance; P4M: DN) after linear or nonlinear transformation. Between these sensors, there is a certain difference between the spectral values in the same band, and this difference may be enlarged or reduced after VI calculations. This leaves the question as to whether there is any difference between the VI products obtained by the above two methods. To answer this question, this study considered two UAV multispectral minisensors, Sequoia and P4M. The research was conducted based on the consistency of spectral values, consistency of VI products, and accuracy of VI products, as these assessment methods have been widely used in different sensors [26–29].

Zhang et al. [30] compared the reflectance and normalized difference vegetation index (NDVI) of the medium spatial resolution satellite Sentinel-2A with those of Landsat-8. The results showed that the Sentinel-2A surface reflectance was greater than the Landsat-8 surface reflectance for all bands except the green, red, and the broad Sentinel-2A near-infrared bands. The Sentinel-2A surface NDVI was greater than the Landsat-8 surface NDVI. Ahmadian et al. [31] estimated the physiological and physical parameters of crops by using the VIs of Landsat8 OLI and Landsat-7 ETM+. The results showed that Landsat-8 OLI was better at capturing small variability in the VIs, making it more suitable for use in the estimation of crop physiological parameters. Roy et al. [32] compared Landsat-8 and Landsat-7 in terms of reflectance and NDVI. The results showed that the reflectance and NDVI of Landsat-8 were both greater than those of Landsat-7. In order to accurately distinguish cassava and sugarcane in images, Phongaksorn et al. [33] compared the reflectance and NDVI of Landsat-5 and THEOS. The results showed that THEOS can better distinguish the two crops. These previous studies show the value of studying different sensor platforms in terms of reflectance and VI in order to evaluate their performance. While there are many other relevant studies focusing on satellite-based sensors [34–38], only a few studies have considered UAV-based minisensors.

Bueren et al. [39] compared four optical UAV-based sensors (RGB camera, near-infrared camera, MCA6 camera, and STS spectroradiometer) to evaluate their suitability for agricultural applications. The STS spectrometer and the multispectral camera MCA6 were found to deliver spectral data that can match the spectral measurements of an Analytical Spectral Devices handheld spectroradiometer (ASD) at ground level when compared over all waypoints. Bareth et al. [40] compared the Cubert UHD185 Firefly and Rikola hyperspectral camera (RHC) to introduce their performance in precision agriculture.

The results showed that they both worked well, and the flight campaigns successfully delivered hyperspectral data. Nebiker et al. [41] compared three sensors (Canon s110 NIR, multi-SEPC 4C Prototype, and multi-SEPC 4C commercial) to investigate their characteristics and performance in agronomical research. The investigations showed that the SEPC 4C (multi-SEPC 4C Prototype and multi-SEPC 4C commercial) matched very well with ground-based field spectrometer measurements, while the Canon s110 NIR expressed significant biases. Deng et al. [42] systematically compared the vegetation observation capabilities of MCA and Sequoia based on reflectance and VI. It was found that the reflectance of the MCA camera had higher accuracy in the near-infrared band, and the reflectance accuracy of the Sequoia camera was more stable in each band. The MCA camera can obtain an NDVI product with a higher accuracy after using a more precise nonlinear calibration method.

In recent years, UAV minisensors have begun to show an end-to-end (user to product) development trend, which simplifies the data processing and VI calculation, thus giving users the best sense of use. It is necessary to further explore the performance of different UAV multispectral minisensors based on previous research. To address this, our main objective of this paper was to experimentally evaluate different UAV multispectral minisensors and compare them in terms of consistency. To meet the main objective, we focused on (1) analyzing the consistency of spectral values, (2) analyzing the consistency of VI products, and (3) assessing the accuracy of NDVI products between two sensors. This research will suggest whether vegetation observations from different sensors complement each other or not, thereby further broadening their application in different fields.

2. Materials and Methods

2.1. Study Area

The study area is located in Fangshan District, Beijing, China (39°33′34.93″ N, 115°47′40.97″ E), which has a warm temperate humid monsoon climate (Figure 1). It covers an area of 0.03 km^2 with flat terrain and 95 m average altitude. The annual average temperature is 11.6 °C, and the annual average precipitation is 602.5 mm. There are a variety of surface types in the area, mainly grassland.

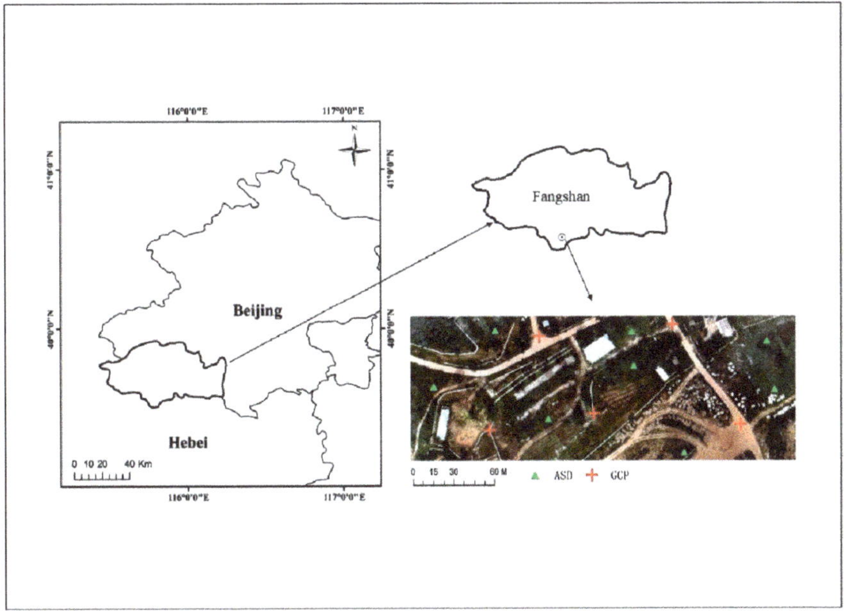

Figure 1. Location of study area.

2.2. Multispectral Sensors and UAV Platforms

Two types of multispectral sensors were compared in this experiment. The Sequoia camera [43] has a total of five imaging sensors, including four multispectral sensors and one RGB sensor (Figure 2). The spectral response function of Sequoia is shown as the solid line in Figure 3, which was provided by the manufacturer. The focal length of the Sequoia camera is 3.98 mm, the image size is 1280 × 960 pixels, and the sensor size is 4.8 mm × 3.6 mm. It is equipped with a sunshine sensor that can record the illumination information of each image, facilitating the calibration of multispectral images. The self-provided calibration panel can be used for radiometric calibration, and the reflectance data can be obtained directly.

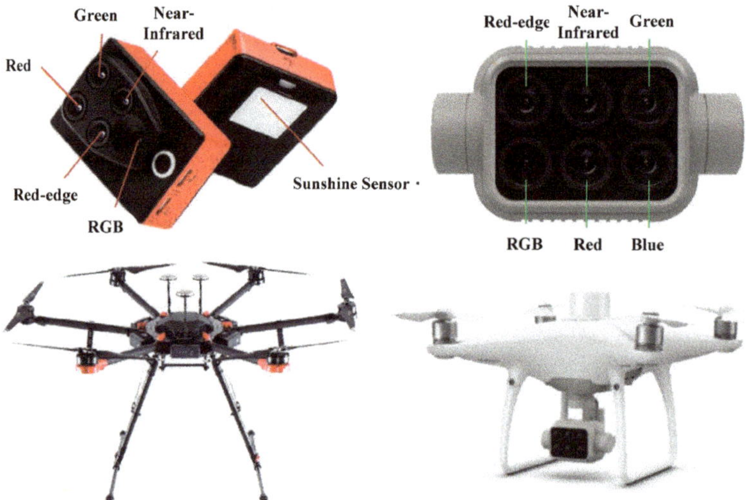

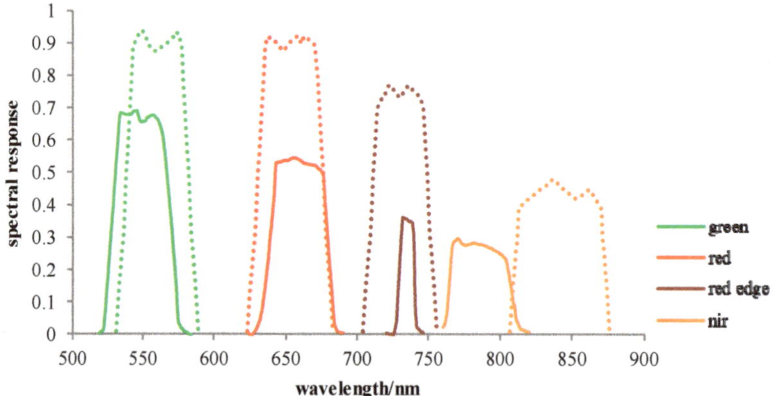

Figure 2. UAV platforms and multispectral sensors (Sequoia and P4M). The Sequoia sensor is carried on the EM6-800 hexarotor UAV (left), and the P4M uses its own aircraft (right).

Figure 3. Spectral response functions of Sequoia (solid lines) and P4M (dashed lines).

The other multispectral sensor considered in the study is the P4M (Figure 2). The P4M camera [44] has a total of six imaging sensors, including five multispectral sensors and one RGB sensor. The spectral response function of P4M is shown as the dashed line in Figure 3, which was provided by the manufacturer.

The focal length of the P4M camera is 5.74 mm, the image size is 1600 × 1300 pixels, and the sensor size is 4.87 mm × 3.96 mm. The P4M camera is also equipped with a sunshine sensor, but the reflectance data cannot be obtained directly. Table 1 shows the band information for Sequoia and P4M.

Table 1. Spectral band information for the Parrot Sequoia (Sequoia) and DJI Phantom 4 Multispectral (P4M). Their approximately equivalent bands (green, red, red edge, and near-infrared) were compared in this study.

Sequoia			P4M		
Band	Central Wavelength (nm)	Wavelength Width (nm)	Band	Central Wavelength (nm)	Wavelength Width (nm)
-	-	-	blue	450	32
green	550	40	green	560	32
red	660	40	red	650	32
red edge	735	10	red edge	730	32
near-infrared	790	40	near-infrared	840	52

During the data collection, the sensors were carried on different UAV platforms. The Sequoia was mounted on a hexarotor UAV called EM6-800 which has the advantages of low cost and high stability. Its payload is 800 g, and the maximum flight time is 40 min; under the maximum load of 1.2 kg, its maximum flight time is 25 min [45]. The UAV is equipped with an onboard flight controller, which includes a compass; an inertial unit; and gyroscopic, barometric, and global positioning system sensors. The onboard flight controller can be used to control the flight missions through flight route and measurement point settings. It can also record and access the data obtained by the mounted sensors for postprocessing. Unlike the Sequoia, the P4M has its own UAV platform, so it can complete the data collection task independently without the help of other aircraft. It has a takeoff weight of 1487 g, and the average flight time is 27 min.

2.3. Data Collection

2.3.1. Sequoia and P4M Data

The UAV flight was conducted during sunny and clear sky (without clouds) conditions from 11:00 to 13:00 on 22 August 2019. During data collection, the Sequoia sensor was mounted on the EM6-800 hexarotor UAV, while the P4M was mounted on its own aircraft. For the Sequoia, a calibration target provided by the manufacturer was recorded to perform radiometric calibration in postprocessing.

In this experiment, the two sensors acquired a total of three sets of image data. The Sequoia images were collected while flying at 56 m height with 5 cm resolution and 100 m height with 10 cm resolution. The P4M images were collected while flying at 100 m height with 5 cm resolution. All of those flights were started within a one-hour period (11:27 to 12:22) to maintain similar illumination among each set of images. These images were acquired with 80% overlap. Sequoia acquired a total of 1715 images and P4M acquired 960 images. The specific parameter settings for the sensors are shown in Table 2.

Table 2. Parameters of sensors used for comparison. During the data collection process, the P4M camera failed to successfully acquire image data with a resolution of 10 cm. The 10 cm image data used in the comparative experiment was obtained by resampling the 5 cm image.

Sensor	Date	Time	Altitude (m)	Solar Zenith (°)	Solar Azimuth (°)	Resolution (m)
P4M	2019.8.22	11:27	100	29.8031	155.00	0.05
Sequoia	2019.8.22	11:59	56	27.9838	170.80	0.05
Sequoia	2019.8.22	12:22	100	27.7342	182.63	0.10

2.3.2. ASD Data

We used the FieldSpec HandHeld 2 field spectroradiometer produced by Analytical Spectral Devices to measure the ground object spectral data. The coordinates and photos of the measured points were collected for ground object identification and visual interpretation of images. The spectroradiometer can perform continuous spectrum measurement in the wavelength range of 325–1075 nm, with the spectral resolution <3.0 nm at 700 nm, wavelength accuracy of ±1 nm, and field angle of 25°. It can measure the reflection, transmission, radiance, or irradiance in real time and obtain the continuous spectral curve of the measured object. The spectrum measurement was carried out in sunny and cloudless weather and the time was between 11:00 to 13:00. During the UAV flight, synchronous ground observation was carried out to ensure that the solar elevation angle, zenith angle, and weather conditions measured by the ground object spectrum were consistent with the UAV data. During data collection, the surveyors wore black clothing to absorb sunlight and reduce spectral interference. The spectrum measurement was carried out under natural light conditions, the spectroradiometer was held vertically downward at 1 m above the ground, and the sensor covered about 0.68 m^2 ground area. To improve the accuracy of measurement, each ground point was repeatedly measured (ten times), and then the average value was taken. The spectroradiometer was calibrated every 10 min to reduce the interference of weather change on the spectrum measurement. Considering the small area, similar vegetation species, and relatively uniform ground surface in the study area, a total of eight ground points were selected (Figure 1). These points were selected randomly. Finally, according to Equation (1), the radiance of the ground object was converted into reflectance using the calibration coefficient provided by the reference plate.

$$R_i = \frac{\int_{\lambda_{imin}}^{\lambda_{imax}} R_\lambda C_\lambda \mathrm{d}\lambda}{\int_{\lambda_{imin}}^{\lambda_{imax}} C_\lambda \mathrm{d}\lambda} \tag{1}$$

where R_i is the reflectance of band i ($i = 1, 2, 3, 4$), λ_{imax} and λ_{imin} are the maximum and minimum values of wavelength i, C_λ is the transmittance of wavelength, and R_λ is the reflectance of wavelength λ.

The reflectance data can be used to calculate the true NDVI of the ground point (ASD-NDVI). The area of each ground point is about 0.68 m^2. Therefore, when comparing the NDVI obtained by the two sensors with the true NDVI (ASD-NDVI), 272 pixels were selected and the average NDVI was taken from the 5 cm resolution image. Similarly, when comparing the NDVI obtained by the two sensors with the true NDVI (ASD-NDVI), 68 pixels were selected and the average NDVI was taken from the 10 cm resolution image.

2.3.3. GCP

Five ground control points (GCPs) were evenly established on the field using printed white crosses to ensure the overlap between the Sequoia and P4M imagery at different times (Figure 1). The GCP and ASD coordinates were measured with 0.025 m horizontal accuracy and 0.035 m vertical accuracy. A geodetic dual-frequency global navigation satellite system (GNSS) receiver was used in a rapid-static manner (approximately 4 min for each measurement) using the relative positioning approach from a master station located at a point with known coordinates.

2.4. Methodology

This research methodology involved data acquisition (Section 2.3), preprocessing (Section 2.4), VI selection (Section 2.4), ROI selection (Section 2.4), and analysis (Section 3). The details are described after methodology chart (Figure 4).

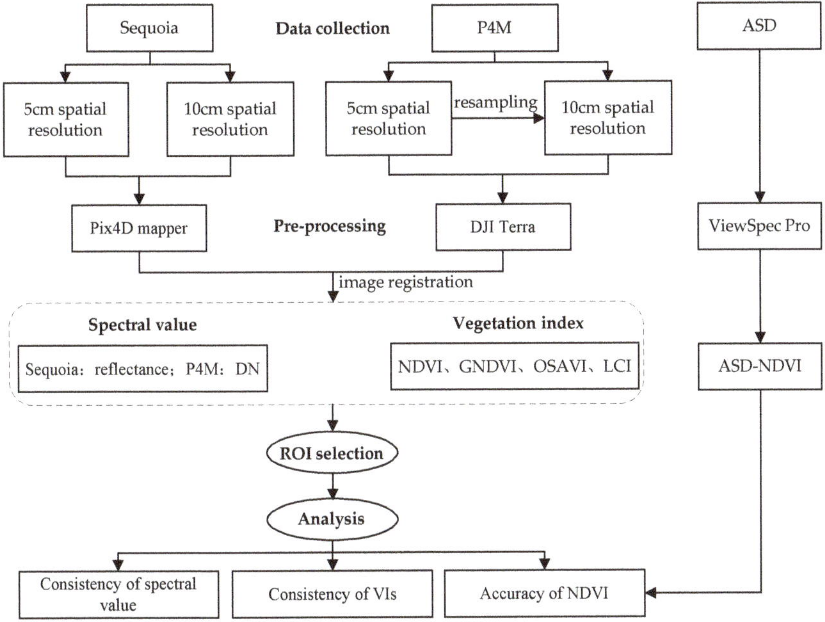

Figure 4. Methodological flowchart of the research.

2.4.1. Image Resampling

To standardize the spatial resolution of images acquired by different sensors, it is necessary to resample images that are very suitable for experimental comparison [46]. In order to avoid the contingency of the experimental results and to ensure the maneuverability of the UAV flight process, we compared the images of Sequoia and P4M with different spatial resolutions (5 and 10 cm). Therefore, we used ENVI software to resample the P4M images with the spatial resolution of 5 cm to obtain images with the spatial resolution of 10 cm. The pixel aggregate method was adopted in the resampling process [47].

2.4.2. Image Preprocessing

For preprocessing, Sequoia and P4M images were imported into Pix4D mapper [48] and DJI Terra software, respectively. Different steps of initial processing were followed, including point cloud processing, 3D model construction, feature extraction, feature construction, and orthophoto generation. As the Sequoia images can be used to directly obtain the reflectance data of the study area after processing, the VIs were calculated using reflectance data from VI equations. As the P4M images can be used to directly obtain the VIs, there was no need to get reflectance data for these images. Then, the processed images were imported into ENVI software to clip, match, and select different ROIs in a single band for comparison.

Figure 5 shows the processed 5 cm spatial resolution image. From Figure 5, it can be seen that there is a slight difference between a and b. For example, the building in the bottom left corner appears white in the P4M image but red in Sequoia, the stones on the right are white in P4M but red in Sequoia, and some roads which are white in P4M are yellow in Sequoia. These differences may be caused by the saturation of the red band in the Sequoia sensor. There are also some small differences between c and d. The Sequoia-derived NDVI (Sequoia-NDVI) is greater than the P4M-derived NDVI (P4M-NDVI); the range of Sequoia-NDVI is −0.19 to 0.93, and the range of P4M-NDVI is −0.43 to 0.85. There are some errors of Sequoia-NDVI in the visual range: Some buildings in the bottom left corner have the

NDVI of about 0.7 (yellow part), but the building does not correspond to such a large NDVI value in reality.

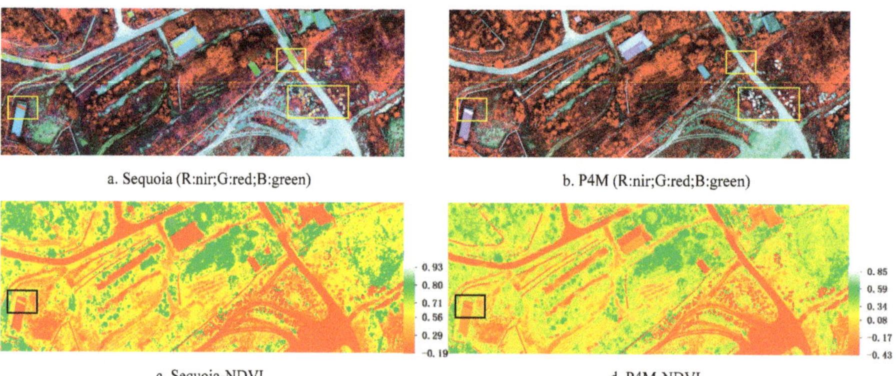

Figure 5. Image pairs of Sequoia and P4M (5 cm) after data processing: (**a**,**b**) two false color composites formed by the combination of near-infrared, red, and green bands; (**c**,**d**) normalized difference vegetation index (NDVI) products of the two sensors. The left side corresponds to the Sequoia camera, and the right side corresponds to the P4M camera. The yellow squares indicate the difference between Sequoia-derived RGB and P4M-derived RGB; the black squares indicate the difference between Sequoia-NDVI and P4M-NDVI.

Figure 6 shows the processed 10 cm spatial resolution images of Sequoia. Compared with the 5 cm spatial resolution results, both the false color RGB images (Figure 5a) and NDVI products (Figure 5c) are different. In the 10 cm resolution RGB image, the red part of the building in the bottom left corner still exists, but it is significantly smaller than in the 5 cm resolution image; the stones on the right are shown in white instead of red, and the road is shown in white instead of yellow. The problem of red band saturation does not seem to be obvious in 10 cm resolution images. Similarly, the NDVI value of the building in the bottom left corner seems normal.

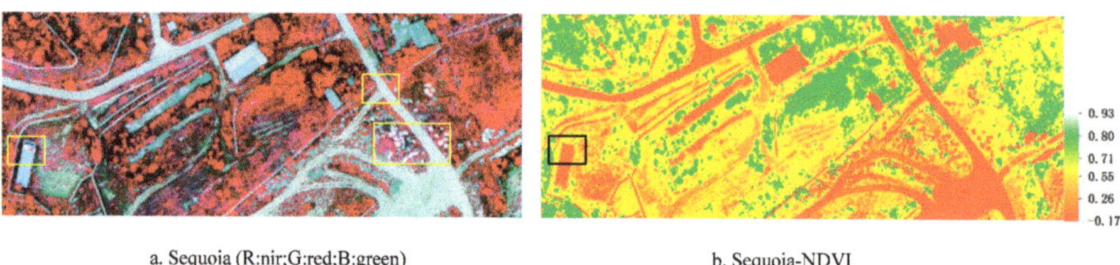

Figure 6. Image pairs of Sequoia (10 cm) after data processing: (**a**) the false color composite formed by the combination of near-infrared, red, and green bands; (**b**) normalized difference vegetation index (NDVI) product of Sequoia.

2.4.3. ROI Selection

For the comparison of different sensors, the most common method is to compare the spectrum information or VI of each corresponding band by statistical regression [49]. In this experiment, four commonly used VIs were compared between Sequoia and P4M. In the spectrum information comparison, the reflectance of Sequoia cannot be directly compared with the DN value of P4M, so we used linear regression to characterize the spectral difference of these two sensors. Additionally, eight ground points were also measured by ASD, but the number of points was too limited to establish a fitting relationship between the ASD and the sensor. Therefore, we compared the NDVI between ASD and sensor point by point. For this experiment, we selected some ROIs in the same location (between images of two sensors) and then compared the average value in each ROI [50]. The images of the study area were divided into 10 × 8 grids on average, and a ROI was selected from each grid. A total of 80 homogeneous ROIs (including vegetation and nonvegetation) were selected in the experiment. The selected ROIs were in flat terrain, properly sized, homogeneous, and almost identical (no other object features were included) [51]. The relation function between Sequoia and P4M was fitted using ordinary least square (OLS) regression. The goodness of fit was defined by the correlation coefficient (also written as R^2) [52]. Root-mean-square error (RMSE) was used to measure the deviation degree between the Sequoia-NDVI or P4M-NDVI and ASD-NDVI, as shown in Equation (2).

$$\text{RMSE} = \sqrt{\frac{\sum_{i}^{n} (y_i - y_i')^2}{n}} \qquad (2)$$

where y_i is the ASD-NDVI, y_i' is the average Sequoia-NDVI or P4M-NDVI, and n is the total number of ground points ($n = 8$).

2.4.4. VI Selection

VIs can reflect the growth status of vegetation. Different VIs may have certain differences in reflecting vegetation characteristics [53]. In vegetation studies, among all the possible existent VIs, NDVI, green normalized difference vegetation index (GNDVI), optimal soil-adjusted vegetation index (OSAVI), and leaf chlorophyll index (LCI) are commonly used. These four VIs were compared in this experiment, as shown in Table 3. NDVI is currently the most widely used VI in the world. In agriculture, NDVI is one of the most important tools for crop yield estimation, biomass estimation, and so on [54]. Using the unique response characteristics of vegetation in the near-infrared band, NDVI combines the spectral values of the red band and near-infrared band to quantitatively describe the vegetation coverage in the study area.

Table 3. Four vegetation indices (VIs) used for the research.

VI	Formula	Reference
Normalized Difference Vegetation Index	(NIR − R)/(NIR + R)	[57]
Green Normalized Difference Vegetation Index	(NIR − G)/(NIR + G)	[55]
Optimal Soil-Adjusted Vegetation Index	(NIR − R)/(NIR + R + 0.16)	[58]
Leaf Chlorophyll Index	(NIR − RE)/(NIR + R)	[56]

NIR, R, G, RE: Reflectance of near-infrared, red, green, and red edge bands.

Compared with NDVI, GNDVI is more sensitive to the change in vegetation chlorophyll content [55]. It combines the spectral values of the green band and the near-infrared band. OSAVI can reduce the interference of soil and vegetation canopy [15]. It also combines the spectral values of the red band and the near-infrared band. LCI is a sensitive indicator of chlorophyll content in leaves and is less affected by scattering from the leaf surface and internal structure variation [56]. It combines the spectral values of the red band, red edge band, and the near-infrared band. Different VIs were selected so that their VI equations contained different bands (Figure 3).

3. Results

3.1. Consistency of Spectral Values

In order to get better experiment results, we compared the images with 5 and 10 cm spatial resolution using the scatter plots of the Sequoia and P4M spectral values for the approximately equivalent spectral bands (green, red, red edge, and near-infrared). In the experiment, Sequoia used the spectral reflectance and P4M used the DN value of the image.

In the first experiment (5 cm spatial resolution), the spectral values of Sequoia and P4M were highly correlated (Figure 7). The two sensors showed a high correlation in the approximately equivalent four bands, and the correlation coefficient of the fitting function was not less than 0.90. The two sensors had the highest correlation in the red band (R^2 = 0.9709), followed by the green band (R^2 = 0.9699) and the red edge band (R^2 = 0.9208); the correlation for the near-infrared band was lower than those of the other three bands (R^2 = 0.9042). It was seen that the spectral values of Sequoia and P4M had an excellent correlation in the green and red bands, and the R^2 was greater than 0.96. Meanwhile, the correlation was low in the red edge and the near-infrared bands, and the R^2 was slightly less than 0.92. Thus, these results showed that spectral values of these two sensors had a high correlation in the green and red bands and a low correlation in the red edge and near-infrared bands.

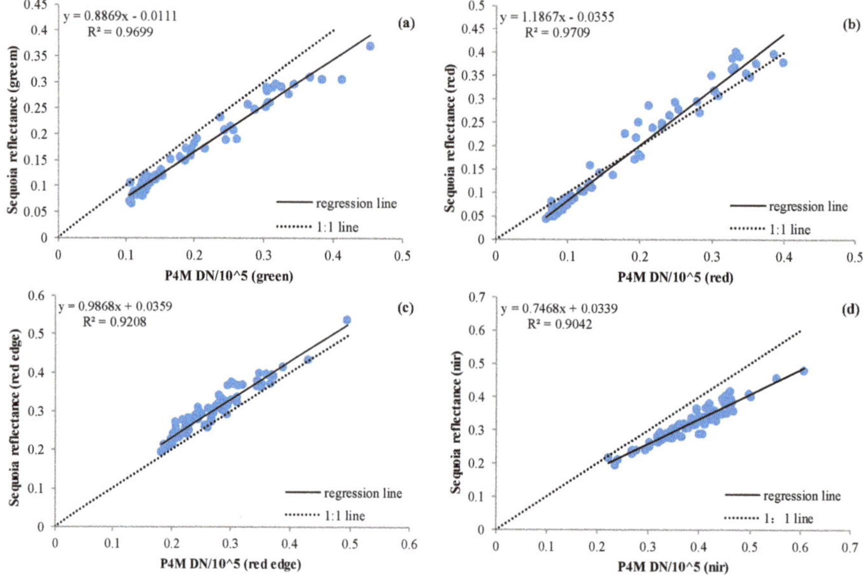

Figure 7. Scatter plots of Sequoia and P4M spectral values (5 cm) in the green band (**a**), red band (**b**), red edge band (**c**) and near infrared band (**d**). The solid lines show OLS regression of the Sequoia and the P4M data, and the dotted lines are 1:1 lines for reference.

In the second experiment (10 cm spatial resolution), the spectral values of Sequoia and P4M were also well correlated (Figure 8). In the four bands of these sensors, the correlation coefficient of the fitting equation was not less than 0.91, showing a strong correlation. As seen in the 5 cm spatial resolution results, the two sensors had the highest correlation in the red band (R^2 = 0.9793), followed by the green band (R^2 = 0.9727) and the red edge band (R^2 = 0.9436); the correlation for the near-infrared band was lower than those of the other three bands (R^2 = 0.9199). The spectral values of Sequoia and P4M were highly correlated in the green and red bands, and the R^2 was greater than 0.97. Similarly, the correlations between the two sensors in the red edge and the near-infrared bands were low, and the R^2 was slightly less than 0.94. These results also showed that spectral values of these two sensors had a high correlation in the green and red bands and a weak correlation in the red edge and near-infrared bands.

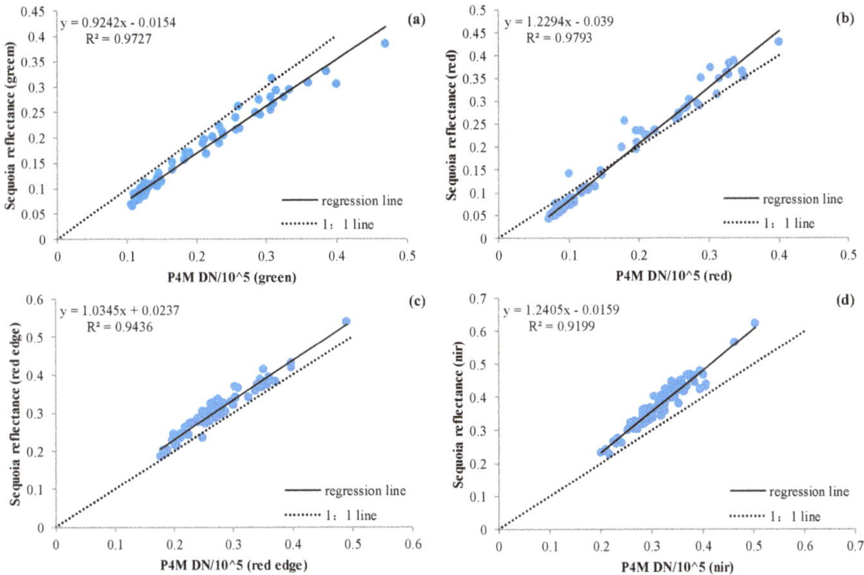

Figure 8. Scatter plots of Sequoia and P4M spectral values (10 cm) in the green band (**a**), red band (**b**), red edge band (**c**) and near infrared band (**d**).

In short, the spectral values of Sequoia and P4M were highly correlated in both the green and red bands ($R^2 > 0.96$), but the correlation was slightly lower in the red edge and the near-infrared bands ($R^2 < 0.96$). The correlation of spectral values for the two sensors at 10 cm spatial resolution was slightly higher than that of 5 cm. Thus, if we are interested in using both images at the same time, the 10 cm spatial resolution image may be the better choice. Although these two sensors were highly correlated, there was also a slight difference, which may be caused by a variety of mixing factors, including the difference in spectral response function (Figure 3). Among the compared bands, the center wavelength and the wave width of these two sensors were both close in the green and red bands. Although the center wavelength of these two sensors was close in the red edge band, there was a big difference in the wave width. In the near-infrared band, the center wavelength and the wave width of the two sensors were significantly different. This explains how the spectral values differ between Sequoia and P4M.

3.2. Consistency of VI Products

The VI products of Sequoia and P4M were highly correlated (Figure 9). Four VIs were compared in this paper, namely NDVI, GNDVI, OSAVI, and LCI, as shown in Figure 9. The results on the left were obtained with 5 cm spatial resolution image, and those on the right were obtained with 10 cm spatial resolution. The black dotted lines are the 1:1 reference lines, and the solid lines are the fitting functions of these sensor-derived VIs (using OLS regression).

Among the four VIs, NDVI had the highest correlation, followed by OSAVI, GNDVI, and LCI. In the comparison of 5 cm spatial resolution images, the correlation of NDVI was the highest ($R^2 = 0.9863$), followed by OSAVI ($R^2 = 0.9859$), while GNDVI and LCI were lower (GNDVI: $R^2 = 0.9595$; LCI: $R^2 = 0.9516$). In the comparison of 10 cm spatial resolution images, the correlation of NDVI was still the highest ($R^2 = 0.9842$), followed by OSAVI ($R^2 = 0.9806$), while GNDVI and LCI were lower (GNDVI: $R^2 = 0.9518$; LCI: $R^2 = 0.9546$).

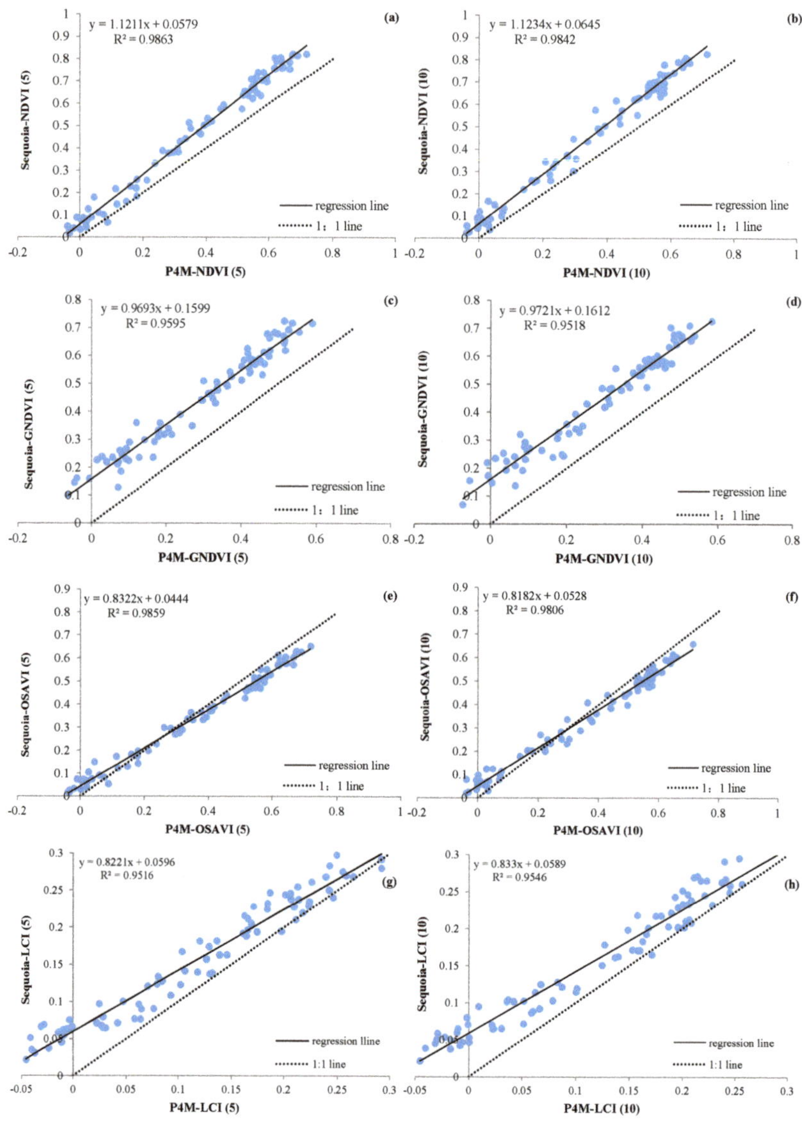

Figure 9. Scatter plots of Sequoia and P4M vegetation indices, which corresponded to the NDVI, GNDVI, OSAVI and LCI with 5cm spatial resolution (**a, c, e** and **g**) and 10cm (**b, d, f** and **h**).

Sequoia-NDVI had better result than P4M-NDVI (most of the scattered points were distributed above the 1:1 line, and a very small part of the scattered points were located on or below the 1:1 line). The legend in Figure 5 also shows that Sequoia-NDVI was slightly higher than P4M-NDVI; the fitting results of GNDVI and LCI were similar to NDVI, and there were also some differences. In both resolutions, Sequoia-GNDVI was higher than P4M-GNDVI (all scattered points were distributed above the 1:1 line), but the distributions of points were more dispersed than those of NDVI. The fitting result of OSAVI was different from those of the previous three indices. In both resolutions, Sequoia-OSAVI was only partially higher than P4M-OSAVI (the scattered points were evenly distributed above, below, and on the 1:1 line).

Table 4 shows Sequoia and P4M VI transformation functions derived by OLS regression of the data shown in Figure 9. The transformation functions for the 5 and 10 cm spatial resolution images are listed separately, where S represents the VI of Sequoia and P represents the VI of P4M. Both sensors had good consistency in those four indices.

Table 4. Sequoia and P4M VI transformation functions derived by OLS regression of the data illustrated in Figure 9.

VIs	N	5 cm Function	R^2	10 cm Function	R^2
NDVI	80	S = 1.1211 × P + 0.0579	0.9863	S = 1.1234 × P + 0.0645	0.9842
GNDVI	80	S = 0.9693 × P + 0.1599	0.9595	S = 0.9721 × P + 0.1612	0.9518
OSAVI	80	S = 0.8322 × P + 0.0444	0.9859	S = 0.8182 × P + 0.0528	0.9806
LCI	80	S = 0.8221 × P + 0.0596	0.9516	S = 0.8330 × P + 0.0589	0.9546

S: Sequoia, P: P4M.

NDVI, GNDVI, OSAVI, and LCI were used for different combinations of surface reflectivity, so their values were partly determined by the reflectance of the green, red, red edge, and near-infrared bands. There was a certain difference in the spectral response functions of the sensors, which led to slight differences between the VI products. Although users cannot directly obtain the reflectance from P4M image, they can still obtain high-quality VI products. It was seen that P4M, which integrates aircraft, cameras, and data processing software, optimizes the user's experience and improves the working efficiency by providing good VI products.

3.3. Accuracy of NDVI

Both the Sequoia-NDVI and P4M-NDVI had high accuracy, not only with a small deviation from ASD-NDVI but also with a good correlation (Figure 10). Two sets of spatial resolution data (5 and 10 cm) are compared in Figure 10: the left part shows the fitting scattered points of Sequoia-NDVI and ASD-NDVI, while the right part shows the fitting scattered points of P4M-NDVI and ASD-NDVI (blue dots correspond to 5 cm resolution and orange triangles correspond to 10 cm resolution). The Sequoia-NDVI was highly consistent with ASD-NDVI, and the correlation was high. In the comparative study of 5 cm spatial resolution images, RMSE = 0.0622 and R^2 = 0.8523; in 10 cm spatial resolution images, RMSE = 0.0684 ad R^2 = 0.8497. Similar to Sequoia, P4M-NDVI was also highly consistent with ASD-NDVI, maintaining a good correlation. In the comparative study, RMSE = 0.0886 and R^2 = 0.8785 for 5 cm spatial resolution images, while RMSE = 0.0842 and R^2 = 0.8785 for 10 cm spatial resolution images. This indicates that both Sequoia and P4M can provide NDVI products with high accuracy. Furthermore there was no big difference between the VI products obtained from these sensors.

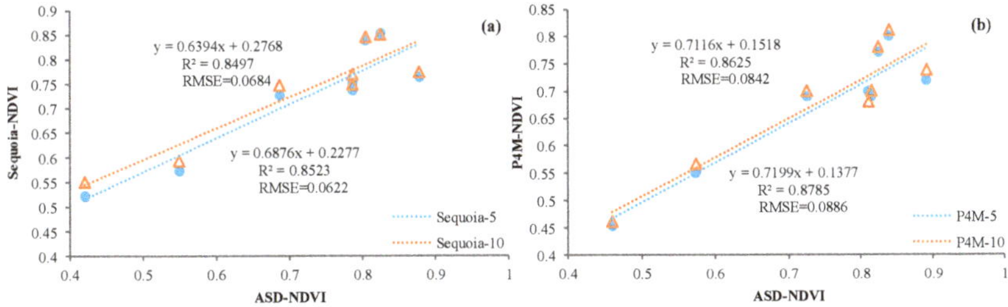

Figure 10. Scatter plots of Sequoia-NDVI (**a**) and P4M-NDVI (**b**) with ASD-NDVI. The blue dotted lines show OLS regression of Sequoia-NDVI (P4M-NDVI) and ASD-NDVI data with 5 cm resolution. The orange dotted lines show OLS regression of Sequoia-NDVI (P4M-NDVI) and ASD-NDVI data with 10 cm resolution.

4. Discussion

4.1. Differences between Sequoia and P4M

Different sensors may have different spectral response functions [59], and such differences will cause systematic deviations in the spectral values of the images. The consistency of spectral values between the two sensors studied showed a clear difference in the near-infrared band (Figures 7 and 8). The reason for this might be due to their different spectral response functions (Figure 3). Compared with the spectral response function of Sequoia, the spectral range of P4M in the near-infrared band was wider than that of Sequoia (Sequoia: 40 nm; P4M: 52 nm), and the positions of center wavelength in the near-infrared band were also different (Sequoia: 790 nm; P4M: 840 nm). The results also showed that there was some difference in the red edge band. Although the center wavelengths of these two sensors were close in the red edge band (Sequoia: 735 nm; P4M: 730 nm), there were big differences in the wave width (Sequoia: 10 nm; P4M: 32 nm). In contrast, the center wavelength and the wave width of these two sensors were both close in the green (Sequoia: 550 nm and 40 nm; P4M: 560 nm and 32 nm) and red bands (Sequoia: 660 nm and 40 nm; P4M: 650 nm and 32 nm). The different spectral response functions may explain the difference in the spectral values between Sequoia and P4M.

In addition, other factors such as the spectral reflection characteristics of the ground object, the nonuniformity of the ground surface, the observation time, and the solar elevation angle also increased the randomness and uncertainty of this systematic deviation [60–62]. In this experiment, the ground surface of the study area was uniform, and the observation time was similar for both sensors, as was the solar elevation angle. Therefore, the reason for the difference in spectral values was probably related to the reflection characteristics of ground target features. The spectral value of a single pixel may be influenced by both the spectral response function and the reflection characteristics of the target feature. Therefore, some objects in the image having high reflection characteristics in a specific spectral band may often be more affected by the difference in spectral response function.

The acquisition of the VI usually requires a series of conversion processes on the spectral values; thus, if there is a deviation in the spectral values, the VI may also be affected. In analysis of the consistency of the VIs between the two sensors, the four VI products of Sequoia and P4M were found to be highly correlated, but there were still some differences. These differences may have a great relationship with the differences in spectral values between the sensors. The reason for this difference in spectral values is also the same as explained above (due to spectral response function). Therefore, this difference between VIs may be caused by the spectral response function and the reflection characteristics of the target features. As we know, the VI is obtained by combining the spectral values of different bands, so using different combination methods of spectral values may also affect the quality of the VI.

4.2. Sensitivity of VIs to Spectral Deviation

The calculation of the VI involved spectral values of different spectral bands. NDVI, GNDVI, OSAVI, and LCI were compared in this study, and their calculation included the spectral values of red, green, red edge, and near-infrared bands. Therefore, small changes in the spectral values of each band may have a relatively big impact on the VI results. In addition, the band combination method may also change the sensitivity of the VI to small changes in the spectral values. The experimental results showed that although the spectral values of the Sequoia and P4M were significantly different in the near-infrared band, this difference did not show a significant impact on the VI products. The correlation coefficients of the VI products obtained by these two sensors were greater than 0.95. The NDVI products of the two sensors were also compared with the ASD-NDVI. The results showed that Sequoia-NDVI and P4M-NDVI both have high accuracy. The normalized calculation method of VI eliminated the influence of the difference in spectral values to a certain extent, thus reducing the sensitivity of the VI to such spectral deviations [63].

Poncet et al. [64] found that the error of VIs was correlated with different radiometric calibration methods. In this experiment, for Sequoia, we used a calibration target provided by the manufacturer to perform radiometric calibration in postprocessing. The calibration method might have affected the reflectance, which would have affected the VI.

4.3. Selection of Optimal Spatial Scale

The pixel is the smallest unit that constitutes the remote sensing digital image. It is an important symbol to reflect the features of the image and can be used to characterize the ground conditions in the study area. The pixel size determines the spatial resolution of a digital image and amount of information it can contain. After resampling from high spatial resolution to low resolution, the resultant image (low spatial resolution) will lose spectral information and spectral variation [65]. With the increase of remote sensing image scale, the spectral features of several different ground objects may appear simultaneously in a single pixel, resulting in the generation of a mixed pixel. At this time, the signal intensity of the ground object features in the pixel tends to be stable, and the pixel signals received by different sensors will tend to be similar.

The correlation between the spectral values of Sequoia and P4M in each band (green, red, red edge, and near-infrared) seemed to have a certain relationship with the image scale. When the image scale was small (5 cm), the correlation was low; when the image scale was large (10 cm), the correlation was high (Table 5). Fawcett et al. [66] found that the NDVI consistency of a multispectral sensor was similar at different spatial resolutions. Our results also show that the NDVI consistencies of Sequoia and P4M at 5 and 10 cm resolutions are similar (5 cm: R^2 = 0.9863; 10 cm: R^2 = 0.9863).

Table 5. Sequoia and P4M spectral value transformation functions and the values of their correlation coefficients (R^2).

Band	5 cm		10 cm	
	Function	R^2	Function	R^2
green	S = 0.8869 × P − 0.0111	0.9699	S = 0.9242 × P − 0.0154	0.9727
red	S = 1.1867 × P − 0.0355	0.9709	S = 1.2294 × P − 0.0390	0.9793
red edge	S = 0.9868 × P + 0.0359	0.9208	S = 1.0345 × P + 0.0237	0.9436
near-infrared	S = 0.7468 × P + 0.0339	0.9042	S = 1.2405 × P − 0.0159	0.9199

S: Sequoia, P: P4M.

4.4. Limitations

The study was carried out on a single date in a single study area with uniform vegetation species; it would be better if different study areas with different vegetation species in different periods were used. When assessing the suitability of UAV sensors in determining VIs, it would be important to include agricultural land, preferably with different nutrient treatments or crop species. Thus, in the future, it would be better to include agricultural land with crop species while doing VI-related research. Similarly, to assess the accuracy of NDVI, eight ASD points were used, as the terrain was uniform; still, better results could be obtained if more points were used. The use of more than two multispectral minisensors could be more meaningful to analyze the consistency of spectral values, consistency of VI products, and accuracy of NDVI. Therefore, detailed research is needed in the future to obtain improved results and conclusions.

5. Conclusions

Different UAV multispectral minisensors have been developed for applications in various fields, but their experimental performance and consistency need to be determined before their application. As a preliminary work towards consistency evaluation, different UAV images from Sequoia and P4M sensors with multispectral bands were acquired and preprocessed for ROI creation and VI calculation. The main objective of this research was to experimentally evaluate different UAV multispectral minisensors and compare them in terms of consistency. Using a combined method of consistency of spectral values, consistency of VI products, and accuracy of NDVI, we came to the following conclusions: First, the data acquisition capability of the Sequoia is similar to that of the P4M; both the spectral values and VIs of the two sensors have good correlation ($R^2 > 0.90$). Second, the VI products obtained from both sensors have good precision, and they are suitable for vegetation remote sensing monitoring. Third, both sensors have similar characteristics, and they may be used interchangeably for large area coverage with high spatial resolution and for daily time series science and applications.

Author Contributions: H.L., T.F., and L.D. designed and developed the research idea. H.Z. conducted the field data collection. H.L. and T.F. processed all remaining data. H.L. performed the data analysis and wrote the manuscript. H.L., T.F., P.G., and L.D. contributed to results and data interpretation, discussion, and revision of the manuscript. All the authors revised and approved the manuscript. All authors have read and agreed to the published version of the manuscript.

Funding: This research was funded by the National Key Research & Development Program of China (No. 2018YFC0706004).

Acknowledgments: The authors are very thankful for the valuable support of Zou Han Yue, Qiao Dan Yu, and Chen Yong.

Conflicts of Interest: The authors declare no conflict of interest. The funders had no role in the design of the study; in the collection, analyses, or interpretation of data; in the writing of the manuscript, or in the decision to publish the results.

References

1. Berni, J.; Zarco-Tejada, P.J.; Suarez, L.; Fereres, E. Thermal and Narrowband Multispectral Remote Sensing for Vegetation Monitoring From an Unmanned Aerial Vehicle. *IEEE T. Geosci. Remote.* **2009**, *47*, 722–738. [CrossRef]

2. Iizuka, K.; Itoh, M.; Shiodera, S.; Matsubara, T.; Dohar, M.; Watanabe, K. Advantages of unmanned aerial vehicle (UAV) photogrammetry for landscape analysis compared with satellite data: A case study of postmining sites in Indonesia. *Cogent Geosci.* **2018**, *4*, 1498180. [CrossRef]

3. Matese, A.; Toscano, P.; Di Gennaro, S.; Genesio, L.; Vaccari, F.; Primicerio, J.; Belli, C.; Zaldei, A.; Bianconi, R.; Gioli, B. Intercomparison of UAV, Aircraft and Satellite Remote Sensing Platforms for Precision Viticulture. *Remote Sens.* **2015**, *7*, 2971–2990. [CrossRef]

4. Kelcey, J.; Lucieer, A. Sensor Correction And Radiometric Calibration Of A 6-Band Multispectral Imaging Sensor For Uav Remote Sensing. *ISPRS—Int. Arch. Photogramm. Remote Sens. Spat. Inf. Sci.* **2012**, *XXXIX-B1*, 393–398. [CrossRef]

5. Dash, J.P.; Watt, M.S.; Pearse, G.D.; Heaphy, M.; Dungey, H.S. Assessing very high resolution UAV imagery for monitoring forest health during a simulated disease outbreak. *ISPRS J. Photogramm.* **2017**, *131*, 1–14. [CrossRef]

6. Lu, B.; He, Y. Species classification using Unmanned Aerial Vehicle (UAV)-acquired high spatial resolution imagery in a heterogeneous grassland. *ISPRS J. Photogramm.* **2017**, *128*, 73–85. [CrossRef]

7. Puliti, S.; Ene, L.T.; Gobakken, T.; Næsset, E. Use of partial-coverage UAV data in sampling for large scale forest inventories. *Remote Sens. Environ.* **2017**, *194*, 115–126. [CrossRef]

8. Bendig, J.; Yu, K.; Aasen, H.; Bolten, A.; Bennertz, S.; Broscheit, J.; Gnyp, M.L.; Bareth, G. Combining UAV-based plant height from crop surface models, visible, and near infrared vegetation indices for biomass monitoring in barley. *Int. J. Appl. Earth Obs.* **2015**, *39*, 79–87. [CrossRef]

9. Candiago, S.; Remondino, F.; De Giglio, M.; Dubbini, M.; Gattelli, M. Evaluating Multispectral Images and Vegetation Indices for Precision Farming Applications from UAV Images. *Remote Sens.* **2015**, *7*, 4026–4047. [CrossRef]

10. Feng, Q.; Liu, J.; Gong, J. UAV Remote Sensing for Urban Vegetation Mapping Using Random Forest and Texture Analysis. *Remote Sens.* **2015**, *7*, 1074–1094. [CrossRef]

11. Kamble, B.; Kilic, A.; Hubbard, K. Estimating Crop Coefficients Using Remote Sensing-Based Vegetation Index. *Remote Sens.* **2013**, *5*, 1588–1602. [CrossRef]

12. Liu, J.; Pattey, E.; Jégo, G. Assessment of vegetation indices for regional crop green LAI estimation from Landsat images over multiple growing seasons. *Remote Sens. Environ.* **2012**, *123*, 347–358. [CrossRef]

13. Motohka, T.; Nasahara, K.N.; Oguma, H.; Tsuchida, S. Applicability of Green-Red Vegetation Index for Remote Sensing of Vegetation Phenology. *Remote Sens.* **2010**, *2*, 2369–2387. [CrossRef]

14. Lunetta, R.S.; Knight, J.F.; Ediriwickrema, J.; Lyon, J.G.; Worthy, L.D. Land-cover change detection using multi-temporal MODIS NDVI data. *Remote Sens. Environ.* **2006**, *105*, 142–154. [CrossRef]

15. Huete, A.R. A soil-adjusted vegetation index (SAVI). *Remote Sens. Environ.* **1988**, *25*, 295–309. [CrossRef]

16. Ye, H.; Huang, W.; Huang, S.; Cui, B.; Dong, Y.; Guo, A.; Ren, Y.; Jin, Y. Recognition of Banana Fusarium Wilt Based on UAV Remote Sensing. *Remote Sens.* **2020**, *12*, 938. [CrossRef]

17. Ashapure, A.; Jung, J.; Chang, A.; Oh, S.; Maeda, M.; Landivar, J. A Comparative Study of RGB and Multispectral Sensor-Based Cotton Canopy Cover Modelling Using Multi-Temporal UAS Data. *Remote Sens.* **2019**, *11*, 2757. [CrossRef]

18. Iizuka, K.; Kato, T.; Silsigia, S.; Soufiningrum, A.Y.; Kozan, O. Estimating and Examining the Sensitivity of Different Vegetation Indices to Fractions of Vegetation Cover at Different Scaling Grids for Early Stage Acacia Plantation Forests Using a Fixed-Wing UAS. *Remote Sens.* **2019**, *11*, 1816. [CrossRef]

19. Lima-Cueto, F.J.; Blanco-Sepúlveda, R.; Gómez-Moreno, M.L.; Galacho-Jiménez, F.B. Using Vegetation Indices and a UAV Imaging Platform to Quantify the Density of Vegetation Ground Cover in Olive Groves (Olea Europaea L.) in Southern Spain. *Remote Sens.* **2019**, *11*, 2564. [CrossRef]

20. Freeman, D.; Gupta, S.; Smith, D.H.; Maja, J.M.; Robbins, J.; Owen, J.S.; Peña, J.M.; de Castro, A.I. Watson on the Farm: Using Cloud-Based Artificial Intelligence to Identify Early Indicators of Water Stress. *Remote Sens.* **2019**, *11*, 2645. [CrossRef]

21. Dash, J.; Pearse, G.; Watt, M. UAV Multispectral Imagery Can Complement Satellite Data for Monitoring Forest Health. *Remote Sens.* **2018**, *10*, 1216. [CrossRef]

22. Albetis, J.; Duthoit, S.; Guttler, F.; Jacquin, A.; Goulard, M.; Poilvé, H.; Féret, J.; Dedieu, G. Detection of Flavescence dorée Grapevine Disease Using Unmanned Aerial Vehicle (UAV) Multispectral Imagery. *Remote Sens.* **2017**, *9*, 308. [CrossRef]

23. Gevaert, C.M.; Suomalainen, J.; Tang, J.; Kooistra, L. Generation of Spectral–Temporal Response Surfaces by Combining Multispectral Satellite and Hyperspectral UAV Imagery for Precision Agriculture Applications. *IEEE J.-Stars.* **2015**, *8*, 3140–3146. [CrossRef]

24. Suárez, L.; Zarco-Tejada, P.J.; Berni, J.A.J.; González-Dugo, V.; Fereres, E. Modelling PRI for water stress detection using radiative transfer models. *Remote Sens. Environ.* **2009**, *113*, 730–744. [CrossRef]

25. Ahmed, O.S.; Shemrock, A.; Chabot, D.; Dillon, C.; Williams, G.; Wasson, R.; Franklin, S.E. Hierarchical land cover and vegetation classification using multispectral data acquired from an unmanned aerial vehicle. *Int. J Remote Sens. Unmanned Aer. Veh. Environ. Appl.* **2017**, *38*, 2037–2052. [CrossRef]

26. Ke, Y.; Im, J.; Lee, J.; Gong, H.; Ryu, Y. Characteristics of Landsat 8 OLI-derived NDVI by comparison with multiple satellite sensors and in-situ observations. *Remote Sens. Environ.* **2015**, *164*, 298–313. [CrossRef]

27. Cheng, Y.; Gamon, J.; Fuentes, D.; Mao, Z.; Sims, D.; Qiu, H.; Claudio, H.; Huete, A.; Rahman, A. A multi-scale analysis of dynamic optical signals in a Southern California chaparral ecosystem: A comparison of field, AVIRIS and MODIS data. *Remote Sens. Environ.* **2006**, *103*, 369–378. [CrossRef]

28. Soudani, K.; François, C.; le Maire, G.; Le Dantec, V.; Dufrêne, E. Comparative analysis of IKONOS, SPOT, and ETM+ data for leaf area index estimation in temperate coniferous and deciduous forest stands. *Remote Sens. Environ.* **2006**, *102*, 161–175. [CrossRef]

29. Goward, S.N.; Davis, P.E.; Fleming, D.; Miller, L.; Townshend, J.R. Empirical comparison of Landsat 7 and IKONOS multispectral measurements for selected Earth Observation System (EOS) validation sites. *Remote Sens. Environ.* **2003**, *88*, 80–99. [CrossRef]

30. Zhang, H.K.; Roy, D.P.; Yan, L.; Li, Z.; Huang, H.; Vermote, E.; Skakun, S.; Roger, J. Characterization of Sentinel-2A and Landsat-8 top of atmosphere, surface, and nadir BRDF adjusted reflectance and NDVI differences. *Remote Sens. Environ.* **2018**, *215*, 482–494. [CrossRef]

31. Ahmadian, N.; Ghasemi, S.; Wigneron, J.; Zölitz, R. Comprehensive study of the biophysical parameters of agricultural crops based on assessing Landsat 8 OLI and Landsat 7 ETM+ vegetation indices. *Gisci. Remote Sens.* **2016**, *53*, 337–359. [CrossRef]

32. Roy, D.P.; Kovalskyy, V.; Zhang, H.K.; Vermote, E.F.; Yan, L.; Kumar, S.S.; Egorov, A. Characterization of Landsat-7 to Landsat-8 reflective wavelength and normalized difference vegetation index continuity. *Remote Sens. Environ.* **2016**, *185*, 57–70. [CrossRef] [PubMed]

33. Phongaksorn, N.; Tripathi, N.K.; Kumar, S.; Soni, P. Inter-Sensor Comparison between THEOS and Landsat 5 TM Data in a Study of Two Crops Related to Biofuel in Thailand. *Remote Sens.* **2012**, *4*, 354–376. [CrossRef]

34. Runge, A.; Grosse, G. Comparing Spectral Characteristics of Landsat-8 and Sentinel-2 Same-Day Data for Arctic-Boreal Regions. *Remote Sens.* **2019**, *11*, 1730. [CrossRef]

35. Chastain, R.; Housman, I.; Goldstein, J.; Finco, M.; Tenneson, K. Empirical cross sensor comparison of Sentinel-2A and 2B MSI, Landsat-8 OLI, and Landsat-7 ETM+ top of atmosphere spectral characteristics over the conterminous United States. *Remote Sens. Environ.* **2019**, *221*, 274–285. [CrossRef]

36. Flood, N. Comparing Sentinel-2A and Landsat 7 and 8 using surface reflectance over Australia. *Remote Sens.* **2017**, *9*, 659. [CrossRef]

37. Mandanici, E.; Bitelli, G. Preliminary comparison of sentinel-2 and landsat 8 imagery for a combined use. *Remote Sens.* **2016**, *8*, 1014. [CrossRef]

38. Claverie, M.; Vermote, E.F.; Franch, B.; Masek, J.G. Evaluation of the Landsat-5 TM and Landsat-7 ETM+ surface reflectance products. *Remote Sens. Environ.* **2015**, *169*, 390–403. [CrossRef]

39. von Bueren, S.K.; Burkart, A.; Hueni, A.; Rascher, U.; Tuohy, M.P.; Yule, I. Deploying four optical UAV-based sensors over grassland: Challenges and limitations. *Biogeosciences* **2015**, *12*, 163–175. [CrossRef]

40. Bareth, G.; Aasen, H.; Bendig, J.; Gnyp, M.L.; Bolten, A.; Jung, A.; Michels, R.; Soukkamäki, J. Low-weight and UAV-based hyperspectral full-frame cameras for monitoring crops: Spectral comparison with portable spectroradiometer measurements. *Photogramm.-Fernerkund.-Geoinf.* **2015**, *2015*, 69–79. [CrossRef]

41. Nebiker, S.; Lack, N.; Abächerli, M.; Läderach, S. Light-weight multispectral UAV sensors and their capabilities for predicting grain yield and detecting plant diseases. *Int. Arch. Photogramm. Remote Sens. Spat. Inf. Sci.* **2016**, *41*. [CrossRef]

42. Deng, L.; Mao, Z.; Li, X.; Hu, Z.; Duan, F.; Yan, Y. UAV-based multispectral remote sensing for precision agriculture: A comparison between different cameras. *ISPRS J. Photogramm.* **2018**, *146*, 124–136. [CrossRef]

43. Parrot. Available online: https://support.parrot.com/us/support/products/parrot-sequoia (accessed on 23 July 2020).

44. P4 Multispectral. Available online: https://www.dji.com/nl/p4-multispectral?site=brandsite&from=nav (accessed on 23 July 2020).

45. EM6-800. Available online: http://www.easydrone.com.cn (accessed on 23 July 2020).

46. Gurjar, S.B.; Padmanabhan, N. Study of various resampling techniques for high-resolution remote sensing imagery. *J. Indian Soc. Remote.* **2005**, *33*, 113–120. [CrossRef]

47. Keys, R. Cubic convolution interpolation for digital image processing. *IEEE Trans. Acoust. Speech Signal Process.* **1981**, *29*, 1153–1160. [CrossRef]

48. Allred, B.; Eash, N.; Freeland, R.; Martinez, L.; Wishart, D. Effective and efficient agricultural drainage pipe mapping with UAS thermal infrared imagery: A case study. *Agr. Water Manag.* **2018**, *197*, 132–137. [CrossRef]

49. Xu, H.; Zhang, T. Assessment of consistency in forest-dominated vegetation observations between ASTER and Landsat ETM+ images in subtropical coastal areas of southeastern China. *Agr. For. Meteorol.* **2013**, *168*, 1–9. [CrossRef]

50. Saunier, S.; Goryl, P.; Chander, G.; Santer, R.; Bouvet, M.; Collet, B.; Mambimba, A.; Kocaman Aksakal, S. Radiometric, Geometric, and Image Quality Assessment of ALOS AVNIR-2 and PRISM Sensors. *IEEE T. Geosci. Remote.* **2010**, *48*, 3855–3866. [CrossRef]

51. Xiao-ping, W.; Han-qiu, X. Cross-Comparison between GF-2 PMS2 and ZY-3 MUX Sensor Data. *Spectrosc. Spect. Anal.* **2019**, *39*, 310–318.

52. Halliday, M.A.K.; Hasan, R. *Cohesion in English*; Taylor & Francis: New York, NY, USA, 2014.

53. Boyte, S.P.; Wylie, B.K.; Rigge, M.B.; Dahal, D. Fusing MODIS with Landsat 8 data to downscale weekly normalized difference vegetation index estimates for central Great Basin rangelands, USA. *Gisci. Remote Sens.* **2017**, *55*, 376–399. [CrossRef]

54. Rouse, J.W., Jr.; Haas, R.H.; Schell, J.A.; Deering, D.W. *Monitoring Vegetation Systems in the Great Plains with Erts*; Nasa Special Publication: Washington, DC, USA, 1973; pp. 309–317.

55. Gitelson, A.A.; Kaufman, Y.J.; Merzlyak, M.N. Use of a green channel in remote sensing of global vegetation from EOS-MODIS. *Remote Sens. Environ.* **1996**, *58*, 289–298. [CrossRef]

56. Pu, R.; Gong, P.; Yu, Q. Comparative analysis of EO-1 ALI and Hyperion, and Landsat ETM+ data for mapping forest crown closure and leaf area index. *Sensors* **2008**, *8*, 3744–3766. [CrossRef]

57. Haboudane, D. Hyperspectral vegetation indices and novel algorithms for predicting green LAI of crop canopies: Modeling and validation in the context of precision agriculture. *Remote Sens. Environ.* **2004**, *90*, 337–352. [CrossRef]
58. Zarco-Tejada, P.J.; Miller, J.R.; Morales, A.; Berjón, A.; Agüera, J. Hyperspectral indices and model simulation for chlorophyll estimation in open-canopy tree crops. *Remote Sens. Environ.* **2004**, *90*, 463–476. [CrossRef]
59. Ghimire, P.; Lei, D.; Juan, N. Effect of Image Fusion on Vegetation Index Quality—A Comparative Study from Gaofen-1, Gaofen-2, Gaofen-4, Landsat-8 OLI and MODIS Imagery. *Remote Sens.* **2020**, *12*, 1550. [CrossRef]
60. Flood, N. Continuity of reflectance data between Landsat-7 ETM+ and Landsat-8 OLI, for both top-of-atmosphere and surface reflectance: A study in the Australian landscape. *Remote Sens.* **2014**, *6*, 7952–7970. [CrossRef]
61. Barsi, J.A.; Lee, K.; Kvaran, G.; Markham, B.L.; Pedelty, J.A. The spectral response of the Landsat-8 operational land imager. *Remote Sens.* **2014**, *6*, 10232–10251. [CrossRef]
62. Wang, Z.; Coburn, C.A.; Ren, X.; Teillet, P.M. Effect of soil surface roughness and scene components on soil surface bidirectional reflectance factor. *Can. J. Soil Sci.* **2012**, *92*, 297–313. [CrossRef]
63. Wang, Z.; Liu, C.; Huete, A. From AVHRR-NDVI to MODIS-EVI: Advances in vegetation index research. *Acta Ecol. Sin.* **2003**, *23*, 979–987.
64. Poncet, A.M.; Knappenberger, T.; Brodbeck, C.; Fogle, M.; Shaw, J.N.; Ortiz, B.V. Multispectral UAS data accuracy for different radiometric calibration methods. *Remote Sens.* **2019**, *11*, 1917. [CrossRef]
65. Woodcock, C.E.; Strahler, A.H. The factor of scale in remote sensing. *Remote Sens. Environ.* **1987**, *21*, 311–332. [CrossRef]
66. Fawcett, D.; Panigada, C.; Tagliabue, G.; Boschetti, M.; Celesti, M.; Evdokimov, A.; Biriukova, K.; Colombo, R.; Miglietta, F.; Rascher, U. Multi-Scale Evaluation of Drone-Based Multispectral Surface Reflectance and Vegetation Indices in Operational Conditions. *Remote Sens.* **2020**, *12*, 514. [CrossRef]

remote sensing

Article

Determination of Appropriate Remote Sensing Indices for Spring Wheat Yield Estimation in Mongolia

Battsetseg Tuvdendorj [1,2], Bingfang Wu [1,2,*], Hongwei Zeng [1,2], Gantsetseg Batdelger [3] and Lkhagvadorj Nanzad [2,3,4]

[1] State Key Laboratory of Remote Sensing Science, Aerospace Information Research Institute, Chinese Academy of Sciences, Beijing 100101, China; baku@radi.ac.cn (B.T.); zenghw@radi.ac.cn (H.Z.)
[2] University of Chinese Academy of Sciences, Beijing 100049, China; lkhagvaa@radi.ac.cn
[3] National Remote Sensing Center, Information and Research Institute of Meteorology, Hydrology, and Environment (IRIMHE), Ulaanbaatar 15160, Mongolia; gantsetseg@irimhe.namem.gov.mn
[4] Key Laboratory of Digital Earth Science, Aerospace Information Research Institute, Chinese Academy of Sciences, Beijing 100194, China
* Correspondence: wubf@radi.ac.cn; Tel.: +86-10-64855689; Fax: +86-10-64858721

Received: 6 September 2019; Accepted: 29 October 2019; Published: 1 November 2019

Abstract: In Mongolia, the monitoring and estimation of spring wheat yield at the regional and national levels are key issues for the agricultural policy and food management as well as for the economy and society as a whole. The remote sensing data and technique have been widely used for the estimation of crop yield and production in the world. For the current research, nine remote sensing indices were tested that include normalized difference drought index (NDDI), normalized difference water index (NDWI), vegetation condition index (VCI), temperature condition index (TCI), vegetation health index (VHI), normalized multi-band drought index (NMDI), visible and shortwave infrared drought index (VSDI), and vegetation supply water index (VSWI). These nine indices derived from MODIS/Terra satellite have so far not been used for crop yield prediction in Mongolia. The primary objective of this study was to determine the best remote sensing indices in order to develop an estimation model for spring wheat yield using correlation and regression method. The spring wheat yield data from the ground measurements of eight meteorological stations in Darkhan and Selenge provinces from 2000 to 2017 have been used. The data were collected during the period of the growing season (June–August). Based on the analysis, we constructed six models for spring wheat yield estimation. The results showed that the range of the root-mean-square error (RMSE) values of estimated spring wheat yield was between 4.1 (100 kg ha^{-1}) to 4.8 (100 kg ha^{-1}), respectively. The range of the mean absolute error (MAE) values was between 3.3 to 3.8 and the index of agreement (d) values was between 0.74 to 0.84, respectively. The conclusion was that the best model would be (R^2 = 0.55) based on NDWI, VSDI, and NDVI out of the nine indices and could serve as the most effective predictor and reliable remote sensing indices for monitoring the spring wheat yield in the northern part of Mongolia. Our results showed that the best timing of yield prediction for spring wheat was around the end of June and the beginning of July, which is the flowering stage of spring wheat in this study area. This means an accurate yield prediction for spring wheat can be achieved two months before the harvest time using the regression model.

Keywords: MODIS; northern Mongolia; remote sensing indices; spring wheat; yield estimation

---

1. Introduction

Food security is an important topic for every country in the world [1]. Accurate and timely estimation of the spring wheat yield on regional and national scales is becoming absolutely essential for developing countries like Mongolia. In particular, crop yield estimation and the monitoring of crop production can provide fundamental information for crop producers, decision-makers in planning harvest and for agricultural development overall [2]. The agriculture sector is the second contributor to the Mongolian economy after mining [3]. However, only 13% of agricultural production is sourced from crops, mostly spring wheat, the remaining 87% is from the livestock [4] since the Mongolian climate is more suitable for extensive grazing, which covers more than 80% of the total land area. The spring wheat is below 1% of the total land area and around 1.35 million hectares of the total land is suitable for crop cultivation [5]. The northern part of Mongolia has the most favorable natural conditions and a more suitable area for rain-fed crops [6]. Hence, most of the spring wheat is grown in the northern provinces due to above-average precipitation. However, precipitation can only support the basic water requirement of spring wheat, and little variation in precipitation would cause a big fluctuation in crop yield. The vegetation cover, crop yields, and their growth are highly dependent on the amount of precipitation and the related soil moisture [7,8]. Mongolia has an extreme continental climate, with a short growing season, high evaporation, and low precipitation, which pose serious limitations for the Mongolian agriculture development. Because of the high altitude, our country's climate is much colder than other countries in the same latitude. More than 80% of total spring wheat cultivation is rain-fed and only 5000 hectares is irrigated for spring wheat in Mongolia [4]. Therefore, agricultural production is particularly sensitive to climate variability and climatic conditions make agriculture very challenging. Due to the impacts of climate change, more extreme and continued droughts have occurred in many parts of Mongolia and have directly affected the vegetation and crop growth, biodiversity and socioeconomics in Mongolia [9]. Nanzad et al. [10] found that about 41–57% of Mongolia has been ravaged by mild to severe droughts for many of the last 17 years. A consecutive severe drought in 2002, 2005, 2007, 2010, 2013, and 2015 lowered spring wheat production severely as shown in Figure 1 and spring wheat had to be imported as local production declined due to weather conditions [4].

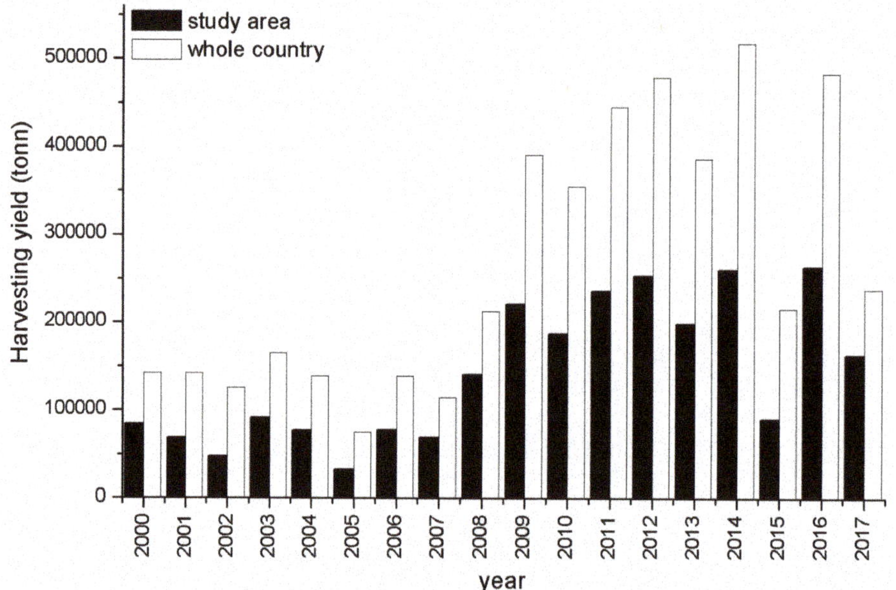

Figure 1. Whole Mongolian grain production with study area (National Statistics Office, 2019).

Weather information is normally used to forecast crop yield, but there is a lack of continuous measurement among others due to cost factors. Using Earth observation satellite imagery for monitoring temporal and spatial variation, combined with the point observation as a co-monitoring has advantages.

Furthermore, satellite imagery is produced at a lower cost than the traditional way and is more easily accessible for use [11,12]. The use of remote sensing data helps to assess crop conditions in different fields at regional and whole country levels, even in remote areas, as it gives a timely and accurate measurement. Therefore, there have been many attempts in the applications of remote sensing in crop yield estimation, monitoring and mapping and most of these work streams indicated that remote sensing technology was prospective and promising [13–21]. A number of field studies have shown that models based on remote sensing data enable to estimate crop yield in many countries. Usually, remote sensing derived indices are connected to crop yield using empirical regression-based models [22,23]. During the past decades, remote sensing has been broadly used in forecasting crop yield. The Advanced Very-High-Resolution Radiometer (AVHHR) is the most popular sensor, the most widely used in terms of crop monitoring and yield forecasting since the early 1980s for a large scale [24,25]. In recent years, satellite-derived data such as Moderate Resolution Imaging Spectroradiometer (MODIS), Landsat, and Sentinel data were used for the yield prediction and monitoring and meaningful results have been obtained [11,12,26–28]. Lewis et al., [29] used AVHRR-NDVI data for maize production forecasts and correlated results showed that forecasts could be obtained one month before the harvest. In Spain, Vicente-Serrana et al. [30] combined AVHHR-NDVI data and drought indices and were able to predict wheat and barley yield four months before harvest. Moreover, Peterson [12] found the best timing to predict crop yield was from two to four months before the harvesting using NDVI, EVI, and NDWI of MODIS for different crops in Africa. Recently, some remote sensing indices such as the normalized multiband drought index (NMDI), vegetation supply water index (VSWI), and visible and shortwave infrared drought index (VSDI) were utilized in a number of studies for drought and crop monitoring and crop yield estimation according to previous studies [31–34]. The more promising method is using crop growth modeling that incorporates updated crop biophysical parameters such as leaf area index (LAI) and a fraction of absorbed photosynthetically active radiation (fPAR) retrieved from satellite imagery and by using survey information of crops throughout the growing season in local to regional areas. For example, Huang et al. [35] found that more accurate county-level winter wheat estimation was obtained using the WOFOST-PROSAIL model. Furthermore, many researchers have developed crop growth models to estimate crop yields [36–40]. However, the crop growth models require more specific information, such as daily weather data, soil properties, and crop growth determining factors, which would make analytical costs excessive. It is obvious that no general indicator can be used to predict crop yields in all regions. The applicability of the indicator will vary with the region, crop type, and crop growth stage.

Some recent studies in Mongolia were conducted to monitor the cropland cover changes, to assess land degradation for the agricultural region. Erdenee et al. [5] have used Landsat TM and ETM data in the detection of changes cropland over Tsagaannuur, Selenge provinces from 1989 and 2000. Otgonbayar et al. [41] investigated to prepare a cropland suitability map of Mongolia using Landsat and MODIS (MOD13, MOD15, and MOD17). Furthermore, Enkhjargal et al. [42] used MODIS and SPOT time-series remotely sensed data from 2000–2013 to estimate long-term soil moisture content in agricultural regions of Mongolia. Ariya [43] used Landsat images from 2000 and 2015 to assess land degradation for the agricultural area of Mongolia. Nevertheless, to date, no studies of crop yield estimation using remote sensing indicators have been done yet in Mongolia. Recently, remote sensing indicators are employed to monitor the drought across the pasture lands of Mongolia [44] and provide valuable information for drought management and reduction. Compared to drought, more attention should be paid to crop yield, while these indices were not getting enough attention in the field of crop yield in Mongolia. The small variation of precipitation would cause the big fluctuation of crop yield so that it is very important to forecast spring wheat yield early for food security in Mongolia. Although spring wheat accounts for a small proportion of Mongolia's land area, as the size of the spring wheat field is large enough that it provides the possibility of predicting spring wheat yields based on remote sensing technology in Mongolia.

Therefore, our analysis focuses on Selenge and Darkhan provinces of Mongolia due to the availability of high-quality spring wheat yield data for those regions. It is the first attempt to estimate the spring wheat yield using space observation technology in Mongolia. The main objectives of this study were as follows: (i) to evaluate the potential of using remote sensing nine indices to estimate spring wheat yield; (ii) to choose the more suitable remote sensing indices for predicting spring wheat yield; (iii) to identify the best timing and more accurate model to estimate spring wheat yield in Northern Mongolia.

2. Study Area and Data

2.1. Study Area

Mongolia is divided into five different agro-ecological regions, which reflect distinct geographical patterns of agricultural production and climate. The study was conducted in Darkhan and Selenge provinces, which is located in Selenge-Onon agro-ecological region (N48–51 °C and E104–108 °C) (Figure 2). Selenge and Darkhan provinces are the principle cropping area for and account for more than 50% of national total grain production (Figure 1). In this region, most of the spring wheat cultivated area is rain-fed cropland. Thus, determining the fluctuation of spring wheat yield is highly dependent on the weather condition. The region averages between 90 to 110 frost-free days and has annual mean precipitation between 250 and 400 mm. In addition, crop growing duration is short (90–140 days) in this region, and depends on location and altitude. Approximately 90% of the nationwide precipitation is lost to evapotranspiration, which is associated with a continental climate. The remaining 10% of the total precipitation has been unable to evaporate, and 37% contribute to soil and underground reserves and streams, while 63% is surface runoff. In other words, 3–4% of total precipitation becomes potentially available as a water resource in the form of soil moisture or groundwater [6]. Additionally, the mean annual temperature is between 0.0 °C and 2.5 °C with cold temperature in January to –20 °C and warm temperature in July to 19 °C. Average elevations in this region have ranged between 1500 and 2000 m [45].

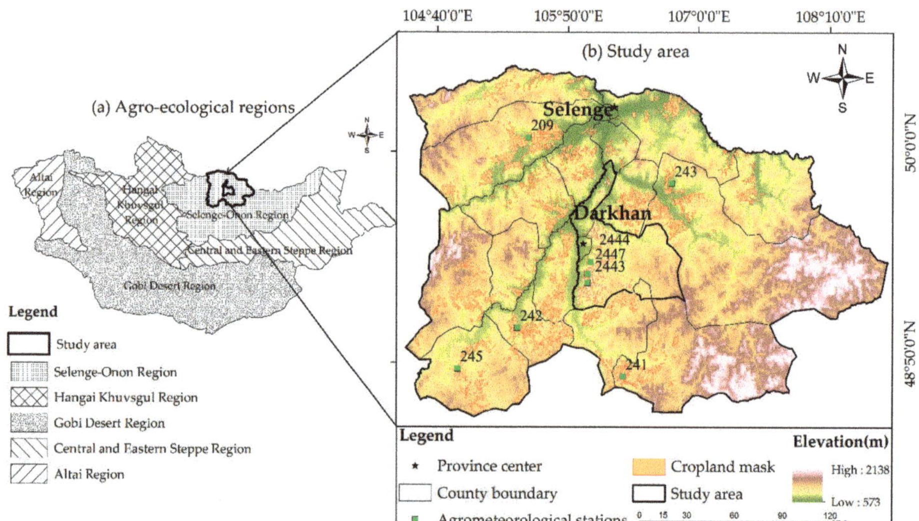

Figure 2. Study area: (**a**) Agro-ecological regions of Mongolia; (**b**) Digital Elevation Model data (DEM), Spatial distribution of agrometeorological stations in Selenge and Darkhan provinces and cropland.

2.2. MODIS Data and Processing

This study applied the use of 1 km spatial resolution, daily intervals MODIS/MOD1B time-series data to evaluate spring wheat yield. The MODIS data provided a high temporal resolution and wide coverage areas but a low spatial resolution. MODIS time-series data during the growing season (June to August) were obtained for the study area from the Atmosphere Archive and Distribution System (LAADS) of (NASA) for 2000–2017. Available online https://ladsweb.modaps.eosdis.nasa.gov (accessed on 20-10-2019). The study area covered over three tiles of granules of MODIS data such as an H24V03, H24V04, and H25V04 (H is horizontal and V is vertical coordinate). All downloaded MODIS 18 years' images were re-projected from sinusoidal to the Albers equal area projection. The reflectance bands (NIR, red, blue, and SWIR) were calculated using MOD021KM level 1b calibrated data. In the processing section, all the collected images were re-projected, mosaicked and calibrated for atmospheric and geometric correction using ENVI IDL software. In this study, the nine vegetation and drought indices (NDVI, NMDI, NDWI, VCI, TCI, VHI, NDDI, VSDI, and VSWI) are calculated from cloud-free and corrected reflectance bands.

2.3. Crop Data

In this study, we have utilized annual spring wheat yield data for eight agrometeorological stations from 2000 to 2017 in Northern Mongolia (Table 1). The sowing stage of spring wheat is generally in the first decade of May. All the crops present an important vegetative development in the June–August period and spring wheat harvest occurs generally in September. The statistics of spring wheat yield were obtained from the agrometeorological division of Information and Research Institute of Meteorology, Hydrology, and Environment (IRIMHE) in Mongolia. Agrometeorological stations measure the crop phenology stage, growth condition and damage, crop density and height for every 10 days from May to September. Finally, the spring wheat yield sown from sampling surveys at the end of the growing season was also measured. Spring wheat yield was collected in 50×50 cm plots, in four repeated samplings at each agrometeorological station. From the homologies sample plots, four samples of spring wheat were taken through crop cutting from a different area at equal distance and their average was taken to minimize random errors. Before thrashing and weighting the spring wheat grain yield, the sample plot was placed in the oven for 5–10 min and at $20°C–25°C$ to easily split the grain yield and the straw. The final spring wheat grain weight of each station was converted to 100kg ha^{-1} unit. As shown in (Figure 3), the phenological stages of the normal growth cycle of spring wheat and mean climate variables in the study area from May to September (2000–2017). In order to obtain the crop remote sensing indices values correctly, we had to solve image masking for regional spring wheat yield estimation. Applying a cropland mask to select remote sensing indices values as input to a crop yield model significantly improves the accuracy of the crop yield estimation [18,46]. A copy of the crop cover mask was obtained from the land cover map of Mongolia, provided by Elbegjargal et al. [47], and used to reduce the influence of non-agricultural areas on the remote sensing indices signal [48]. Finally, all areas with non-agricultural land were masked out and the regional annual yield was estimated and mapped for only cropland areas yield estimated maps were produced by model (M4) in the study area from 2000 to 2017.

Table 1. Information of meteorological stations with location (spring wheat yield available stations).

N	Province Name	Station Name	Station ID	Latitude	Longitude	Crop Type
1	Selenge	Tsagaannuur	209	50.0886	105.3252	wheat
2	Selenge	Baruunkharaa	241	48.7856	106.2624	wheat
3	Selenge	Orkhon	242	49.0039	105.4208	wheat
4	Selenge	Eruu	243	49.6842	106.6008	wheat
5	Selenge	Orkhontuul	245	48.7208	105.0358	wheat
6	Darkhan	Tsaidam	2443	49.3221	105.9826	wheat
7	Darkhan	6th Brigad	2444	49.3602	106.0782	wheat
8	Darkhan	Altangadas	2447	49.2328	105.9466	wheat

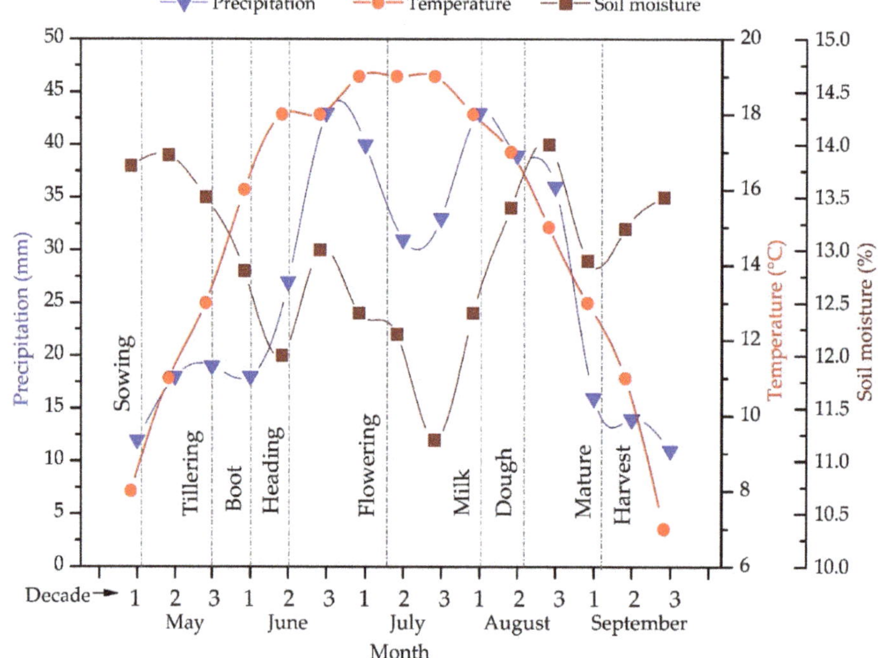

Figure 3. Growing stage of spring wheat and climate variables distribution in the study area.

3. Methodologies

The purpose of this study is to create a predictive measure of spring wheat yield computed from satellite data. The methodology has three main parts. Firstly, we calculate the nine remote sensing indices using MODIS data. Secondly, nine remote sensing indices are used for the relationship between actual spring wheat yield as input for the testing of the estimation model of spring wheat yield. Thirdly, we develop spring wheat yield regression models and map estimated spring wheat yield for the regional level to the 18 years. The general flowchart of this research, the processing method, and the individual steps are illustrated in Figure 4.

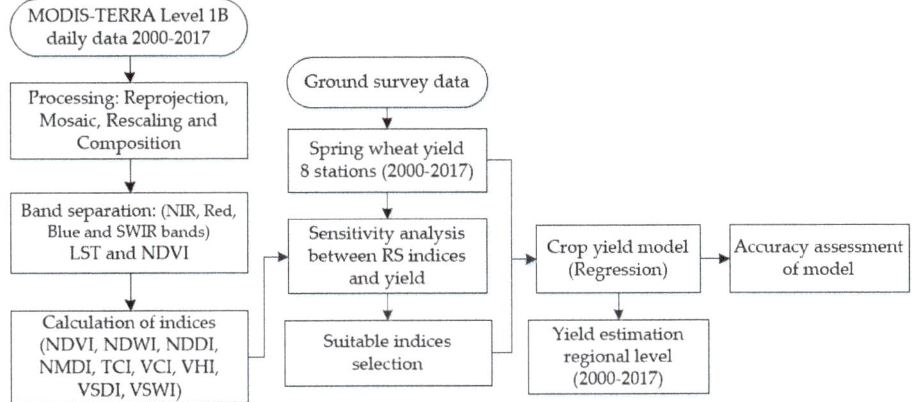

Figure 4. Flowchart of processing method.

3.1. Calculation of Remote Sensing Nine Indices:

Crop yield is markedly influenced by the growth conditions in each crop stage. Nine typical vegetation and drought indices were computed and compared with spring wheat yield. These indices are commonly used for crop yield estimation, drought monitoring, and only require optical spectral bands for calculation. For example, the NDVI has been most commonly used for vegetation monitoring, crop yield assessment and forecasting [27,49,50]. Kogan [51] has been developed the VCI to study the response of vegetation to drought conditions worldwide. Unganai et al. [52] conducted that the vegetation condition index (VCI) derived from AVHHR-NDVI correlated significantly with maize yield in Zimbabwe. The growth condition and crop yield are strongly correlated at the pixel level. Each pixel's value of remote sensing indices was directly taken at point locations of the eight stations. We examined the relationship between NDVI, NDWI, NMDI, TCI, VCI, VHI, NDDI, VSDI, and VSWI with actual spring wheat yield during the growing season (June–August) for 2000–2017 across Northern Mongolia. The 10 day and monthly nine remote sensing indices (NDVI, NMDI, NDWI, VCI, TCI, VHI, NDDI, VSDI, and VSWI), which are derived from transformations of the red, NIR, blue, and SWIR spectral bands, was used to continuous time series of data the represented the spring wheat growth indices temporal curve for each pixel in the study area. We used Table 2 for remote sensing indices calculation. Active plants are intensely absorbed by red and blue bands and reflected by the near-infrared band (NIR) [53]. Depending on the type of plant and the crop yield stage, the absorption of red and blue bands and the degree of the NIR band is different. When vegetation is stressed by lack of water, and also at the end of the growing period, the chlorophyll absorption declines and the ratio of NIR to RED or visible reflectance decreases. Therefore, vegetation index can be defined as the difference between RED and NIR bands reflections measured from satellites. The SWIR-1 and SWIR-2 bands are sensitive to the soil and vegetation moisture content. Hence, we calculated nine remote sensing indices utilized NIR, red, blue, and SWIR bands for spring wheat yield monitoring. Furthermore, we determined which indices from nine remotely sensed indices were more suitable to estimate spring wheat yields. To reduce the impact of atmosphere and cloud, the 10-day remote sensing-based indices derived from the daily data using maximum value composition (MVC) [54] were considered. Besides, we have retrieved monthly remote sensing indices from 10 days MVC indices by a simple moving average method. Particularly, each monthly remote sensing index was calculated by averaging three temporally 10 days MVC indices.

Table 2. Equations of tested nine remote sensing indices.

N	Remote Sensing Based Indices	Equation	References
1	Normalized Difference Vegetation Index	$NDVI = \frac{NIR-RED}{NIR+RED}$	[55]
2	Normalized Difference Water Index	$NDWI = \frac{NIR-SWIR}{NIR+SWIR}$	[56]
3	Vegetation Condition Index	$VCI = \frac{(NDVIj-NDVImin)}{(NDVImax+NDVImin)} \times 100\%$	[57]
4	Temperature Condition Index	$TCI = \frac{Tmax-Tj}{Tmax-Tmin} \times 100\%$	[57]
5	Vegetation Health Index	$VHI = a \times VCI + (1-a) \times TCI$	[57]
6	Normalized Multi-Band Drought Index	$NMDI = \frac{NIR-(SWIR1-SWIR2)}{NIR-(SWIR1+SWIR2)}$	[34]
7	Vegetation Supply Water Index	$VSWI = Ts/NDVI$	[58]
8	Normalized Difference Drought Index	$NDDI = \frac{(NDVI-NDWI)}{(NDVI+NDWI)}$	[59]
9	Visible and Shortwave Infrared Drought Index	$VSDI = 1 - [(SWIR - BLUE) + (RED - BLUE)]$	[33]

NIR–near-infrared band, Band2 (841–876 nm); RED–red band, Band1 (620–670 nm); SWIR1 - shortwave infrared band, Band6 (1628–1652 nm); SWIR2–shortwave infrared band, Band7 (2105–2155 nm); Blue–Band 3 (459–479 nm); Tj, Tmax and Tmin–surface temperature (current, maximum, and minimum).

3.2. Sensitive Analysis between Remote Sensing Indicators and Crop Yield

In this study, Pearson's correlation coefficient (R) between remote sensing indices and spring wheat yield was calculated for every 10 days and every month of the growing season from June to August for the northern part of Mongolia for 2000–2017 using following (Equation (1)). Pearson's correlation coefficient (R) represents the degree and direction of the linear regression between two continuous variables that are measured on the equal interval. The range of values for the R is from −1 to 1 (R > 0 it is a positive linear relationship, R = 0 it is indicated that there is no relationship and R < 0 it is a negative linear relationship).

$$R = \frac{\sum_{i=1}^{n}(x_i - \overline{x})(y_i - \overline{y})}{\sqrt{\sum_{i=1}^{n}(x_i - \overline{x})^2 \sum_{i=1}^{n}(y_i - \overline{y})^2}}, \tag{1}$$

where x_i and y_i presents remote sensing indices and the value of spring wheat yield at different time periods, n is the number of samples, $\overline{x}$ and $\overline{y}$ are the average values of x_i and y_i.

3.2.1. Crop Yield Estimation Model

The process of crop production is directly influenced by many biological, physiological and biophysical laws that are directly responsible for plants, and theoretically, it is possible to model the whole process of vegetative growth by mathematically describing the physical processes of these processes. Nowadays, dynamic modeling techniques are based on simple statistical equations, based on complex differential equations systems, in modeling the events of agroecosystems. In this study, using the remote sensing nine indices, we tested and attempted to develop the spring wheat yield estimation model. The relationship between spring wheat yield and remote sensing indices was observed through the linear regression model, where the independent variable was represented by remote sensing nine indices and the dependent variable was spring wheat. The estimation model is multilinear and includes slope and an interception constant coefficient. The empirical regression methods based on spectral indices have commonly used for modeling crop yield in many studies [49,60,61]. Bolton et al. [62] found a good linear correlation between different crop yields with MODIS-NDVI, EVI, and NDWI data at the county levels. Sui et al. [11] developed the estimation model for winter wheat production based on the environmental factors derived from satellite at a regional level, with errors of <12% for winter wheat yield, respectively. To determine when the correlation between remote sensing indices and spring wheat yield is strongest, we estimated the coefficient of determination (R^2) using data from every 10 days and monthly in the growing period of spring wheat. In order to make the final predicted results more accurate and stable, we considered it necessary to select the critical growth stage and remote sensing indices from all indices. We used a stepwise regression model in order to select the best candidate indices for yield estimation using SPSS 12 software. Stepwise regression provides a strong mean between the one or more independent variables and a dependent variable that conforms to the general equation for a multidimensional flat.

$$\hat{Y} = b_0 + b_1 X_1 + b_2 X_2 + \ldots + b_n X_n, \tag{2}$$

where, $\hat{Y}$ is the dependent variable and predicted spring wheat yield, $X_1, X_2 \ldots X_n$ are the independent variables and MODIS remote sensing indices, and $b_0, b_1, b_2 \ldots b_n$ are the regression coefficients and n is the number of independent variables.

3.2.2. Model Performance Evaluation

Then we selected the month that produced the highest coefficients to the determination to develop multilinear regression models based on all 18 years of data. Generally, the comprehensive method to validate models is to correlate the measured values against the predicted values [48]. We used model fitting and performance statistics such as the coefficient of determination (R^2), root mean square error (RMSE), mean absolute error (MAE), bias and an index of agreement (d) agreement.

$$RMSE = \sqrt{\frac{\sum_{i=1}^{n}\left(\hat{Y}-Y\right)^2}{n}}, \tag{3}$$

$$MAE = \frac{1}{n}\sum_{i=1}^{n}\left|\hat{Y}-Y\right|, \tag{4}$$

$$d = 1 - \frac{\sum_{i=1}^{n}\left(Y_i - \hat{Y}_i\right)^2}{\sum_{i=1}^{n}\left(\left|\hat{Y}_i - \overline{Y}\right| + \left|Y_i - \overline{Y}\right|\right)^2}, \tag{5}$$

where Y is observed values, $\hat{Y}$ is modeled values, $\overline{Y}$ is an average of observed values, and n is a number of yield and RS data.

4. Results

4.1. Temporal Climate Variables and Remote Sensing Indices Profiles for Spring Wheat

Figure 5 describes annual spring wheat yield with the amount of precipitation and average temperature throughout each growing season from 2000 to 2017. The temperature change, precipitation, and soil moisture have a significant impact on wheat yield. Mainly in the period May–September which accounts for 85% of annual precipitation. Especially in June–August, about 50%–60% of the annual precipitation occurs in Mongolia [63]. According to the [64] results show that the break of rainy season caused a similar reduction in soil moisture around mid-July in Mongolia. The highest mean precipitation of the growing months at the eight meteorological stations was 352 mm in 2013, 309 mm in 2009, 293 mm in 2008, 290 mm in 2012 and lowest annual precipitation occurred in 143 mm in 2002, 172 mm in 2001, and 199 mm in 2005 for the study area. In summer, especially growing season temperature and precipitation were negatively correlated. From (Figure 5) the lowest yield of spring wheat harvested in 2002(4.2kg ha^{-1}) and the highest yield in 2014(21.9kg ha^{-1}) in these two provinces. The main reason is that high temperatures and low precipitation led to soil moisture deficits, which is a significant impact on wheat yield. The climate of Mongolia is dry and semi-arid, and the growing vegetation cover and crop yields and development are highly dependent on the amount of precipitation and the related soil moisture [7,8].

Processed growing season (June–August) Moderate Resolution Imaging Spectroradiometer (MODIS-Terra) daily reflectance bands for study area for the years to 2000 to 2017. Based on the corrected surface reflectance NIR, red, blue, SWIR bands and LST of MODIS data, we have computed nine remote sensing indices and listed in (Table 2). We extracted time series values of NDVI, NDVI,

NMDI, NDWI, VCI, TCI, VHI, NDDI, VSDI, and VSWI interpolated 10 day and monthly intervals in growing season (June–August) from 2000 to 2017. As shown in (Figure 6), the long-term annual remote sensing nine indices variables for spring wheat yield in Northern Mongolia.

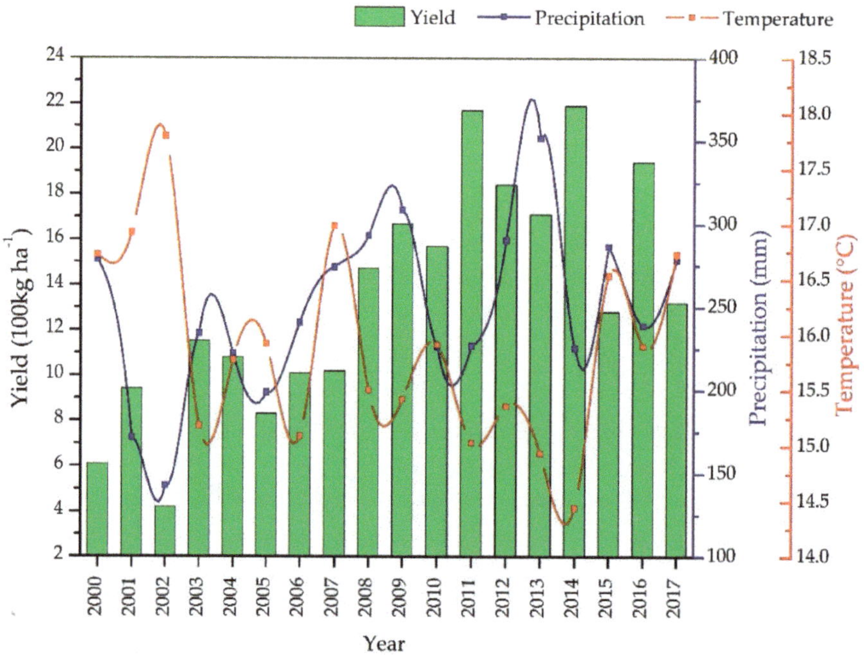

Figure 5. Long-term trend of average spring wheat yield and climate variables (2000–2017).

4.2. Sensitivity Analysis between Remote Sensing Indicators and Crop Yield

The Pearson's correlation analysis examined between multi-year spring wheat yield data of eight stations and NDVI, NMDI, NDWI, TCI, VCI, VHI, NDDI, VSDI, and VSWI for June-August from 2000 to 2017, for a total of about 135 data points. To test the regional effectiveness of remote sensing indices and determine the best index during the growing period (June to August) for spring wheat yield estimation, we compared the 10 days and monthly remote sensing indices with the spring wheat yield. In general, in terms of the sown stage in the middle of May, emergency in early June, flowering in late June and early July, milk, and dough in August, maturity in early September, and harvesting in late September in the study area (each growing stage shown in Figure 3). For each of the remote sensing indices, the correlation values and significant *p*-values were produced by eight stations from 2000 to 2017 (Table 3).

The results show that most of the indices had found a higher correlation with spring wheat yield in June and July. This period covers the heading and flowering phenological stage of spring wheat. To see all of the correlations between June remote sensing indices and spring wheat yield in more detail, refer to Figure 7. About 50–60% of the annual precipitation occurred in the growing period, especially most of the precipitation occurred in June and the beginning of July. Because of that reason, there is sufficient moisture and suitable weather condition. Slightly lower correlation was found in August since crops are harvested in September. We can see that results of correlation (R) among the nine indices, NDDI, VSDI, VSWI, and crop yield were negatively correlated with spring wheat yield, while other NDVI, VCI, TCI, VHI, NDWI, and NMDI were positively correlated. The highest correlation coefficient (R) values of between 10 day remote sensing 9 indices with spring wheat yield were NDVI

(0.51) in first 10 days of July, NMDI (0.18) in first 10 days of August, NDWI (0.48) in first 10 days of June, TCI (0.57) in third 10 days of June, VCI (0.31) in first 10 days of July, VHI (0.46) in third 10 days of June, NDDI (−0.38) in second 10 days of July, VSDI (−0.56) in first 10 days of July and August, and VSWI (−0.36) in first 10 days of June, respectively. These were statistically significant values with a yield of $p < 0.05$. Indices including NDDI, VSDI, and VSWI were negatively correlated with spring wheat yield, this implies when the values of these indices increase, the spring wheat yield decreased.

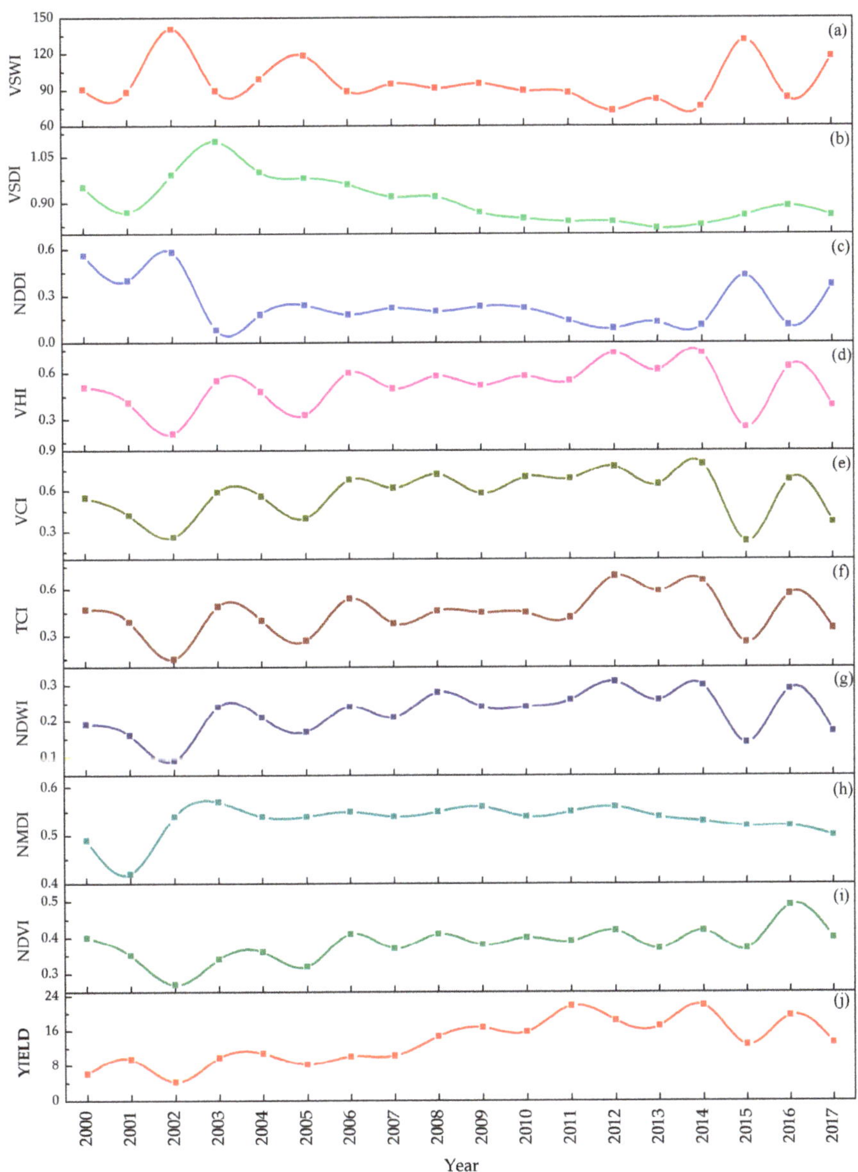

Figure 6. Annual growth season (June–August) average VSWI (**a**), VSDI (**b**), NDDI (**c**), VHI (**d**), VCI (**e**), TCI (**f**), NDWI (**g**), NMDI (**h**), NDVI (**i**), and Spring wheat yield (**j**).

Table 3. Multi-year correlation between 10-day indices and spring wheat yield in June to August (2000–2017).

Month	Decade	Index								
		NDVI	**NMDI**	**NDWI**	**TCI**	**VCI**	**VHI**	**NDDI**	**VSDI**	**VSWI**
	First 10 days	0.29	0.05	0.48	0.31	0.2	0.34	−0.36	−0.51	−0.36
June	Second 10 days	0.32	0.13	0.4	0.24	0.16	0.32	−0.23	−0.44	−0.31
	Third 10 days	0.31	0.02	0.46	0.57	0.06	0.46	−0.27	−0.52	−0.22
	First 10 days	0.51	0.01	0.39	0.25	0.31	0.41	−0.26	−0.56	−0.33
July	Second 10 days	0.3	0.18	0.39	0.24	0.24	0.32	−0.38	−0.52	−0.3
	Third 10 days	0.27	0.1	0.34	0.18	0.27	0.21	−0.35	−0.54	−0.21
	First 10 days	0.16	0.18	0.28	0.01	0.17	0.12	−0.37	−0.56	−0.13
August	Second 10 days	0.31	0.17	0.27	0.08	0.25	0.35	−0.14	−0.14	−0.22
	Third 10 days	0.3	-0.07	0.42	0.26	0.28	0.35	−0.33	−0.11	−0.35

Significant at highlighted values = $p < 0.05$, Number of samples = 126 − 134 ((Due to the clouds or error data of satellite images there were some 10days of some indices are missed).

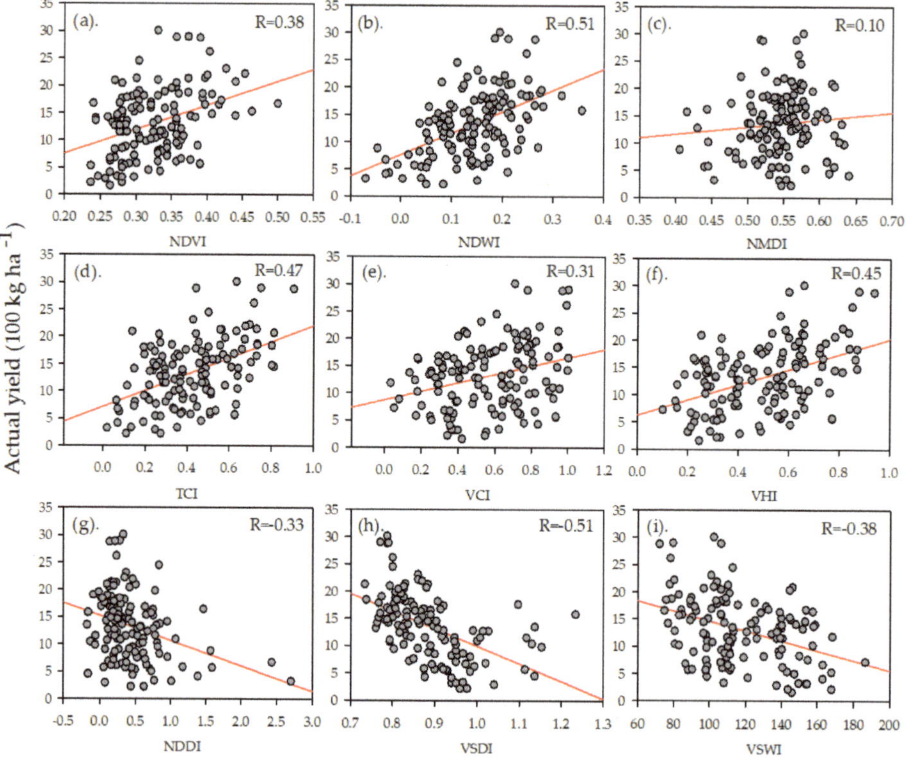

Figure 7. The correlations between spring wheat yield with June (**a**) NDVI, (**b**) NDWI, (**c**) NMDI, (**d**) TCI, (**e**) VCI, (**f**) VHI, (**g**) NDDI, (**h**) VSDI, and (**i**)VSWI. Number of samples N = 135.

Also, we did a correlation between monthly nine remote sensing indices and spring wheat yield at each station in the monitoring period. The results show that the absolute value of correlation coefficients (R) of each month as shown in Table 4. These were statistically significant $p < 0.01$ with yield, except NMDI.

Table 4. Multi-year correlation between monthly indices and spring wheat yield in June- August (2000–2017).

N	Index								
	NDVI	**NMDI**	**NDWI**	**TCI**	**VCI**	**VHI**	**NDDI**	**VSDI**	**VSWI**
June	0.38***	0.1	0.51***	0.47***	0.31***	0.45***	−0.33***	-0.51***	−0.38***
July	0.47***	0.12	0.4***	0.29***	0.35***	0.35***	−0.39***	-0.57***	−0.3***
August	0.28***	0.15	0.35***	0.17*	0.38***	0.32***	−0.37***	-0.33***	−0.25**

Significant at *= $p < 0.05$, ** = $p < 0.01$, *** = $p < 0.001$; Number of samples=134.

The relationship between NMDI and spring wheat yield showed was the lowest results (0.10–0.15), which is indicating that this index not suitable for spring wheat yield estimation in this region. From Table 4 we can see that monthly NDWI was the highest correlated indices with yield in June (0.51) and monthly VSDI was the highest correlated with yield in July (−0.57) and these were statistically significant $p < 0.001$, respectively. It indicates the soil and crops moisture and water content are most important for crop yield.

4.3. Yield Estimation Model

In this study, we tested nine remote sensing indices to develop the best and most accurate estimation models for spring wheat yield for Northern Mongolia. The yield was estimated at station level based on remote sensing indices for the spring wheat-growing season (June to August) and ground crop yield data. Each model has used eight stations crop yield data and nine remote sensing indices (10 daily and monthly) from 2000 to 2017. We used stepwise regression as a technique for choosing independent nine remote sensing indices for a multiple linear regression equation from a list of candidate indices. The results of regression analysis and best-fitted models are summarized in (Table 5).

Table 5. Best-fit estimation models for spring wheat yield during the growing period (June–August).

Month	Model	Equations	R^2	SEM	p-Value
June	Model 1	$y = 44.837 - 41.661 \times \text{VSDI}_{63} + 37.745 \times \text{NDWI}_6$	0.53	4.6	<0.001
	Model 2	$y = 35.041 - 36.358 \times \text{VSDI}_{63} + 24.621 * \text{NDWI}_6 + 13.668 \times \text{VHI}_{63}$	0.57	4.3	<0.001
July	Model 3	$y = 44.721 - 39.502 * \text{VSDI}_7 + 21.296 \times \text{NDWI}_{71}$	0.44	4.6	<0.001
	Model 4	$y = 34.492 - 37.189 \times \text{VSDI}_{63} + 28.571 \times \text{NDWI}_6 + 19.61 \times \text{NDVI}_{71}$	0.55	4.3	<0.001
August	Model 5	$y = 52.224 + 12.774 \times \text{NDWI}_{81} - 46.207 \times \text{VSDI}_{81}$	0.39	4.9	<0.001
	Model 6	$y = 62.527 + 25.492 \times \text{NDWI}_{81} - 50.254 \times \text{VSDI}_{81} - 23.552 \times \text{NDVI}_{81}$	0.44	4.8	<0.001

Number of samples (N = 135), VSDI63-third 10 days of June, VSDI; NDWI6–June, NDWI; VSDI7–July, VSDI; NDVI71-first 10 days, NDVI; VSDI81, NDWI81, and NDVI81- first 10 days of August, VSDI, NDWI, and NDVI.

The models had R^2 values ranging from 0.39 to 0.57 and all models had statistically significant $p < 0.001$, respectively. The models with high R^2 values, low RMSE, and MAE values indicate the best model for spring wheat yield estimation. The highest R^2 values were 0.57 in June and 0.55 in July. Final all models include NDWI and VSDI (VHI and NDVI in some models) from June to August were good predictors of spring wheat yield.

In order to test the estimate performance of the method, we used the coefficient of determination (R^2), root means square error (RMSE), mean absolute error (MAE), bias and index of agreement (d) to evaluate the estimated spring wheat yield in regional level. We compared the predicted yield with the actual yield of eight stations for 2000–2017, the results showed in (Figure 8).

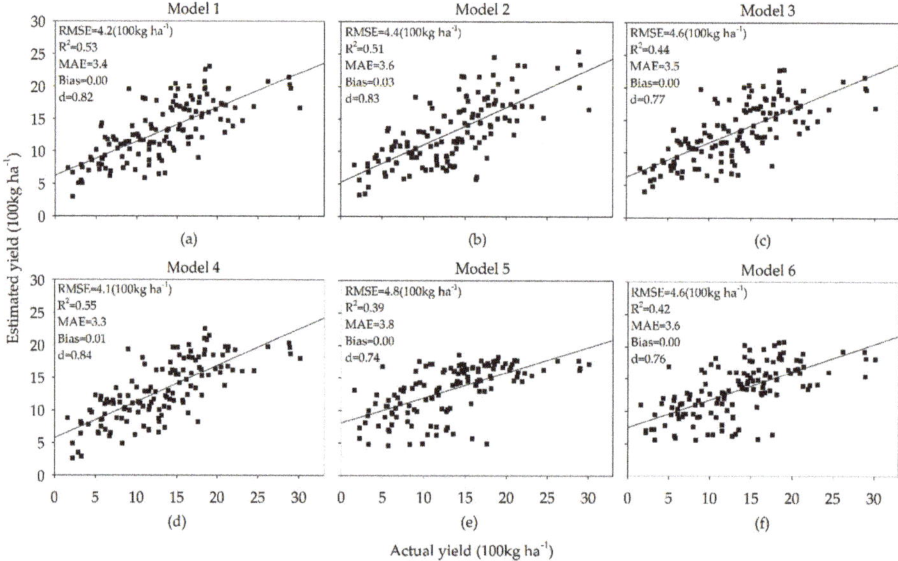

Figure 8. Comparison between actual spring wheat yield with estimated spring wheat yield for each model; model 1 (**a**), model 2 (**b**), model 3 (**c**), model 4 (**d**), model 5 (**e**), model 6 (**f**). Each point indicates the estimated yield versus the actual yield for a single station and year.

The best timing and more accurate model for spring wheat yield estimation was found at the end of June and beginning of July. Model 4 was selected as the best estimation model for spring wheat in Northern Mongolia. Model 4 has combined variables of VSDI, NDWI, and NDVI, and index of agreement (d) value was 0.84, the relationship between estimated and actual yield was $R^2 = 0.55$, mean absolute error was MAE = 3.3 and root mean square error was RMSE = 4.1(100 kg ha^{-1}), respectively.

4.4. Evaluation of Spring Wheat Yield at the Regional Scale

Using a cropland mask, the calibrated model was applied in the study area. The spring wheat yield was evaluated based on the best model 4 from 2000 to 2017 in the Northern part of Mongolia (Selenge and Darkhan provinces) and the results are shown in (Figure 9). The temporal and spatial spring wheat yield maps produced based on the best model using NDWI (June), VSDI (third decade of June), and NDVI (first decade of July) indices, at the regional level have been generated in cropland area from 2000 to 2017. The spatial patterns of estimated spring wheat yield ranged from 1.3–35(100kg ha^{-1}). In generated spring wheat yield maps, green colors indicating the normal and favorable condition and highest values of crop yield; red color shows unfavorable conditions and less amount of crop yield, respectively. From this result, we can see that spring wheat yield was low in drought years (2000–2002, 2005, 2006, 2015, and 2017). The irrigation system is limited in Mongolia, most of the spring crops at their vegetative and reproductive stages suffer water stress due to recurrent drought. Drought stress influences the water supply to vegetation and reduces accumulated biomass and production of crops [65].

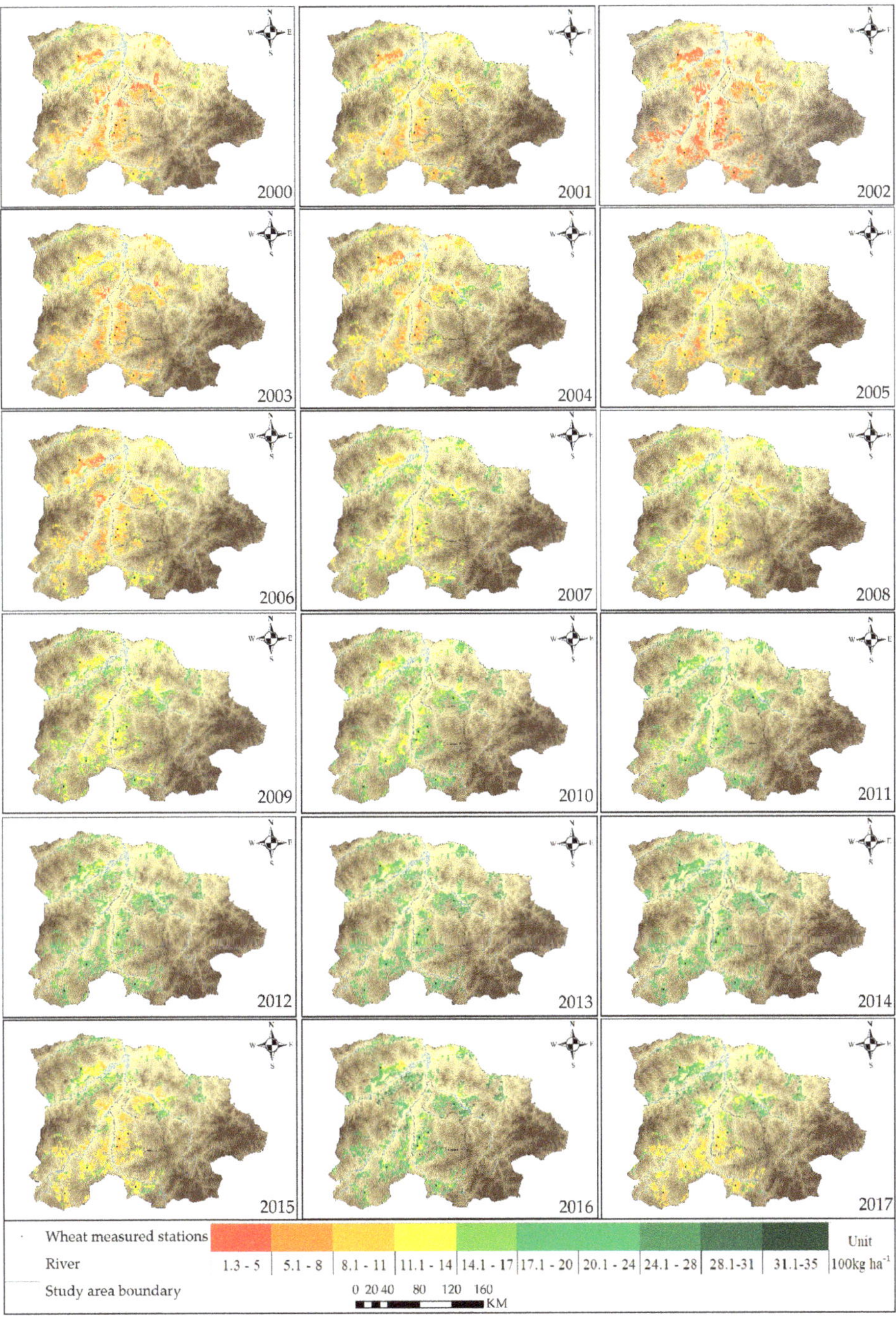

Figure 9. Estimated temporal and spatial spring wheat yield map at the regional level.

5. Discussions

The study paves a new way for crop monitoring in Northern Mongolia. We have explored nine remote sensing indices in decade and month intervals. We found that there are 4 indices (VHI, VSDI, NDWI, and NDVI) that are more relevant than other indices for spring wheat yield estimation. This study has found that the NDWI and VSDI are the best indices for Mongolian crop monitoring. The NDWI was mainly indicated as an effective tool for water stress, soil, and vegetation moisture conditions and water content in vegetative areas, which was determined by the NIR and SWIR bands [11,12,33,48,62]. The supply of moisture to the north-central cropping region of Mongolia comes out as the main factor that clearly demonstrates the results of findings [66]. Water deficiency causes a physiological disorder that can inhibit cell division and differentiation, leading to the reduction of plant size and yield [67]. Our best results were obtained through Model 4 that showed $R^2 = 0.55$, respectively. The results of the relationship between indices showed that MSAVI obtained the best wheat yield estimation model ($R^2 = 0.63$), which was slightly higher than our result. Indices (SAVI, MSAVI, NDVI, and EVI) selected for wheat yield estimation in irrigated Indus Basin of Pakistan [68] are the difference from ours, because at, irrigation area, water is not stress issue. Water is a stressed factor in rainfed spring wheat in Mongolia. Dempewolf et al. [49] developed the wheat yield forecasting model in Punjab province of Pakistan using time-series MODIS and Landsat derived vegetation indices (NDVI, WDRVI, EVI2, SANDVI). The final results show that a forecasted wheat yield was within 0.2% and 11.5% of actual values, which was lower than our result. Bolton and Friedl et al. [62] compared the accuracy of different indices (EVI2, NDVI, and NDWI) in different zones for maize and soybean yield in the Central United States, and the EVI2 obtained best accuracy result with $R^2 = 0.73$, respectively. This result was higher than our result.

We also find the highest correlation between indices with spring wheat yield and peaked at the flowering stage. The peak period of vegetative for spring wheat yield is June and July in this region. This implies that vegetation has its strongest response to moisture availability during this period. The growth condition of spring wheat during flowering stage might have more yield information than other stages at all growth stages, which means the developed regression model can be predicted two months before harvesting the crop and the correlation of estimated and actual yield from heading to flowering periods is higher than other crop growth stages. These results are in agreement with previous studies showing this to be the most suitable time to predict yield [27,48,60,61,69].

Furthermore, we find that later June is the most critical time for spring wheat yield formation. Indices in later June are used in every equation. We examined the relationship between actual spring wheat yield with 10 days and monthly indices for the regression model in the growing period (June–August). Similarly, we tried out the relationship between accumulated and relative values of nine indices with spring wheat yield from 2000 to 2017. The results of the correlation between accumulated and relative values of remote sensing nine indices with actual yield were lower than 10 day and monthly indices value. Juan Sui et al. [11] developed the dry aboveground mass and wheat yield estimation model using several remote sensing indices derived from MODIS and Himawari-8 sensor. A dry aboveground mass and yield errors of <10% and 12% were reported in Hengshui city of Hebei province, which was slightly lower than our results. Lopresti et al. [27] performed based on time-series MODIS-NDVI data for wheat yield estimation model obtained a higher correlation with estimated and actual yield ($R^2 = 0.75$), which was higher than our results. Also, several regression models for crop yield estimation based on MODIS, NDVI, are presented [18,48,61]. Moriondo et al. [61] have carried out the NDVI data to estimate wheat yield in the Grosseto and Foggia provinces of Italy. The results of the correlation coefficient between simulated and actual yield were 0.73–0.77, with corresponding RMSE were 0.44Mg/ha and 0.47Mg/ha, respectively.

The impacts of global warming are already confronted in Mongolia, visible from records between 1940 and 2013 from 48 meteorological stations. According to Dagvadorj et al. [70], the temperature has increased by 2.07°C compare to mean. Due to the impacts of climate change, more extreme and continued droughts have occurred in many parts of Mongolia and which directly affect the vegetation and crop growth, biodiversity and socioeconomics in Mongolia [9]. Reportedly, [10,44,71] 2000–2002, 2004, 2005, 2007, and 2009 years were extremely affected by mild to severe drought and slight drought-hit Mongolia in 2003 and 2011. An additional notable finding of this study is that the spatial regional spring wheat yield distributions shown that the spring wheat yield was high in 2011, 2012, 2013, and 2016 and was low from 2000 to 2003 and 2015. It was statistically significant ($p < 0.001$), respectively and confirmed our result that during the drought years' spring wheat yield was low. Perhaps, in the year 2003, we had the highest precipitation in our monitoring period. The amount of precipitation, soil type, soil moisture, and changes in air temperature have a significant impact on wheat yield. Particularly, drought and soil moisture deficit influence the most reduced crop yield and vegetation size [72]. Thus, from our results, we recommend developing an irrigation system for spring wheat cultivation and increase the number of crop yield observation samples in this region. These results obviously show the promising application of NDWI and VSDI data in crop yield assessment at relatively cheap cost and timely.

6. Conclusions

In Mongolia, the application of remote sensing methodology in agricultural policy and practices is in its nascent stage. This was the first time a multi-regression model based on remote sensing indicators was used to estimate crop yield in Northern Mongolia which is the main spring wheat-producing region. For this purpose, the best and most suitable indices were first defined through the testing of correlations between the nine indices and the actual spring wheat yield. Our results show that NDVI, NDWI, VCI, TCI, VHI, and NMDI indices with spring wheat yield were positively correlated (0.47, 0.51, 0.38, 0.47, 0.45, and 0.15), respectively and NDDI, VSWI, and VSDI with spring wheat yield were negatively correlated (−0.39, −0.57, and −0.38), respectively. Furthermore, the results confirmed the importance of the integration of both satellite and ground data for crop yield estimation. Consequently, we selected the NDWI, VSDI, and NDVI as the most suitable indices out of the nine indices, which are NDVI, NDWI, VCI, TCI, VHI, and NMDI. The highest negatively and positively correlated indices are a combination of NIR, red, blue, and SWIR bands. SWIR and red bands are found more sensitive to moisture variation and water stress of crops and soils [33]. Among nine indices, NDWI (0.51) in June and VSDI (−0.57) in July show the highest correlated indices with actual spring wheat yield, which indicates that the soil and crop moisture, as well as water content, are very important factors for crop yield.

As next step, we refined and developed the regression model using the above three selected indices in order to estimate crop yield. In total, six models were elaborated to be used for the growing period which is June to August. Timeline observation showed a higher correlation between indices and spring wheat yield during the flowering stage in June and July. Therefore, it is suggested that a suitable time to estimate spring wheat yield is at this stage of one to two months before the harvest. Whereas the results for the month of August showed a lower correlation indicating the lateness to estimate. The best results were obtained through Model 4 that used a combination of indicators from the period of third 10 days of June of VSDI, average of June NDWI and first 10 days of July of NDVI. Therefore, the Model 4 is the most effective predictor for crop yield monitoring in the northern part of Mongolia.

In this paper, we could estimate only 74% of the actual yield. This was due to several reasons that possibly could be ascribed to the different spatial resolution between MODIS data (1 km) and the ground measured spring wheat yield data (station-based measurements). Also, phenomena such as the different soil structures and the amount of precipitation have a big influence on the yield. However, the application of remote sensing regression model results enormously enrich the ground station collected data by providing large scale, region-wide data for the decision-makers to better manage food security challenges. In the future, work needs to be carried out to apply more consistently high-resolution images, such as Landsat and Sentinel for more accurate estimation of crop yield. In general, a comprehensive and systematic use of remote sensing technology in the agriculture sector of

Mongolia is to be considered, including broader policy for research and development, the introduction of the latest technology and equipment and targeted capacity-building activities.

Author Contributions: B.T. contributed to the research experiments, methodology, software, data analysis, and the writing-original draft of the manuscript; B.W. conceived the experiments, and was responsible for the research analysis; H.Z. contributed to experimental design and the manuscript revision; G.B. contributed to the collection of crop yield data and the manuscript revision; L.N. contributed the manuscript revision.

Funding: This paper was funded by the National Natural Science Foundation of China (41561144013, 41601464), National Key R&D Program of China (2016YFA0600304), the International Partnership Program of Chinese Academy of Sciences (131C11KYSB20160061,121311KYSB20170004), the Queensland-Chinese Academy of Sciences (Q-CAS) Collaborative Science Fund (grant number: 131211KYSB20170008 & 2017000257).

Acknowledgments: We are very grateful to colleagues of the Information and Research Institute of Meteorology, Hydrology, and Environment (IRIMHE) of Mongolia. Special thanks to Elbegjargal Nasanbat for his comments and suggestions. We thank the anonymous reviewers for reviewing the manuscript and providing comments to improve the manuscript.

Conflicts of Interest: The authors declare no conflict of interest.

References

1. Rosegrant, M.W. Global Food Security: Challenges and Policies. *Science* **2003**, *302*, 1917–1919. [CrossRef]
2. Becker-Reshef, I.; Vermote, E.; Lindeman, M.; Justice, C. A generalized regression-based model for forecasting winter wheat yields in Kansas and Ukraine using MODIS data. *Remote Sens. Environ.* **2010**, *114*, 1312–1323. [CrossRef]
3. Batchuluun, A.; Lin, J.Y. An Analysis of Mining Sector Economics in Mongolia. *Glob. J. Bus. Res.* **2010**, *4*, 81–93.
4. Coslet, C.; Palmeri, F.; Sukhbaatar, J.; Batjargal, E.; Wadhwa, A. Special Report FAO/WFP Crop and Livestock Assessment Mission to Mongolia. 2017. Available online: http://www.wfp.org/food-security/reports/ (accessed on 28 October 2017).
5. Erdenee, B.; Tana, G.; Tateishi, R. Cropland information system in Mongolia using remote sensing and geographical information system: Case study in Tsagaannuur, Selenge aimag. *Int. J. Geomat. Geosci.* **2010**, *1*, 577–586.
6. Altansukh, N. Mongolia: Country Report the FAO International Technical Conference on Plant Geneticresources. Available online: http://www.fao.org/fileadmin/templates/agphome/documents/PGR/SoW1/east/MONGOLIA.pdf (accessed on 28 October 2017).
7. Bao, G.; Qin, Z.; Bao, Y.; Zhou, Y.; Li, W.; Sanjjav, A. NDVI-Based Long-Term Vegetation Dynamics and Its Response to Climatic Change in the Mongolian Plateau. *Remote Sens.* **2014**, *6*, 8337–8358. [CrossRef]
8. Gantsetseg, B.; Ishizuka, M.; Kurosaki, Y.; Mikami, M. Topographical and hydrological effects on meso-scale vegetation in desert steppe, Mongolia. *J. Arid Land* **2017**, *9*, 132–142. [CrossRef]
9. Batima, P.; Natsagdorj, L.; Gombluudev, P.; Erdenetsetseg, B. Observed climate change in Mongolia. *AIACC Work. Pap.* **2005**, *12*, 1–25.
10. Nanzad, L.; Zhang, J.; Tuvdendorj, B.; Nabil, M.; Zhang, S.; Bai, Y. NDVI anomaly for drought monitoring and its correlation with climate factors over Mongolia from 2000 to 2016. *J. Arid Environ.* **2019**, *164*, 69–77. [CrossRef]
11. Sui, J.; Qin, Q.; Ren, H.; Sun, Y.; Zhang, T.; Wang, J.; Gong, S. Winter wheat production estimation based on environmental stress factors from satellite observations. *Remote Sens.* **2018**, *10*, 962. [CrossRef]
12. Petersen, L.K. Real-time prediction of crop yields from MODIS relative vegetation health: A continent-wide analysis of Africa. *Remote Sens.* **2018**, *10*, 1726. [CrossRef]
13. Atzberger, C. Advances in Remote Sensing of Agriculture: Context Description, Existing Operational Monitoring Systems and Major Information Needs. *Remote Sens.* **2013**, *5*, 949–981. [CrossRef]
14. Dong, T.; Shang, J.; Qian, B.; Liu, J.; Chen, J.M.; Jing, Q.; McConkey, B.; Huffman, T.; Daneshfar, B.; Champagne, C.; et al. Field-Scale Crop Seeding Date Estimation from MODIS Data and Growing Degree Days in Manitoba, Canada. *Remote Sens.* **2019**, *11*, 1760. [CrossRef]
15. He, T.; Xie, C.; Liu, Q.; Guan, S.; Liu, G. Evaluation and Comparison of Random Forest and A-LSTM Networks for Large-scale Winter Wheat Identification. *Remote Sens.* **2019**, *11*, 1665. [CrossRef]

16. Kogan, F. World droughts in the new millennium from AVHRR-based vegetation health indices. *Eos* **2002**, *83*, 557. [CrossRef]

17. Kogan, F.N. Operational space technology for global vegetation assessment. *Bull. Am. Meteorol. Soc.* **2001**, *82*, 1949–1964. [CrossRef]

18. Ren, J.; Chen, Z.; Zhou, Q.; Tang, H. Regional yield estimation for winter wheat with MODIS-NDVI data in Shandong, China. *Int. J. Appl. Earth Obs. Geoinf.* **2008**, *10*, 403–413. [CrossRef]

19. Sakamoto, T.; Gitelson, A.A.; Arkebauer, T.J. Near real-time prediction of U.S. corn yields based on time-series MODIS data. *Remote Sens. Environ.* **2014**, *147*, 219–231. [CrossRef]

20. Singh, R.P.; Roy, S.; Kogan, F. Vegetation and temperature condition indices from NOAA AVHRR data for drought monitoring over India. *Int. J. Remote Sens.* **2003**, *24*, 4393–4402. [CrossRef]

21. Yeom, J.; Jung, J.; Chang, A.; Ashapure, A.; Maeda, M.; Maeda, A.; Landivar, J. Comparison of Vegetation Indices Derived from UAV Data for Differentiation of Tillage Effects in Agriculture. *Remote Sens.* **2019**, *11*, 1548. [CrossRef]

22. Guindin-Garcia, N.; Gitelson, A.A.; Arkebauer, T.J.; Shanahan, J.; Weiss, A. An evaluation of MODIS 8- and 16-day composite products for monitoring maize green leaf area index. *Agric. For. Meteorol.* **2012**, *161*, 15–25. [CrossRef]

23. Rembold, F.; Atzberger, C.; Savin, I.; Rojas, O. Using Low-Resolution Satellite Imagery for Yield Prediction and Yield Anomaly Detection. *Remote Sens.* **2013**, *5*, 1704–1733. [CrossRef]

24. Malingreau, J.P. Global vegetation dynamics: Satellite observations over Asia. *Int. J. Remote Sens.* **1986**, *7*, 1121–1146. [CrossRef]

25. Tucker, C.J.; Vanpraet, C.L.; Sharman, M.J.; Van Ittersum, G. Satellite remote sensing of total herbaceous biomass production in the senegalese sahel: 1980–1984. *Remote Sens. Environ.* **1985**, *17*, 233–249. [CrossRef]

26. Battude, M.; Al Bitar, A.; Morin, D.; Cros, J.; Huc, M.; Sicre, C.M.; Le Dantec, V.; Demarez, V. Estimating maize biomass and yield over large areas using high spatial and temporal resolution Sentinel-2 like remote sensing data. *Remote Sens. Environ.* **2016**, *184*, 668–681. [CrossRef]

27. Lopresti, M.F.; Di Bella, C.M.; Degioanni, A.J. Relationship between MODIS-NDVI data and wheat yield: A case study in Northern Buenos Aires province, Argentina. *Inf. Process. Agric.* **2015**, *2*, 73–84. [CrossRef]

28. Sandford, S.A.; Bernstein, M.P.; Materese, C.K. NASA Public Access. *Astrophys. J. Suppl. Ser.* **2013**, *205*, 1–58.

29. Lewis, J.E.; Rowland, J.; Nadeau, A. Estimating maize production in kenya using ndvi: Some statistical considerations. *Int. J. Remote Sens.* **1998**, *19*, 2609–2617. [CrossRef]

30. Vicente-Serrano, S.; Cuadrat-Prats, J.M.; Romo, A. Early prediction of crop production using drought indices at different time-scales and remote sensing data: Application in the Ebro Valley (north-east Spain). *Int. J. Remote Sens.* **2006**, *27*, 511–518. [CrossRef]

31. Carlson, T.N.; Gillies, R.R.; Perry, E.M. A method to make use of thermal infrared temperature and NDVI measurements to infer surface soil water content and fractional vegetation cover. *Remote Sens. Rev.* **1994**, *9*, 161–173. [CrossRef]

32. Carlson, T.N.; Perry, E.M.; Schmugge, T.J. Remote estimation of soil moisture availability and fractional vegetation cover for agricultural fields. *Agric. For. Meteorol.* **1990**, *52*, 45–69. [CrossRef]

33. Zhang, N.; Hong, Y.; Qin, Q.; Zhu, L. Evaluation of the Visible and Shortwave Infrared Drought Index in China. *Int. J. Disaster Risk Sci.* **2013**, *4*, 68–76. [CrossRef]

34. Wang, L.; Qu, J.J. NMDI: A normalized multi-band drought index for monitoring soil and vegetation moisture with satellite remote sensing. *Geophys. Res. Lett.* **2007**, *34*. [CrossRef]

35. Huang, J.; Ma, H.; Sedano, F.; Lewis, P.; Liang, S.; Wu, Q.; Su, W.; Zhang, X.; Zhu, D. Evaluation of regional estimates of winter wheat yield by assimilating three remotely sensed reflectance datasets into the coupled WOFOST–PROSAIL model. *Eur. J. Agron.* **2019**, *102*, 1–13. [CrossRef]

36. Jones, J.W.; Hoogenboom, G.; Porter, C.H.; Boote, K.J.; Batchelor, W.D.; Hunt, L.A.; Wilkens, P.W.; Singh, U.; Gijsman, A.J.; Ritchie, J.T. The DSSAT cropping system model. *Eur. J. Agron.* **2003**, *18*, 235–265. [CrossRef]

37. Rezzoug, W.; Gabrielle, B.; Suleiman, A.; Benabdeli, K. Application and evaluation of the DSSAT-wheat in the Tiaret region of Algeria. *African J. Agric. Res.* **2008**, *3*, 284–296.

38. Zhuo, W.; Huang, J.; Li, L.; Huang, R.; Gao, X.; Zhang, X.; Zhu, D. Assimilating SAR and Optical Remote Sensing Data into WOFOST Model for Improving Winter Wheat Yield Estimation. In Proceedings of the 2018 7th International Conference on Agro-geoinformatics (Agro-geoinformatics), Hangzhou, China, 6–9 August 2018; pp. 1–5.

39. Ceglar, A.; van der Wijngaart, R.; de Wit, A.; Lecerf, R.; Boogaard, H.; Seguini, L.; van den Berg, M.; Toreti, A.; Zampieri, M.; Fumagalli, D.; et al. Improving WOFOST model to simulate winter wheat phenology in Europe: Evaluation and effects on yield. *Agric. Syst.* **2019**, *168*, 168–180. [CrossRef]

40. Gilardelli, C.; Stella, T.; Frasso, N.; Cappelli, G.; Bregaglio, S.; Chiodini, M.E.; Scaglia, B.; Confalonieri, R. WOFOST-GTC: A new model for the simulation of winter rapeseed production and oil quality. *Field Crop. Res.* **2016**, *197*, 125–132. [CrossRef]

41. Otgonbayar, M.; Atzberger, C.; Chambers, J.; Amarsaikhan, D.; Böck, S.; Tsogtbayar, J. Land Suitability Evaluation for Agricultural Cropland in Mongolia Using the Spatial MCDM Method and AHP Based GIS. *J. Geosci. Environ. Prot.* **2017**, *5*, 238–263. [CrossRef]

42. Natsagdorj, E.; Renchin, T.; De Maeyer, P.; Dari, C.; Tseveen, B. Long-term soil moisture content estimation using satellite and climate data in agricultural area of Mongolia. *Geocarto Int.* **2019**, *34*, 722–734. [CrossRef]

43. Ariya, B. Land Degradation Assessment in an Agricultural Area of Mongolia: Case Study in Orkhon Soum. United Nations University Land Restoration Training Programme [final project]. 2017, pp. 1–20. Available online: http://www.unulrt.is/static/fellows/document/ariya2017.pdf (accessed on 28 October 2017).

44. Chang, S.; Wu, B.; Yan, N.; Davdai, B.; Nasanbat, E. Suitability assessment of satellite-derived drought indices for Mongolian grassland. *Remote Sens.* **2017**, *9*, 650. [CrossRef]

45. Tsegmid, S.; Vorobiev, V. *The National Atlas*; Mongolian People's Republic (in Mongolian). GUGK SSSR GUGK MNR Mosc: Ulaanbaatar, Mongolia, 1990.

46. Kastens, J.; Kastens, T.; Kastens, D.; Price, K.; Martinko, E.; Lee, R. Image masking for crop yield forecasting using AVHRR NDVI time-series imagery. *Remote Sens. Environ.* **2005**, *99*, 341–356. [CrossRef]

47. Elbegjargal, N.; Khudulmur, S.; Tsogtbaatar, J.; Dash, D.; Mandakh, N. *Desertification Atlas of Mongolia*; Institute of Geoecology, Mongolian Academy of Sciences: Ulaanbaatar, Mongolia, 2014.

48. Mkhabela, M.S.; Bullock, P.; Raj, S.; Wang, S.; Yang, Y. Crop yield forecasting on the Canadian Prairies using MODIS NDVI data. *Agric. For. Meteorol.* **2011**, *151*, 385–393. [CrossRef]

49. Dempewolf, J.; Adusei, B.; Becker-Reshef, I.; Hansen, M.; Potapov, P.; Khan, A.; Barker, B. Wheat yield forecasting for Punjab Province from vegetation index time series and historic crop statistics. *Remote Sens.* **2014**, *6*, 9653–9675. [CrossRef]

50. Kogan, F.; Gitelson, A.; Zakarin, E.; Spivak, L.; Lebed, L. AVHRR-based spectral vegetation index for quantitative assessment of vegetation state and productivity: Calibration and validation. *Photogramm. Eng. Remote Sens.* **2003**, *69*, 899–906. [CrossRef]

51. Kogan, F.N. Droughts of the late 1980s in the United States as derived from NOAA polar-orbiting satellite data. *Bull. Am. Meteorol. Soc.* **1995**, *76*, 655–668. [CrossRef]

52. Unganai, L.S.; Kogan, F.N. Drought monitoring and corn yield estimation in southern Africa from AVHRR data. *Remote Sens. Environ.* **1998**, *63*, 219–232. [CrossRef]

53. Gitelson, A.A.; Kaufman, Y.J.; Merzlyak, M.N. Use of a green channel in remote sensing of global vegetation from EOS-MODIS. *Remote Sens. Environ.* **1996**, *58*, 289–298. [CrossRef]

54. Holben, B.N. Characteristics of maximum-value composite images from temporal AVHRR data. *Int. J. Remote Sens.* **1986**, *7*, 1417–1434. [CrossRef]

55. Tucker, C.J. Red and photographic infrared linear combinations for monitoring vegetation. *Remote Sens. Environ.* **1979**, *8*, 127–150. [CrossRef]

56. Gao, B.C. NDWI—A normalized difference water index for remote sensing of vegetation liquid water from space. *Remote Sens. Environ.* **1996**, *58*, 257–266. [CrossRef]

57. Kogan, F.N. Application of vegetation index and brightness temperature for drought detection. *Adv. Space Res.* **1995**, *15*, 91–100. [CrossRef]

58. Jianbo, W.; Jianjun, B.; Lele, L.; Yuan, Y. Vegetation supply water index based on MODIS data Analysis of the in Yunnan in spring of 2012. In Proceedings of the Third International Conference on Agro-Geoinformatics, Beijing, China, 11–14 August 2014; pp. 1–7.

59. Gu, Y.; Brown, J.F.; Verdin, J.P.; Wardlow, B. A five-year analysis of MODIS NDVI and NDWI for grassland drought assessment over the central Great Plains of the United States. *Geophys. Res. Lett.* **2007**, *34*, 06407. [CrossRef]

60. Balaghi, R.; Tychon, B.; Eerens, H.; Jlibene, M. Empirical regression models using NDVI, rainfall and temperature data for the early prediction of wheat grain yields in Morocco. *Int. J. Appl. Earth Obs. Geoinf.* **2008**, *10*, 438–452. [CrossRef]

61. Moriondo, M.; Maselli, F.; Bindi, M. A simple model of regional wheat yield based on NDVI data. *Eur. J. Agron.* **2007**, *26*, 266–274. [CrossRef]

62. Bolton, D.K.; Friedl, M.A. Forecasting crop yield using remotely sensed vegetation indices and crop phenology metrics. *Agric. For. Meteorol.* **2013**, *173*, 74–84. [CrossRef]

63. Nandintsetseg, B.; Shinoda, M. Seasonal change of soil moisture in Mongolia: Its climatology. *Int. J. Clim.* **2011**, *31*, 1143–1152. [CrossRef]

64. Shinoda, M. Phenology of Mongolian Grasslands and Moisture Conditions. *J. Meteorol. Soc. Jpn.* **2007**, *85*, 359–367. [CrossRef]

65. Delfine, S.; Loreto, F.; Alvino, A. Drought-stress Effects on Physiology, Growth and Biomass Production of Rainfed and Irrigated Bell Pepper Plants in the Mediterranean Region. *J. Am. Soc. Hortic. Sci.* **2001**, *126*, 297–304. [CrossRef]

66. Azzaya, D. Agro-Meteorological Assessment of Plant Growth Conditions in the Central Region of Mongolia. Ph.D. Thesis, Mongolian University of Life Sciences, Ulaanbaatar, Mongolia, 1997.

67. Loka, D. Effect of Water-Deficit Stress on Cotton During Reproductive Development. Ph.D. Thesis, University of Arkansas, Fayetteville, NC, USA, 2012.

68. Liaqat, M.U.; Cheema, M.J.M.; Huang, W.; Mahmood, T.; Zaman, M.; Khan, M.M. Evaluation of MODIS and Landsat multiband vegetation indices used for wheat yield estimation in irrigated Indus Basin. *Comput. Electron. Agric.* **2017**, *138*, 39–47. [CrossRef]

69. Royo, C.; Aparicio, N.; Villegas, D.; Casadesus, J.; Monneveux, P.; Araus, J.L. Usefulness of spectral reflectance indices as durum wheat yield predictors under contrasting Mediterranean conditions. *Int. J. Remote Sens.* **2003**, *24*, 4403–4419. [CrossRef]

70. Dagvadorj, D.; Batjargal, Z.; Natsagdorj, L. *Mongolia Second Assessment Report on Climate Change-2014*; Ministry of Environment and Green Development of Mongolia with Financial Support from the GIZ Programme: Ulaanbaatar, Mongolia, 2014.

71. Dorjsuren, M.; Liou, Y.A.; Cheng, C.H. Time-series MODIS and in situ data analysis for Mongolia drought. *Remote Sens.* **2016**, *8*, 509. [CrossRef]

72. Alizadeh, V.; Shokri, V.; Soltani, A.; Yousefi, M.A. Effects of Climate Change and Drought-Stress on Plant Physiology. *Int. J. Adv. Biol. Biomedical Res.* **2014**, *2*, 468–472.

 *remote sensing*

Article

To Blend or Not to Blend? A Framework for Nationwide Landsat–MODIS Data Selection for Crop Yield Prediction

Yang Chen [1,*], Tim R. McVicar [2], Randall J. Donohue [2], Nikhil Garg [1], François Waldner [3], Noboru Ota [4], Lingtao Li [2] and Roger Lawes [4]

[1] CSIRO Data61, Goods Shed North, 34 Village St, Docklands, VIC 3008, Australia; Nikhil.Garg@data61.csiro.au

[2] CSIRO Land and Water, GPO Box 1700, Canberra, ACT 2061, Australia; Tim.Mcvicar@csiro.au (T.R.M.); Randall.Donohue@csiro.au (R.J.D.); Lingtao.Li@csiro.au (L.L.)

[3] CSIRO Agriculture and Food, 306 Carmody Rd, St Lucia, QLD 4067, Australia; Franz.Waldner@csiro.au

[4] CSIRO Agriculture and Food, 147 Underwood Ave, Floreat, WA 6014, Australia; Noboru.Ota@csiro.au (N.O.); Roger.Lawes@csiro.au (R.L.)

* Correspondence: y.chen@csiro.au; Tel.: +61-9518-5990

Received: 1 May 2020; Accepted: 19 May 2020; Published: 21 May 2020

Abstract: The onus for monitoring crop growth from space is its ability to be applied anytime and anywhere, to produce crop yield estimates that are consistent at both the subfield scale for farming management strategies and the country level for national crop yield assessment. Historically, the requirements for satellites to successfully monitor crop growth and yield differed depending on the extent of the area being monitored. Diverging imaging capabilities can be reconciled by blending images from high-temporal-frequency (HTF) and high-spatial-resolution (HSR) sensors to produce images that possess both HTF and HSR characteristics across large areas. We evaluated the relative performance of Moderate Resolution Imaging Spectroradiometer (MODIS), Landsat, and blended imagery for crop yield estimates (2009–2015) using a carbon-turnover yield model deployed across the Australian cropping area. Based on the fraction of missing Landsat observations, we further developed a parsimonious framework to inform when and where blending is beneficial for nationwide crop yield prediction at a finer scale (i.e., the 25-m pixel resolution). Landsat provided the best yield predictions when no observations were missing, which occurred in 17% of the cropping area of Australia. Blending was preferred when <42% of Landsat observations were missing, which occurred in 33% of the cropping area of Australia. MODIS produced a lower prediction error when ≥42% of the Landsat images were missing (~50% of the cropping area). By identifying when and where blending outperforms predictions from either Landsat or MODIS, the proposed framework enables more accurate monitoring of biophysical processes and yields, while keeping computational costs low.

Keywords: MODIS; Landsat; data blending; crop yield prediction; gap-filling

---

1. Introduction

The world's human population is projected to increase by more than 35% by 2050 [1]. To contribute to improved global food security, the next generation of crop models and agricultural decision support tools needs to efficiently and consistently operate across various scales [2]. Accurate nationwide crop yield forecasts may ensure food security to the citizens. More accurate crop yield prediction at the subfield scale can provide farmers with more detailed information for guiding, within the growing season, in-field variable rate applications of fertilizer, herbicides, and pesticides. An efficient approach to monitor crop growth uses satellite observations providing repeated synoptic regional assessments [3–6]. High-temporal-frequency (HTF) observations are required to accurately track crop development [7] and predict yield [8], and high-spatial-resolution (HSR) observations are necessary to capture within-field heterogeneity [9].

There is a trade-off between temporal frequency and spatial resolution [10–12] as no single sensor can regularly image vast areas of the Earth used for nation-wide dryland cropping at a high spatial resolution. Commercial options of products combine HTF and HSR images achieved by increasing the number of satellites in orbit, such as PlanetScope; however, such commercial options are not affordable for national-scale monitoring, especially in a country as large as Australia (7.7 million km^2), with the southern Australian cereal-based agricultural system notionally covering 530,000 km^2 [13]. The Moderate Resolution Imaging Spectroradiometer (MODIS) provides complete coverage of the globe every day at a 250-m spatial resolution from red and near-infrared bands. This resolution constrains the capacity of describing cropping systems, crop growth, and field heterogeneities, especially when fields are small-to-moderate sized and landscapes are fragmented [14,15]. Sensors with higher spatial resolution, such as Sentinel-2 or Landsat, are more suitable for these smaller fields/management units, but their lower temporal frequencies limit their ability to capture rapidly evolving crop processes, especially when factoring in the potential clouds [16]. Lobell et al. [17] used valid Landsat observations (cloud cover <10%) during the growing season to generate optical-based vegetation indices (Vis) and then fitted a multilinear regression between these VIs and a large number of Agricultural Production Systems Simulator (APSIM) simulations for yield prediction. In a country like Australia, most of the arable non-irrigated land grows winter wheat, barley, oats, and canola, and their growth depends on the in-season rainfall; therefore, totally cloud-free time-series observations for their entire growing season are infrequent.

To overcome these challenges, various methods for spatial filtering [18], temporal gap-filling [19–21], and data fusion [22–24] were devised with a varying degree of success. As spatial filtering and temporal gap-filling disregard the spatial and temporal correlation of a pixel, they are highly sensitive to the choice of size/length of the spatial/temporal window [20,25]. Data fusion techniques, on which this article focuses, were shown to improve the temporal resolution of fine-spatial-resolution data by blending observations from sensors with different spatial and temporal characteristics. Prominent examples are the Spatial and Temporal Adaptive Reflectance Fusion Model (STARFM; Gao et al. [12]) and the Enhanced STARFM (ESTARFM; Zhu et al. [11]). However, blending is no "silver bullet", as it often introduces unforeseen spatial and temporal variances [10]. Therefore, it is critical to systematically evaluate the benefits of blending and identify where and when blending helps to improve monitoring.

Blending satellite data with complementary frequency and spatial resolution characteristics, such as MODIS (herein considered as an HTF and low-spatial-resolution (LSR) sensor) and Landsat (herein considered as a low-temporal-frequency (LTF) and HSR sensor), provides a solution of synthetic imagery that is both HTF and HSR [26]. Current literature such as Dong et al. [27] found that using Landsat images provides a higher crop yield prediction accuracy for field scales over MODIS images, and a further improvement can be achieved by combining Landsat and MODIS (L–M) blended data with the incomplete Landsat series. To date, the utility of blended output is not yet tested for regional and national crop yield mapping [28–35] (Table 1); this study fills that niche.

This study quantifies and evaluates the benefits of blending satellite data of different temporal frequencies and spatial resolutions for crop yield prediction. The specific objectives of this study are to (i) estimate yields using MODIS, Landsat, and L–M blended data and then compare the yield prediction at both pixel and field scales, (ii) identify the fraction of missing Landsat data during a growing season considering the 16-day acquisition cycle to determine a threshold where the blended data can improve the prediction, and (iii) quantify the improvements in the yield prediction accuracy based on the threshold.

Table 1. Review of studies that used data fusion approaches to estimate crop yields and other variables highly related to crop yield (e.g., growing season evapotranspiration and crop phenology). Records are sorted chronologically then alphabetically. When more than one site is reported in a single paper, these are identified with an (A) and (B) in the relevant columns as required. R^2: the coefficient of determination; $RMSE$: root-mean-square error; MAE: mean absolute error; $RMAE$: relative MAE; NR: not reported. In Australia, dryland agriculture is practiced across the region known colloquially as the "wheatbelt". STARFM—Spatial and Temporal Adaptive Reflectance Fusion Model; ESTARFM—enhanced STARFM; MODIS—Moderate Resolution Imaging Spectroradiometer; NIR—near infrared; US—United States; NE—New England; NASA—National Aeronautics and Space Administration; L–M—Landsat/MODIS blend; HTF—high temporal frequency; HSR—high spatial resolution.

Reference	Blending Algorithm	Remote Sensing (RS) Data	Crop Type	RS Variables (e.g., Vegetation Index (VI))/Study Period/Region/Study Area (km^2)	RS-Bases Model to Estimate Crop Yields	Key Results and Accuracy
[28]	Spatial and Temporal Adaptive Vegetation Index Fusion Model (STAVIF)	MODIS and Huanjing Satellite Charge-Coupled Device (HJ-CCD)	Winter wheat	Normalized difference vegetation index (NDVI)/2008–2009/Yucheng, Shandong, China/NR	Empirical model	The estimated winter wheat biomass correlated well with observed biomass ($R^2 = 0.88$ and $MAE = 17.2$ kg/ha) using the blended data.
[29]	STARFM	Satellite for Observation of Earth (SPOT) 5 and HJ-1 CCD	Winter wheat	NDVI and ratio vegetation index (RVI) (NIR/Red)/2008–2009/(A) Rugao county, Jiangsu, and (B) Anyang county, Henan, China/(A) 0.36 km^2 and (B) 0.30 km^2	Empirical model	(A) The accumulated NDVI derived from the blended data gave a higher prediction accuracy ($R^2 = 0.67$ and $RMSE = 0.36$ t/ha) for wheat yield at Rugao. (B) The accumulated RVI derived from the blended data produced a higher prediction accuracy ($R^2 = 0.65$ and $RMSE = 0.36$ t/ha) for wheat yield at Anyang.
[31]	STARFM	MODIS and Landsat	Corn and soybean	Evapotranspiration (ET)/2013/Central Valley, California, the US/(A) 0.34 km^2 and (B) 0.21 km^2	Empirical model	The daily ET derived from the blended data produced the $RMAE$ of 19% with the observed ET (mm/day). The spatial pattern of cumulative ET corresponded to the measured yield.
[27]	ESTARFM	MODIS and Landsat	Winter wheat	Green leaf area index (GLAI)/2013/Southwestern Ontario, Canada/225 km^2	Semi-empirical model	The Landsat GLAI (GLAIL) produced an R^2 of 0.77 and RMSE of 2.31 t/ha; the blended GLAI (GLAIF) resulted in an R^2 of 0.71 and $RMSE$ of 1.93 t/ha; the combination of GLAIL and GLAIF led to further improvements ($R^2 = 0.76$ and $RMSE = 1.76$ t/ha).
[30]	ESTARFM	MODIS and Landsat	Corn and soybean	NDVI/2001–2014/Central Iowa, the US/200 km^2	Empirical model	A linear correlation ($R^2 = 0.83$) between remotely sensed green-up dates and the emergence dates reported by NASA.
[32]	STARFM	MODIS and Landsat	Maize	ET/2010–2014/Mead, NE, the US/(A) 0.49 km^2, (B) 0.52 km^2, and (C) 0.65 km^2	Empirical model	The county-level correlation between observed and predicted maize yields improved from 0.47 to 0.93 when aligning the ratio of actual-to-reference ET by emergence date rather than calendar date.

Table 1. *Cont.*

Reference	Blending Algorithm	Remote Sensing (RS) Data	Crop Type	RS Variables (e.g., Vegetation Index (VI))/Study Period/Region/Study Area (km^2)	RS-Bases Model to Estimate Crop Yields	Key Results and Accuracy
[33]	STARFM	MODIS and Landsat	Corn and soybean	NDVI and enhanced vegetation index (EVI2)/2001–2015/Central Iowa, the US/200 km^2	Empirical model	Maximum EVI2 derived from L–M blended data produced the highest R^2 (0.59 and 0.39) and the lowest *RMAE* (6.1% and 9.1%) for corn and soybeans, respectively, compared with using single data source alone.
[34]	A pixel-wise linear regression model	MODIS and Landsat	Alfalfa, barley, maize, peas, durum wheat, spring wheat, and winter wheat	NDVI/2008–2015/Montana, the US/4.13 million ha	Semi-empirical model	A correlation of 0.96 ($R^2 = 0.92$, relative *RMSE* = 37.0%, $p < 0.05$) resulted when comparing the yield prediction using the blended data with the reported crop production data on county level.
[26]	ESTARFM	MODIS and Landsat	Cotton and winter wheat	NDVI/2004–2014/Fergana Valley, Uzbekistan/NR	Semi-empirical model	The R^2 is 0.56 (*RMSE* = 0.63 t/ha) for wheat, and 0.631(*RMSE* = 0.48 t/ha) for cotton, respectively.
[35]	STARFM	MODIS and Landsat	Corn and soybean	GLAI/2015/Southwestern Ontario, Canada/112 km^2	Semi-empirical model	The *RMSE* of yield prediction is 1.46 t/ha (R^2 = 0.56) for corn and 0.86 t/ha (R^2 = 0.54) for soybean using the blended data.
This study	ESTARFM	MODIS and Landsat	Wheat, barley, and canola	NDVI/2009–2015/Australian wheatbelt/~53 million ha	Semi-empirical model	Comparing HTF, HSR data against the blended data for yield prediction at various scales. Identifying a threshold to determine when and where the blended data can improve in the nationwide yield prediction at the 25-m pixel resolution when using multiple spatio-temporal resolution images. Quantifying and evaluating the improvements in the yield prediction accuracy at various scales based on the threshold.

2. Study Area and Data

2.1. Study Area

The southern Australian agricultural system is dominated by dryland agriculture where cereals (e.g., wheat, barley, and oats), oilseeds (e.g., canola), and legumes (e.g., lupins, chickpeas, field peas, and soybeans) are planted, often in rotation with annual pastures and fallows. Australian wheatbelt (Figure 1) spans an extremely variable agroecological environment with respect to the climate across the continent, leading to the high spatial variability in Australian grain production. The precipitation varies enormously across the country (Figure S1, Supplementary Materials); winter (June–August) precipitations are dominant in Western Australia and South Australia, while summer (December–February) precipitations are dominant in Queensland and northern New South Wales. In southern New South Wales and Victoria, total precipitation is more uniformly distributed throughout the seasons where summer precipitation is more intense indicating that winter precipitation is more frequent [36]. Long-term average monthly accumulated precipitation records strongly correlate with the average number of rainy days (Figure S1, Supplementary Materials).

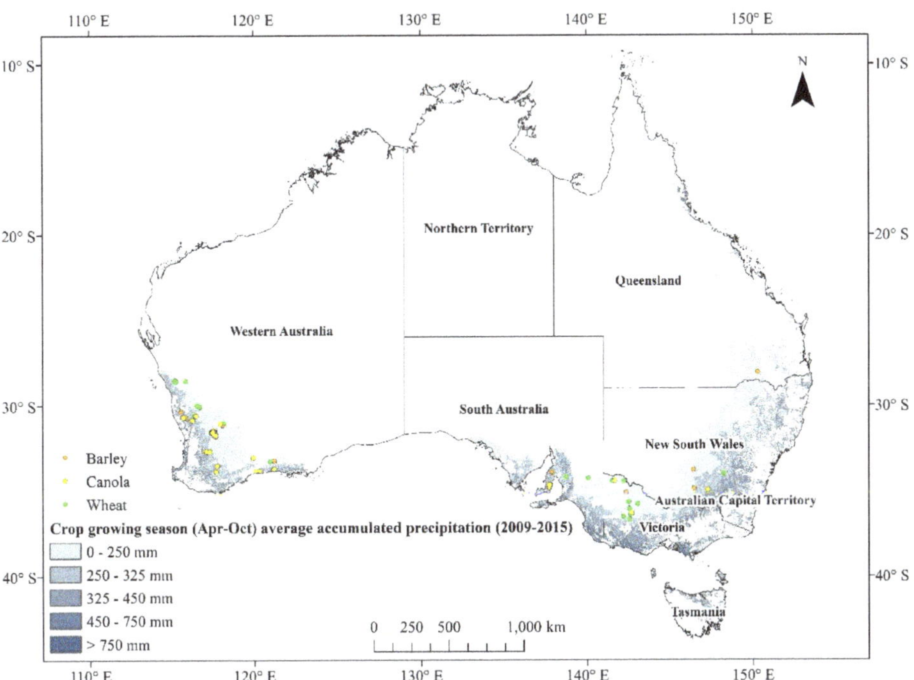

Figure 1. Study sites and the crop growing season (April–October) average accumulated precipitation (mm for the years 2009–2015) across the Australian wheatbelt (~53 million ha) [13]. The precipitation data are sourced from Jeffrey et al. [37].

2.2. Data

2.2.1. Satellite Images and L–M Blended Data

Time series of Landsat-5 Thematic Mapper (TM), Landsat-7 Enhanced Thematic Mapper (ETM)+, and Landsat-8 Operational Land Imager (OLI) and Thermal Infrared Sensor (TIRS) standardized surface reflectance data are sourced from Geoscience Australia (http://geoscienceaustralia.github.io/digitalearthau/index.html), which contains 239 scenes covering the Australian wheatbelt across seven years (2009–2015). These images are nadir-corrected, adjusted for bidirectional reflectance distribution function, and topographically corrected followed by Li et al.'s methods [38], resulting in a 25-m pixel resolution. This is as, in Geoscience Australia's processing stream, they oversample the Landsat imagery to 25 m to allow easier integration with other remote sensing data sources. Time series of normalized difference vegetation index (NDVI) are then computed using the red and near-infrared bands [39]. The 16-day composite of MODIS NDVI (MOD13Q1 v006) data are used due to the low clouds, low view angle, and the highest NDVI value at 250-m spatial resolution. The MODIS data are sourced from United States Geological Survey for the corresponding periods and area (16 tiles) and are then downscaled to 25 m by 25 m spatial resolution using nearest neighbor interpolation [40] for input to the blending algorithm and for consistency to enable further analysis.

The inter-annual variability of precipitation (Figure 2b) illustrates that the probability of rain days is higher during June–August (i.e., tillering to flowering phase), which is a critical period for crop yield prediction [41,42]. Figure 2b also shows the value of 0.58 in July at the 75th quartile, which indicates that the probability of cloud-free Landsat images could be lower than 42% in most areas. Here, 75% of field-scale Landsat missing data occur during the growing season (i.e., 113–289 days of the year (DOY)), and only 25% of the Landsat data are complete sequences (Figure 3).

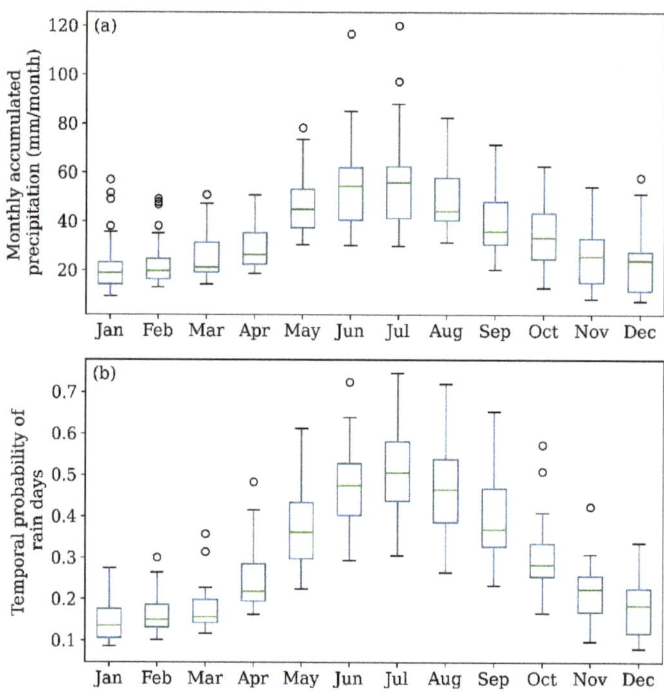

Figure 2. Box–whisker plots of 1901 to 2018 averaged (**a**) monthly accumulated precipitation (mm/month) across the Australian wheatbelt (see Figure 1) and (**b**) monthly probability of rain days (bottom). For both parts, the horizontal line represents the median of the data, the box spans from 25th to 75th quartiles of the data, and the circles past the end of the whiskers are outliers, while the rain day threshold is 0 mm/day.

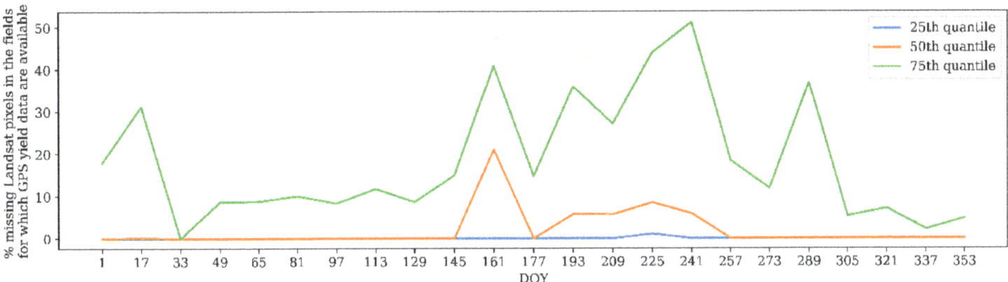

Figure 3. The quantile of Landsat missing pixels in the fields for which observed yield data are available (2009 to 2015).

To blend MODIS and Landsat NDVI data and create the 16-day time series of gap-filled Landsat-like images, we applied ESTARFM [11], which is superior when the spatial variance is dominant [10]. We follow Jarihani et al.'s [22] recommendation to "index then blend", as it yields more accurate results and provides higher computational efficiency than the "blend then index" approach. Figure S2 (Supplementary Materials) illustrates the blending process and the resulting Landsat-like image. As a result of data blending, each pixel is described by 22 observations across the calendar year. For more details about ESTARFM see [11,43].

2.2.2. Yield Data

Australia's major dryland crops are selected for yield estimates and validation, including 35 fields of canola, 123 fields of wheat, and 52 fields of barley (Table S1, Supplementary Materials). The average area of these fields is ~112 ha with a standard deviation of ~69 ha. Using various yield monitoring systems mounted on grain harvesters operated by farmers, these observed data were collected at these sites across the country from 2009 to 2015 (Figure 1). The data obtained by these commercially available yield monitors were used to construct a yield data image at 5-m resolution [44], which was upscaled to 25 m and 250 m to match the Landsat and MODIS resolutions, respectively.

2.2.3. Climate Data

Climate data were sourced from Science Information for Land Owners (SILO) which provides nationwide meteorological variables (e.g., maximum and minimum air temperature, and precipitation) at daily temporal frequencies by interpolating observations made by the Australian Bureau of Meteorology onto a 0.05° by 0.05° grid [37].

3. Methods

3.1. Yield Prediction

We use a semi-empirical model [C-Crop; 6]) because of its low data requirements for calibration across large areas. C-Crop correlates actual grain weight (t/ha) to end-of-season above-ground plant carbon mass (C), and the estimation of C is based on biophysical principles of plant photosynthesis [6]. C_i is estimated using the carbon mass (C_{i-1}) from the previous time step and the current period's allocated net assimilation flux N_i (gCm^{-2}) minus the dead biomass that enters the litter store, and i (1–22) is the model time step (every 16 days in a calendar year).

$$C_i = (1 - \rho)(C_{i-1} + N_i) \tag{1}$$

where ρ (periods^{-1}) is the reciprocal of carbon longevity (i.e., the turn-over rate of plant live carbon into senesced tissue per time step i).

3.1.1. Net Primary Productivity

Net primary productivity (NPP) is the rate of carbon assimilation from atmospheric CO_2 to organic material (biomass) for a given area while accounting for the energy loss due to autotrophic respiration [6,45].

$$N_i = 0.75 \left(G_i \frac{fPAR_i}{0.95} - 16r_{10} \frac{C_{i-1}}{cn} \sigma_i \right) \tag{2}$$

where $\frac{fPAR_i}{0.95}$ is the fraction of total assimilation flux G_i allocated to the above-ground plant biomass (gCm^{-2}) at time step i. The plant maintenance respiration is calculated as a function of leaf nitrogen, air temperature, and previous biomass C_{i-1}. r_{10} is the plant tissue respiration rate at 10 °C; cn stands for plant carbon-to-nitrogen ratio; σ_i is a scalar that modifies the respiration rate according to the daily air temperature [46].

3.1.2. Gross Primary Productivity

The total assimilation flux G_i, = also known as the gross primary productivity (GPP) ($gCm^{-2} \cdot day^{-1}$), can be calculated using the remote sensing-based plant light use efficiency (LUE) approach. The chloroplasts use incoming solar radiation with a spectral range between 400 nm and 700 nm in photosynthesis [47].

$$G_i = PAR_i \times fPAR_i \times RUE_i \tag{3}$$

$$PAR = 2.3(R_O \times \tau_\partial)\rho_{sw} \tag{4}$$

where $(R_O \times \tau_\partial)\rho_{sw}$ represents the shortwave irradiance (R_s), R_O is daily top-of-atmosphere shortwave irradiance ($J/m^2/day$) [48,49], τ_∂ is atmospheric transmissivity calculated using the Bristow–Campbell relationship [50,51] calibrated for Australia, and ρ_{sw} is the ratio of shortwave irradiance at a sloping surface to that at a horizontal surface [52].

$fPAR$ is the portion of PAR that is absorbed by a photosynthetic organism, and it is estimated using a linear relation between $fPAR$ and rescaled NDVI by thresholds (i.e., local minimum and crop-specific maximum NDVIs) [53,54]. LUE is highly linearly related to a diffuse fraction (f_D) and photosynthetic carbon flux [55].

$$LUE_i = 0.024 \times f_{Di} + 0.00061A_x \tag{5}$$

where A_x is the maximum photosynthetic capacity ($\mu molCm^{-2} \cdot s^{-1}$), which is a crop-specific parameter; is 23, 40, and 45 for barley [56], canola [57], and wheat [57], respectively; f_D is the ratio of diffused to total solar irradiance varying from 0.2 (under clear skies) to 1.0 (under overcast skies) [48]. For a full description of C-Crop, see Donohue et al. [6].

3.2. Validation

Two sets of data are used for validation and further analysis, depending on the pixel-level completeness of time-series Landsat NDVIs at 25-m pixel resolution across the growing season (i: 8–19) between April (DOY 113) and October (DOY 304) (Figure S3, Supplementary Materials). Firstly, the coordinates of complete time-series Landsat pixels are used to obtain the resampled MODIS, L–M blended, and observed yield data for validation at the 25-m resolution as the first dataset. Secondly, these complete time-series Landsat pixels are upscaled at 250 m pixel size to extract the MODIS and the L–M blended data with the same pixel size for the validation at the moderate resolution. Thirdly, the pixel-level yield predictions are aggregated for each respective field by averaging the yield values of the pixels within, for field-level validation. Fourthly, and finally, the predicted yields are evaluated using the model coefficient of determination (R^2), the root-mean-square error ($RMSE$), and the mean bias error (MBE). The R^2 describes the proportion of the response variable that can be explained by the model. $RMSE$ gives more weight to the largest errors, and the MBE indicates the systematic error of the model to under or overestimate. These procedures are then repeated for the second dataset created according to the coordinates of incomplete time-series Landsat NDVIs at the 25-m pixel resolution. As C-Crop requires a time series of NDVI, the results are assessed without the incomplete time-series pixels of Landsat.

3.3. Identification of Threshold for When to Blend

The incomplete time-series pixels of Landsat are gap-filled with the L–M blended pixels (L_{LM}). The threshold to indicate when blending is beneficial is identified by quantifying the impact of the fraction of missing data on the prediction accuracy at 25 m. Firstly, we group MODIS and L_{LM} time series in eight groups, based on the fraction of Landsat missing data (<10%, 10–20%, 20–30%, 30–40%, 40–50%, 50–60%, 60–70%, and 70–80%). Then, the accuracy of each group is analyzed by calculating the R^2 and the $RMSE$. The performance of C-Crop using MODIS and L_{LM} is compared by the fraction of missing data in Landsat across time. The threshold can be identified, when the model provides the same R^2 and $RMSE$ using L_{LM} as with MODIS (Figure S3, Supplementary Materials). This threshold determines when, where, or how much L–M blended data improves crop yield prediction when the fraction of missing data in Landsat is lower than the identified threshold.

3.4. Evaluation of the Improvement in Yield Prediction Accuracy

The improvement in prediction accuracy using the identified threshold for multiple spatio-temporal data selection is statistically quantified. More specifically, the threshold is applied to the Landsat observations (2000–2018). Firstly, we compute the temporal probability of optimally using MODIS, Landsat, and L_{LM} images for nationwide crop yield predictions during the past two decades, and then map the results to illustrate the spatial variability of multi-sensor data selection for 25-m pixel-level yield prediction across the wheatbelt. We then evaluate the area percentage of the data sources on a yearly basis and analyze their potential correlation to the annual precipitation (mm/year). Finally, the improvement in the accuracy of predicted yields is evaluated on the field level using MODIS and L_{LM} for Western and eastern Australia, against the reported data [58]. The growing season of 2015 is selected due to the availability of a larger quantity of observed yield data. The incomplete 2015 Landsat series are gap-filled with the blended values corresponding to the threshold value.

4. Results

4.1. Yield Prediction

Exactly 66% of the observed fields have a complete time series of Landsat NDVIs at the pixel level across the growing season. For this dataset, C-Crop performs the best with Landsat images at field ($R^2 = 0.68$; Figure 4a), 250-m ($R^2 = 0.85$; Figure 4d), and 25-m ($R^2 = 0.48$; Figure 4g) pixel resolutions for yield prediction pooled for wheat, barley, and canola. MODIS-based model yields were at least 10% less in terms of the R^2 than when using complete time-series Landsat data. The model performs almost identically when using MODIS and L–M blended data for the predictions at both field and pixel scales. However, it produces the lowest bias for field ($MBE = -0.32$ t/ha; Figure 4c), 250-m ($MBE = -0.23$; Figure 4f), and 25-m ($MBE = -0.21$; Figure 4i) pixels when using the L–M blended data.

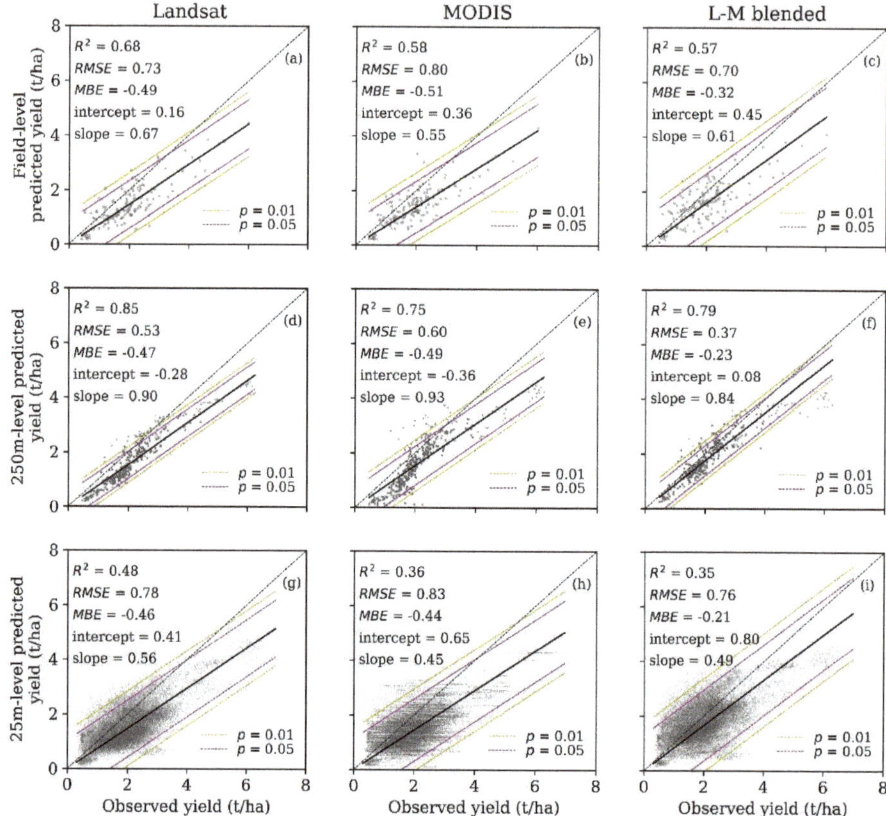

Figure 4. Validation of C-Crop-predicted yield pooled for wheat, barley, and canola using Landsat, MODIS, and L–M blended data where the complete Landsat time series are observed (the fraction of missing Landsat series = 0). From top to bottom, the first (**a–c**), second (**d–f**), and third (**g–i**) rows show the comparison between observed (*x*-axis) and model-predicted yields (*y*-axis) on the field scale (*n* = 139), 250-m pixel level (*n* = 2367), and 25-m pixel level (n = 113,329), respectively, where *n* is the sample size. From left to right, the first (**a,d,g**), second (**b,e,h**), and third (**c,f,i**) columns delineate the validation using Landsat, MODIS, and L–M blended data, respectively. The solid black line is the line of best fit, the purple and the yellow lines represent the upper and lower bounds of the prediction confidence intervals (i.e., $p = 0.01$ and $p = 0.05$), and the black dashed line is the 1:1 line.

Figure 5 shows that 210 fields have incomplete time-series Landsat pixels during the growing season. These series are incomplete due to clouds in some of the Landsat images acquired in the specific growing season when the yield data are observed. Using this set of data, the L–M blended data-based model performs the same as the MODIS-based model when aggregating to 250-m ($R^2 = 0.86$, $RMSE = 0.52$ t/ha, $MBE = -0.40$ t/ha; Figure 5c,d) and field ($R^2 = 0.66$, $RMSE = 0.82$ t/ha, $MBE = -0.49$ t/ha; Figure 5a,b) scales. The L–M blended data-based model shows its advantages at the 25-m pixel level, which explains an extra 7% of the variability in the observed yields when the Landsat time-series pixels are incomplete (Figure 5e,f).

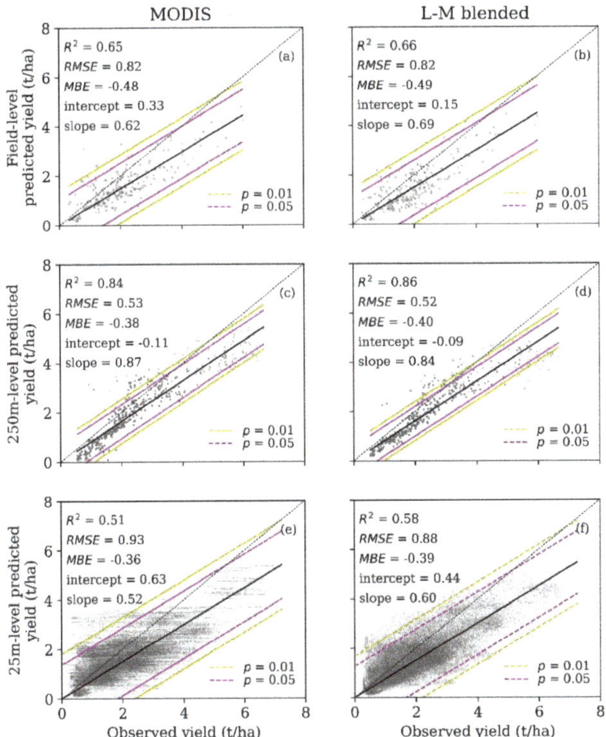

Figure 5. Validation of C-Crop-predicted yield pooled for wheat, barley, and canola using MODIS and L–M blended data when the complete Landsat time series are not available (the fraction of missing Landsat series ≥0.083). From top to bottom, the first (**a,b**), second (**c,d**), and third (**e,f**) rows show the comparison between observed (*x*-axis) and model-predicted yields (*y*-axis) on the field scale (*n* = 210), 250-m pixel level (*n* = 3978), and 25-m pixel level (*n* = 231,667), respectively, where *n* is the sample size. From left to right, the first (**a,c,e**) and second (**b,d,f**) columns delineate the validation using MODIS and L–M blended data, respectively. The solid black line is the line of best fit, the purple and the yellow lines represent the upper and lower bounds of the prediction confidence intervals (i.e., $p = 0.01$ and $p = 0.05$), and the black dashed line is the 1:1 line.

4.2. Identification of the Threshold

Figure 6a shows that the 25-m pixel-level yield prediction accuracy fluctuates between 0.35 and 0.75 using MODIS and L_{LM} data in time and space, where time-series Landsat observations are incomplete. The R^2 derived from MODIS is steady (between 0.4 and 0.5) where the fraction of missing Landsat data ranged between 0.05 and 0.42, which is lower than the R^2 (0.62–0.5) derived from L_{LM} data. The *RMSE* derived from both data sources remains approximately 0.9 t/ha for the same fraction of missing Landsat data (Figure 6b). When more incomplete time-series Landsat data are observed (from 0.42 to 0.75 on both Figure 6a,b), the R^2 based on MODIS increases to around 0.7 and the *RMSE* decreases to 0.4 t/ha (as MODIS is not as cloud-affected as Landsat due to imagery being acquired on more days), whereas the R^2 derived from the L_{LM} fluctuates between 0.6 and 0.4 and the *RMSE* changes between 1.3 and 0.8 t/ha (Figure 6). Given this, up to 42% of missing Landsat data in the growing season is defined as the threshold for when L–M blending is optimal to implement. That is, the 25-m pixel-level yield prediction accuracy can be improved using L_{LM} when the fraction of missing Landsat data at the coordinates is below this threshold. For instance, the L_{LM}-based model increases R^2 by up to 20% when the fraction of the missing Landsat data is below 42% (Figure 6a).

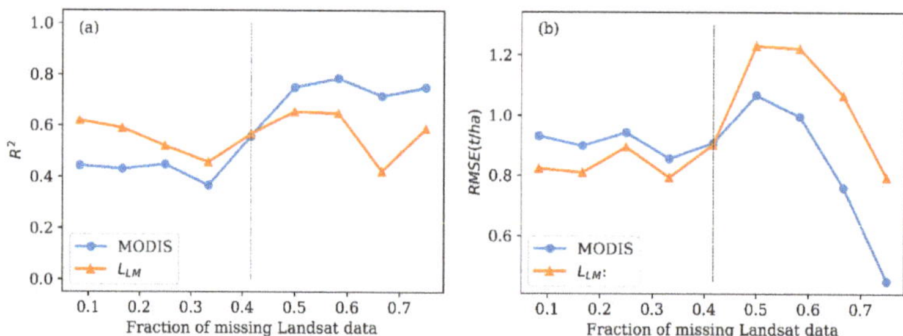

Figure 6. The statistical analysis of missing data in Landsat for 25-m pixel-level yield prediction, by evaluating (**a**) R^2 and (**b**) *RMSE* against the fraction of Landsat missing data during the growing season. L–M blended data were used to fill the gaps (L_{LM}) in the incomplete Landsat series.

4.3. Evaluation of the Improvement in Yield Prediction Accuracy

Given the availability of cloud-free Landsat and MODIS data across the wheatbelt (2000–2018), in concert with the previously determined threshold (Figure 6) and finding (Figure 5), we can demonstrate which imagery (i.e., either MODIS, Landsat, and L_{LM}) is best suitable for 25-m pixel-level crop yield prediction (Figure 7). Figure 7 shows the selection when using multiple satellite products for crop yield estimates across the wheatbelt over the past two decades. For the wheatbelt north of 35° south (S), there is a higher probability of obtaining complete cloud-free Landsat observations over the growing season in the east–west overlapping areas of adjacent Landsat Path-Rows, whilst L_{LM} is optimal elsewhere north of 35° S. South of 35° S, MODIS is optimal, with L_{LM} being optimal in the east–west overlapping areas of adjacent Landsat Path-Rows (in Figure 7).

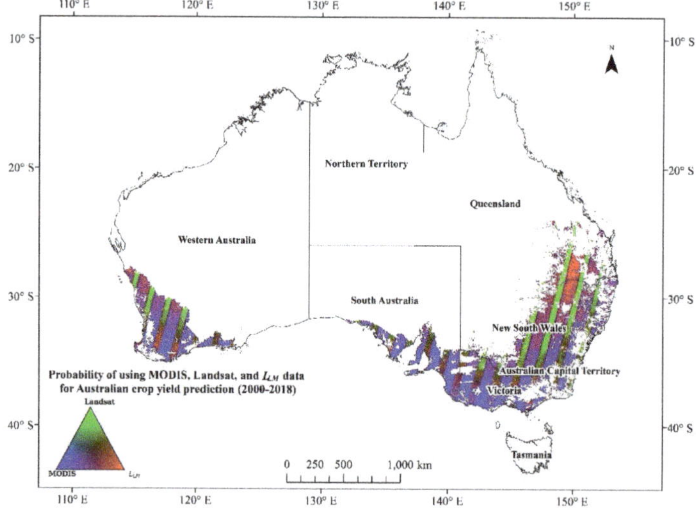

Figure 7. Multi-sensor optimal data selection across the wheatbelt (2000–2018) for 25-m pixel-level crop yield prediction, using the probability of MODIS, Landsat, and L_{LM} images. Blue denotes regions where incomplete Landsat series have a fraction of missing data exceeding 42%, thus indicating where MODIS should be used for yield estimates because it provides more frequent observations than Landsat. Green areas show where adjacent Landsat orbits overlap and, thus, where a complete once every 16-day Landsat series over the whole growing season is available. Areas colored red are those where the fraction of Landsat missing data is below the 0.42 threshold identified previously in Figure 6 when L–M blended data improve the yield prediction accuracy.

For 25-m pixel-level yield prediction, blended data should be preferred on average for 33% of the Australian wheatbelt area (Figure 8a). MODIS and Landsat data remain the preferred data sources in 50% and 17% of cases, respectively. Figure 8b shows that the area percentage is positively correlated with annual precipitation for MODIS and negatively correlated for Landsat and the L-M blended data. MODIS has a larger scatter when annual precipitation is greater than 400 mm/year.

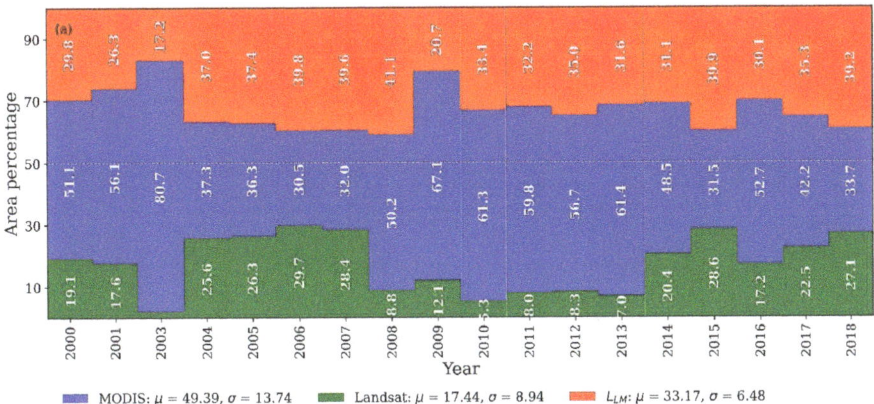

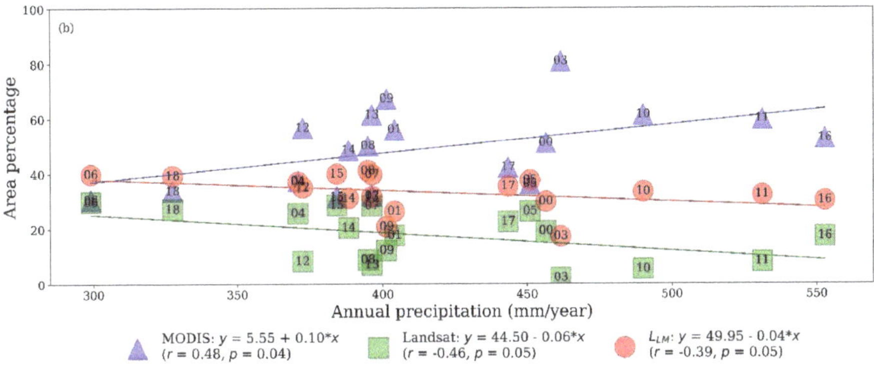

Figure 8. Yearly analysis of multi-sensor data selection for 25-m pixel-level crop yield prediction (2000–2018) by evaluating (**a**) the area percentage and (**b**) its correlation with the annual precipitation (mm/year). The white dashed line shows that the area percentage is 50. μ: mean of population values; σ: standard deviation; r: correlation coefficient. The symbols in (b) are labeled with the last two digits of the year.

Of the 53-million-ha Australia wheatbelt, complete Landsat time series were suitable for 28.6% of that area, MODIS for 31.5%, and L_{LM} for 39.9% during the 2015 growing season (Figure 8a). In this season, there are 104 fields available for assessing the accuracy improvement in yield prediction accuracy for nationwide crop yield prediction (Table S1, Supplementary Materials). Within these observed yield data, 69 fields are located where L–M blended images can improve the prediction accuracy according to the previously defined (see Figure 6) 42% Landsat missing data threshold. Figure 9 shows the predicted yields against the observed values for 63 fields in Western Australia (i.e., WA), and six fields in eastern Australia (i.e., Queensland (QLD), New south Wales (NSW), Victoria (VIC), and South Australia (SA)).

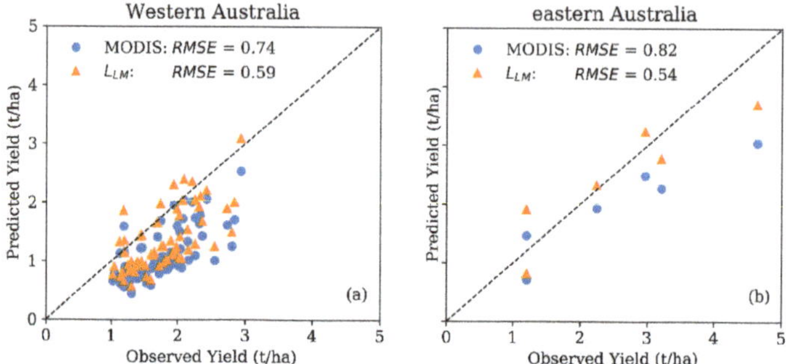

Figure 9. Scattergram for C-Crop-predicted yield against observed values for 2015 at the field level for (**a**) Western Australia ($n = 63$) and (**b**) eastern Australia ($n = 6$) wheatbelt using MODIS and L–M blended data for gap-filling Landsat (L_{LM}) when the fraction of incomplete Landsat series is below 42% for the growing season. The *RMSE* statistics have units of t/ha.

In these areas (see Figure 9), when the fraction of incomplete Landsat series across the growing season of 2015 falls below 42%, using L_{LM} reduced the field-level yield prediction errors by 0.15 t/ha for Western Australia and 0.28 t/ha for eastern Australia. More specifically, the bias of 598 kilotons in grain production over the cropping area of 39,882 km^2 in Western Australia and of 1672 kilotons in grain production over the cropping area of 59,716 km^2 in eastern Australia can be decreased when using the threshold determined herein to decide the areas where the blended data can improve the prediction accuracy.

5. Discussion

Globally, wheat-growing regions are distributed within areas that experience a relatively high frequency of clouds [59]. Persistent cloud cover limits the use of HSR sensors over large spatial extents due to the low frequency of the observations, whereas the other sensors that have HTF are constrained by the LSR. Although blending improves the monitoring of rapidly changing processes such as crop growth, it may introduce unforeseen spatial and temporal variances in the blended images when images are often not observed concurrently [10], and implementing the current suite of algorithms across large areas is currently computationally expensive. Therefore, it is important to systematically quantify and evaluate the utility of blending imagery across time and space before embarking on the development of nationwide blending capabilities.

This study developed a parsimonious approach to identify when and where blending is beneficial for crop yield prediction at the 25-m pixel resolution. When incomplete Landsat series falls below 42% of the possible imagery in the growing season, blending is recommended because it improves the accuracy of the yield estimates. When incomplete Landsat series exceeds 42% of data, the use of HTF LSR data (e.g., MODIS) for the 25-m pixel-level yield prediction is recommended, reducing the computational need for blending. LTF HSR data show benefits, for example, in providing detailed information on plant photosynthesis across space; thus, they should be solely used for crop yield prediction when enough cloud-free images are available throughout the growing season.

Annual precipitation, which is strongly associated with cloudiness [60,61] and, thus, partly governs the amount of missing Landsat data, indicates which data sources should be preferably used (Figure 8). For instance, MODIS should be used for the prediction for greater than 50% of the wheatbelt during wet years (e.g., 2003, 2010, 2011, and 2016). During dry years (e.g., 2006 and 2018), the proportion of the wheatbelt where Landsat is suitable for yield prediction increases by >10% compared to the average proportion (17%). For crop yield prediction, blended data can be optimally applied across approximately 33% of the wheatbelt during dry to normal years.

Crop production fluctuates from year to year due to the rapidly changing patterns of precipitation and temperature [42,62,63]. Over the last 50 years, the amount of precipitation received by the Australian wheatbelt declined [64]. Based on model projections, it was also suggested that the changing climatic conditions will affect both the frequency and intensity of the extreme events such as floods, heatwaves, and droughts, which will impact agricultural production [65,66]. Within the context of the present study, an increase in the number of clear sky days can be beneficial for the use of RS in crop yield forecasting. Moreover, a decline in the number of rain days in recent decades within the wheatbelt area can also be noted (see Figure S4, Supplementary Materials). Here, this decline in the number of rain days is assumed to correlate with the increase in the number of days with clear skies. However, a decrease in the number of rain days does not necessarily imply an absence of clouds, as there could be non-precipitating clouds. However, studies such as Norris et al. [67], using observations and Coupled Model Intercomparison Project Phase 5 (CMIP5) model outputs, showed that, on average, there is a decline in cloud amount across the region from 60° S–60° north (N). This decline was attributed to the poleward shift of mid-latitude storm tracks, among others.

The potential influence of climate change on grain production suggests a decline in world food supply [68] and an increase in the level of exposure of the global population to the risk of hunger [69]. It is, therefore, important to ensure the accuracy and efficiency of yield prediction at anytime, anywhere, and at any scale. A more precise description of grain weight patterns in time and space provides more accurate information for precision agriculture to improve its production and sustainability. The semi-empirical carbon turn-over model used for crop yield prediction is based on historical yield data, and, in the future, relevant drivers (e.g., precipitation and temperature) and their interactions may change under climate change, thus requiring model re-calibration [4,6,70,71]. Our approach was tested with C-Crop but could be extended to other semi-empirical models [72] and models based on machine learning [73].

Development of a national-scale yield prediction system requires time series of HSR and HTF imagery to describe the crop growth; therefore, next-generation HSR and HTF satellite data like Sentinel-2 [74,75] should be tested for nationwide yield estimates post the growing season of 2015. Harmonized Landsat and Sentinel-2 surface reflectance products should be tested for national-scale crop yield prediction [76] when observed yield data for more recent years are available. However, spatio-temporal fusion of MODIS and Landsat data is still necessary for long-term studies that involve historical satellite images collected before 2015 (back to the year 2000, when MODIS was launched). Future studies should also focus on using active RS technologies, such as radio detecting and ranging (Radar) [77,78] and light detection and ranging (LiDAR) [79], for a national yield estimation because of their ability to penetrate clouds and detect a plant's three-dimensional structural characteristics. Data blending between HSR optical imagery and active RS data (e.g., Radar and LiDAR) [80] warrants further study.

6. Conclusions

Sparse time series of satellite remote sensing, due to LTF and/or cloud contamination, represent one of the main barriers limiting accurate crop yield estimation at regional to national scales. Blending of HSR but LTF images with LSR but HTF images was proposed to increase the temporal resolution and maintain spatial details. In this study, the benefits of blending were tested for crop yield prediction across the Australian wheatbelt. We found that, when time series are gap-free, yield prediction from Landsat is the most accurate on both field and pixel scales. When Landsat time series contain <42% missing values during the growing season, blending is recommended for nationwide crop yield prediction at the 25-m pixel resolution. When Landsat time series contain ≥42% missing observations, MODIS outperforms blending. Across Australia, these recommendations would improve yield estimates by 0.15 t/ha for Western Australia and 0.28 t/ha for eastern Australia on average. By identifying where and when to blend, this work paves the way to more accurate monitoring of biophysical processes and yields, while keeping computational costs low.

Supplementary Materials: The following are available online at http://www.mdpi.com/2072-4292/12/10/1653/s1, Table S1: Summary of observed data (the number of field-years) used herein; Figure S1. Long-term (1901–2018) average monthly accumulated precipitation (mm/month) (left) and the long-term average rain days (right) across the wheatbelt for (a, b) Queensland, (c, d) New South Wales, (e, f) Victoria, (g, h) South Australia, (i, j) Western Australia, and (k, l) Tasmania. The precipitation data are sourced from Jeffrey et al. [37], and the wheat belt mask was resampled from native 2-km^2 horizontal resolution onto the precipitation data grid with 5-km^2 horizontal resolution using nearest neighbor [40]. The vertical black lines (a–f) represent one standard deviation of the monthly wheatbelt specific precipitation for each state. The horizontal line (g–l) represents the median of the data, the box spans from 25th to 75th quartiles of the data, and the rain day threshold is 0 mm/day. It should be noted that Northern Territory is not included in the analysis due to its small cropping area; Figure S2. An example of MODIS and Landsat blended NDVI using ESTARFM. The x-axis and y-axis are time (t, DOY) and spatial resolution (m), respectively. Both MODIS and Landsat have valid observations at t_1 and t_2. The image at the center of the top row is a valid observation of MODIS at t_s, and that at the center of the bottom row is Landsat-like blended data at t_s created using the 16-day MODIS composite images on the DOYs 257, 273, and 289, as well as the Landsat images acquired on the DOYs 257 and 289; Figure S3: The implementation flowchart of validation and identification of threshold for when to blend. The shaded areas indicate the original data sources; Figure S4: Long-term (1900–2018) average monthly accumulated precipitation (mm/month) across (a) eastern Australian, (b) Western Australia, and (c) the wheatbelt. The precipitation data are sourced from Jeffrey et al. [37].

Author Contributions: Conceptualization, Y.C., T.R.M., and R.J.D.; methodology, Y.C.; software, N.O. and N.G.; validation, Y.C.; formal analysis, Y.C.; investigation, Y.C. and F.W.; resources, L.L. and N.G.; writing—original draft preparation, Y.C.; writing—review and editing, T.R.M., R.J.D., F.W., N.G., and R.L.; visualization, Y.C. and N.G.; supervision, T.R.M. and R.J.D.; project administration, R.L.; funding acquisition, R.L. All authors read and agreed to the published version of the manuscript.

Funding: The authors acknowledged the support of the Digiscape Future Science Platform, funded by the CSIRO.

Acknowledgments: Appreciation is extended to numerous industry participants who provided training data in the form of yield maps, accessed by the project team. Randall J. Donohue and Tim R. McVicar acknowledge the support of the ARC Center of Excellence for Climate Extremes (Australian Research Council grant CE170100023). We thank the two anonymous reviewers and the Remote Sensing Academic Editor for helpful comments that helped us improve this paper.

Conflicts of Interest: The authors declare no conflicts of interest.

References

1. Cohen, J.E. Human population: The next half century. *Science* **2003**, *302*, 1172–1175. [CrossRef]
2. Jones, J.W.; Antle, J.M.; Basso, B.; Boote, K.J.; Conant, R.T.; Foster, I.; Godfray, H.C.J.; Herrero, M.; Howitt, R.E.; Janssen, S.; et al. Brief history of agricultural systems modeling. *Agric. Syst.* **2017**, *155*, 240–254. [CrossRef] [PubMed]
3. Prasad, A.K.; Chai, L.; Singh, R.P.; Kafatos, M. Crop yield estimation model for Iowa using remote sensing and surface parameters. *Int. J. Appl. Earth Obs. Geoinf.* **2006**, *8*, 26–33. [CrossRef]
4. Doraiswamy, P.C.; Moulin, S.; Cook, P.W.; Stern, A. Crop yield assessment from remote sensing. *Photogramm. Eng. Remote Sens.* **2003**, *69*, 665–674. [CrossRef]
5. Serrano, L.; Filella, I.; Penuelas, J. Remote sensing of biomass and yield of winter wheat under different nitrogen supplies. *Crop. Sci.* **2000**, *40*, 723–731. [CrossRef]
6. Donohue, R.J.; Lawes, R.A.; Mata, G.; Gobbett, D.; Ouzman, J. Towards a national, remote-sensing-based model for predicting field-scale crop yield. *Field Crop. Res.* **2018**, *227*, 79–90. [CrossRef]
7. Myers, E.; Kerekes, J.; Daughtry, C.; Russ, A. Assessing the Impact of Satellite Revisit Rate on Estimation of Corn Phenological Transition Timing through Shape Model Fitting. *Remote Sens.* **2019**, *11*, 2558. [CrossRef]
8. Waldner, F.; Chen, Y.; Horan, H.; Hochan, Z. High temporal resolution of leaf area data improves empirical estimation of grain yield. *Sci. Rep. Press* **2019**, *9*, 1–4. [CrossRef]
9. Sakamoto, T.; Gitelson, A.A.; Arkebauer, T.J. Near real-time prediction of US corn yields based on time-series MODIS data. *Remote Sens. Environ.* **2014**, *147*, 219–231. [CrossRef]
10. Emelyanova, I.V.; McVicar, T.R.; Van Niel, T.G.; Li, L.T.; van Dijk, A.I.J.M. Assessing the accuracy of blending Landsat–MODIS surface reflectances in two landscapes with contrasting spatial and temporal dynamics: A framework for algorithm selection. *Remote Sens. Environ.* **2013**, *133*, 193–209. [CrossRef]
11. Zhu, X.; Chen, J.; Gao, F.; Chen, X.; Masek, J.G. An enhanced spatial and temporal adaptive reflectance fusion model for complex heterogeneous regions. *Remote Sens. Environ.* **2010**, *114*, 2610–2623. [CrossRef]

12. Gao, F.; Masek, J.; Schwaller, M.; Hall, F. On the blending of the Landsat and MODIS surface reflectance: Predicting daily Landsat surface reflectance. *IEEE Trans. Geosci. Remote Sens.* **2006**, *44*, 2207–2218.

13. ABS. *Themes: Land Use on Farms, Australia, Year Ended 30 June 2017*; Australian Bureau of Statistics: Canberra, Australia, 2017. Available online: http://www.abs.gov.au/ausstats/abs@.nsf/mf/4627.0 (accessed on 7 April 2020).

14. Duveiller, G.; Baret, F.; Defourny, P. Crop specific green area index retrieval from MODIS data at regional scale by controlling pixel-target adequacy. *Remote Sens. Environ.* **2011**, *115*, 2686–2701. [CrossRef]

15. Waldner, F.; Defourny, P. Where can pixel counting area estimates meet user-defined accuracy requirements? *Int. J. Appl. Earth Obs. Geoinf.* **2017**, *60*, 1–10. [CrossRef]

16. Whitcraft, A.; Becker-Reshef, I.; Killough, B.; Justice, C. Meeting earth observation requirements for global agricultural monitoring: An evaluation of the revisit capabilities of current and planned moderate resolution optical earth observing missions. *Remote Sens.* **2015**, *7*, 1482–1503. [CrossRef]

17. Lobell, D.B.; Thau, D.; Seifert, C.; Engle, E.; Little, B. A scalable satellite-based crop yield mapper. *Remote Sens. Environ.* **2015**, *164*, 324–333. [CrossRef]

18. Schowengerdt, R.A. *Remote Sensing: Models and Methods for Image Processing*; Elsevier: Amsterdam, The Netherlands, 2006.

19. Kang, S.; Running, S.W.; Zhao, M.; Kimball, J.S.; Glassy, J. Improving continuity of MODIS terrestrial photosynthesis products using an interpolation scheme for cloudy pixels. *Int. J. Remote Sens.* **2005**, *26*, 1659–1676. [CrossRef]

20. Poggio, L.; Gimona, A.; Brown, I. Spatio-temporal MODIS EVI gap filling under cloud cover: An example in Scotland. *ISPRS J. Photogramm. Remote Sens.* **2012**, *72*, 56–72. [CrossRef]

21. Borak, J.S.; Jasinski, M.F. Effective interpolation of incomplete satellite-derived leaf-area index time series for the continental United States. *Agric. For. Meteorol.* **2009**, *149*, 320–332. [CrossRef]

22. Jarihani, A.; McVicar, T.; Van Niel, T.; Emelyanova, I.; Callow, J.; Johansen, K. Blending Landsat and MODIS data to generate multispectral indices: A comparison of "Index-then-Blend" and "Blend-then-Index" approaches. *Remote Sens.* **2014**, *6*, 9213–9238. [CrossRef]

23. Zhang, J. Multi-source remote sensing data fusion: Status and trends. *Int. J. Image Data Fusion* **2010**, *1*, 5–24. [CrossRef]

24. Pohl, C.; Van Genderen, J.L. Review article multisensor image fusion in remote sensing: Concepts, methods and applications. *Int. J. Remote Sens.* **1998**, *19*, 823–854. [CrossRef]

25. Viovy, N.; Arino, O.; Belward, A. The Best Index Slope Extraction (BISE): A method for reducing noise in NDVI time-series. *Int. J. Remote Sens.* **1992**, *13*, 1585–1590. [CrossRef]

26. Löw, F.; Biradar, C.; Dubovyk, O.; Fliemann, E.; Akramkhanov, A.; Narvaez Vallejo, A.; Waldner, F. Regional-scale monitoring of cropland intensity and productivity with multi-source satellite image time series. *GISci. Remote Sens.* **2018**, *55*, 539–567. [CrossRef]

27. Dong, T.; Liu, J.; Qian, B.; Zhao, T.; Jing, Q.; Geng, X.; Wang, J.; Huffman, T.; Shang, J. Estimating winter wheat biomass by assimilating leaf area index derived from fusion of Landsat-8 and MODIS data. *Int. J. Appl. Earth Obs. Geoinf.* **2016**, *49*, 63–74. [CrossRef]

28. Meng, J.; Du, X.; Wu, B. Generation of high spatial and temporal resolution NDVI and its application in crop biomass estimation. *Int. J. Digit. Earth* **2013**, *6*, 203–218. [CrossRef]

29. Wang, L.; Tian, Y.; Yao, X.; Zhu, Y.; Cao, W. Predicting grain yield and protein content in wheat by fusing multi-sensor and multi-temporal remote-sensing images. *Field Crop. Res.* **2014**, *164*, 178–188. [CrossRef]

30. Gao, F.; Anderson, M.C.; Zhang, X.; Yang, Z.; Alfieri, J.G.; Kustas, W.P.; Mueller, R.; Johnson, D.M.; Prueger, J.H. Toward mapping crop progress at field scales through fusion of Landsat and MODIS imagery. *Remote Sens. Environ.* **2017**, *188*, 9–25. [CrossRef]

31. Semmens, K.A.; Anderson, M.C.; Kustas, W.P.; Gao, F.; Alfieri, J.G.; McKee, L.; Prueger, J.H.; Hain, C.R.; Cammalleri, C.; Yang, Y. Monitoring daily evapotranspiration over two California vineyards using Landsat 8 in a multi-sensor data fusion approach. *Remote Sens. Environ.* **2016**, *185*, 155–170. [CrossRef]

32. Yang, Y.; Anderson, M.C.; Gao, F.; Wardlow, B.; Hain, C.R.; Otkin, J.A.; Alfieri, J.; Yang, Y.; Sun, L.; Dulaney, W. Field-scale mapping of evaporative stress indicators of crop yield: An application over Mead, NE, USA. *Remote Sens. Environ.* **2018**, *210*, 387–402. [CrossRef]

33. Gao, F.; Anderson, M.; Daughtry, C.; Johnson, D. Assessing the variability of corn and soybean yields in central Iowa using high spatiotemporal resolution multi-satellite imagery. *Remote Sens.* **2018**, *10*, 1489. [CrossRef]

34. He, M.; Kimball, J.; Maneta, M.; Maxwell, B.; Moreno, A.; Beguería, S.; Wu, X. Regional crop gross primary productivity and yield estimation using fused landsat-MODIS data. *Remote Sens.* **2018**, *10*, 372. [CrossRef]

35. Liao, C.; Wang, J.; Dong, T.; Shang, J.; Liu, J.; Song, Y. Using spatio-temporal fusion of Landsat-8 and MODIS data to derive phenology, biomass and yield estimates for corn and soybean. *Sci. Total Environ.* **2019**, *650*, 1707–1721. [CrossRef] [PubMed]

36. Holper, P.N. *Climate Change, Science Information Paper: Australian Rainfall—Past, Present and Future*; CSIRO: Canberra, Australia, 2011.

37. Jeffrey, S.J.; Carter, J.O.; Moodie, K.B.; Beswick, A.R. Using spatial interpolation to construct a comprehensive archive of Australian climate data. *Environ. Model. Softw.* **2001**, *16*, 309–330. [CrossRef]

38. Li, F.; Jupp, D.L.; Reddy, S.; Lymburner, L.; Mueller, N.; Tan, P.; Islam, A. An evaluation of the use of atmospheric and BRDF correction to standardize Landsat data. *IEEE J. Sel. Top. Appl. Earth Obs. Remote Sens.* **2010**, *3*, 257–270. [CrossRef]

39. Rouse, J.W., Jr.; Haas, R.; Schell, J.; Deering, D. *Monitoring Vegetation Systems in the Great Plains with ERTS*; NASA: Washington, DC, USA, 1974.

40. Sibson, R. A brief description of natural neighbour interpolation. In *Interpreting Multivariate Data*; John Wiley & Sons: New York, NY, USA, 1981; pp. 21–36.

41. Cockram, J.; Jones, H.; Leigh, F.J.; O'Sullivan, D.; Powell, W.; Laurie, D.A.; Greenland, A.J. Control of flowering time in temperate cereals: Genes, domestication, and sustainable productivity. *J. Exp. Bot.* **2007**, *58*, 1231–1244. [CrossRef] [PubMed]

42. Hochman, Z.; Gobbett, D.L.; Horan, H. Climate trends account for stalled wheat yields in Australia since 1990. *Glob. Chang. Biol.* **2017**, *23*, 2071–2081. [CrossRef]

43. Emelyanova, I.V.; McVicar, T.R.; Van Niel, T.G.; Li, L.T.; Van Dijk, A.I.J.M. On blending Landsat-MODIS surface reflectances in two landscapes with contrasting spectral, spatial and temporal dynamics. In *WIRADA Project 3.4: Technical Report*; CSIRO: Water for a Healthy Country Flagship: Canberra, Australia, 2012; p. 72. Available online: https://publications.csiro.au/rpr/pub?list=SEA&pid=csiro:EP128838 (accessed on 7 April 2020).

44. Bramley, R.; Williams, S. *A Protocol for the Construction of Yield Maps from Data Collected Using Commercially Available Grape Yield Monitors*; Cooperative Research Centre for Viticulture: Adelaide, Australia, 2001.

45. Kira, T. Primary production of forests. In *Photosynthesis and Productivity in Different Environments*; Cambridge University Press: Cambridge, UK, 1975.

46. Sitch, S.; Smith, B.; Prentice, I.C.; Arneth, A.; Bondeau, A.; Cramer, W.; Kaplan, J.; Levis, S.; Lucht, W.; Sykes, M.T. Evaluation of ecosystem dynamics, plant geography and terrestrial carbon cycling in the LPJ dynamic global vegetation model. *Glob. Chang. Biol.* **2003**, *9*, 161–185. [CrossRef]

47. McCree, K.J. Test of current definitions of photosynthetically active radiation against leaf photosynthesis data. *Agric. Meteorol.* **1972**, *10*, 443–453. [CrossRef]

48. Roderick, M.L. Estimating the diffuse component from daily and monthly measurements of global radiation. *Agric. For. Meteorol.* **1999**, *95*, 169–185. [CrossRef]

49. Iqbal, M. *An Introduction to Solar Radiation*; Elsevier: Amsterdam, The Netherlands, 2012.

50. Bristow, K.L.; Campbell, G.S. On the relationship between incoming solar radiation and daily maximum and minimum temperature. *Agric. For. Meteorol.* **1984**, *31*, 159–166. [CrossRef]

51. McVicar, T.R.; Jupp, D.L. Estimating one-time-of-day meteorological data from standard daily data as inputs to thermal remote sensing based energy balance models. *Agric. For. Meteorol.* **1999**, *96*, 219–238. [CrossRef]

52. Wilson, J.P.; Gallant, J.C. *Terrain analysis: Principles and Applications*; John Wiley & Sons: New York, NY, USA, 2000.

53. Verger, A.; Baret, F.; Camacho, F. Optimal modalities for radiative transfer-neural network estimation of canopy biophysical characteristics: Evaluation over an agricultural area with CHRIS/PROBA observations. *Remote Sens. Environ.* **2011**, *115*, 415–426. [CrossRef]

54. Li, W.; Weiss, M.; Waldner, F.; Defourny, P.; Demarez, V.; Morin, D.; Hagolle, O.; Baret, F. A generic algorithm to estimate LAI, FAPAR and FCOVER variables from SPOT4_HRVIR and Landsat sensors: Evaluation of the consistency and comparison with ground measurements. *Remote Sens.* **2015**, *7*, 15494–15516. [CrossRef]

55. Donohue, R.J.; Hume, I.; Roderick, M.; McVicar, T.R.; Beringer, J.; Hutley, L.; Gallant, J.C.; Austin, J.; van Gorsel, E.; Cleverly, J. Evaluation of the remote-sensing-based DIFFUSE model for estimating photosynthesis of vegetation. *Remote Sens. Environ.* **2014**, *155*, 349–365. [CrossRef]

56. Tambussi, E.; Nogues, S.; Ferrio, P.; Voltas, J.; Araus, J. Does higher yield potential improve barley performance in Mediterranean conditions? A case study. *Field Crop. Res.* **2005**, *91*, 149–160. [CrossRef]

57. Jensen, C.; Mogensen, V.; Mortensen, G.; Andersen, M.N.; Schjoerring, J.; Thage, J.; Koribidis, J. Leaf photosynthesis and drought adaptation in field-grown oilseed rape (*Brassica napus* L.). *Funct. Plant. Biol.* **1996**, *23*, 631–644. [CrossRef]

58. ABARES. *Australian Agricultural Overview*; Australian Bureau of Agricultural and Resource Economics and Sciences: Canberra, Australia, 2018; p. 26.

59. Wilson, A.M.; Jetz, W. Remotely sensed high-resolution global cloud dynamics for predicting ecosystem and biodiversity distributions. *PLoS Biol.* **2016**, *14*, e1002415. [CrossRef]

60. Jovanovic, B.; Collins, D.; Braganza, K.; Jakob, D.; Jones, D.A. A high-quality monthly total cloud amount dataset for Australia. *Clim. Chang.* **2011**, *108*, 485–517. [CrossRef]

61. Portmann, R.W.; Solomon, S.; Hegerl, G.C. Spatial and seasonal patterns in climate change, temperatures, and precipitation across the United States. *Proc. Natl. Acad. Sci. USA* **2009**, *106*, 7324–7329. [CrossRef]

62. Ludwig, F.; Milroy, S.P.; Asseng, S. Impacts of recent climate change on wheat production systems in Western Australia. *Clim. Chang.* **2009**, *92*, 495–517. [CrossRef]

63. Dreccer, M.F.; Fainges, J.; Whish, J.; Ogbonnaya, F.C.; Sadras, V.O. Comparison of sensitive stages of wheat, barley, canola, chickpea and field pea to temperature and water stress across Australia. *Agric. For. Meteorol.* **2018**, *248*, 275–294. [CrossRef]

64. Cai, W.; Cowan, T. Dynamics of late autumn rainfall reduction over southeastern Australia. *Geophys. Res. Lett.* **2008**, *35*. [CrossRef]

65. Van Dijk, A.I.; Beck, H.E.; Crosbie, R.S.; de Jeu, R.A.; Liu, Y.Y.; Podger, G.M.; Timbal, B.; Viney, N.R. The millennium drought in southeast Australia (2001–2009): Natural and human causes and implications for water resources, ecosystems, economy, and society. *Water Resour. Res.* **2013**, *49*, 1040–1057. [CrossRef]

66. Kiem, A.S.; Johnson, F.; Westra, S.; van Dijk, A.; Evans, J.P.; O'Donnell, A.; Rouillard, A.; Barr, C.; Tyler, J.; Thyer, M. Natural hazards in Australia: Droughts. *Clim. Chang.* **2016**, *139*, 37–54. [CrossRef]

67. Norris, J.R.; Allen, R.J.; Evan, A.T.; Zelinka, M.D.; O'Dell, C.W.; Klein, S.A. Evidence for climate change in the satellite cloud record. *Nature* **2016**, *536*, 72. [CrossRef]

68. Rosenzweig, C.; Parry, M.L. Potential impact of climate change on world food supply. *Nature* **1994**, *367*, 133–138. [CrossRef]

69. Parry, M.L.; Rosenzweig, C.; Iglesias, A.; Livermore, M.; Fischer, G. Effects of climate change on global food production under SRES emissions and socio-economic scenarios. *Glob. Environ. Chang.* **2004**, *14*, 53–67. [CrossRef]

70. Doraiswamy, P.; Hatfield, J.; Jackson, T.; Akhmedov, B.; Prueger, J.; Stern, A. Crop condition and yield simulations using Landsat and MODIS. *Remote Sens. Environ.* **2004**, *92*, 548–559. [CrossRef]

71. Ferencz, C.; Bognar, P.; Lichtenberger, J.; Hamar, D.; Tarcsai, G.; Timar, G.; Molnar, G.; Pásztor, S.; Steinbach, P.; Szekely, B. Crop yield estimation by satellite remote sensing. *Int. J. Remote Sens.* **2004**, *25*, 4113–4149. [CrossRef]

72. Chen, Y.; Donohue, R.J.; McVicar, T.R.; Waldner, F.; Mata, G.; Ota, N.; Houshmandfar, A.; Dayal, K.; Lawes, R.A. Nationwide crop yield estimation based on photosynthesis and meteorological stress indices. *Agric. For. Meteorol.* **2020**, *284*, 107872. [CrossRef]

73. Kamir, E.; Waldner, F.; Hochman, Z. Estimating wheat yields in Australia using climate records, satellite image time series and machine learning methods. *ISPRS J. Photogramm. Remote Sens.* **2020**, *160*, 124–135. [CrossRef]

74. Battude, M.; Al Bitar, A.; Morin, D.; Cros, J.; Huc, M.; Sicre, C.M.; Le Dantec, V.; Demarez, V. Estimating maize biomass and yield over large areas using high spatial and temporal resolution Sentinel-2 like remote sensing data. *Remote Sens. Environ.* **2016**, *184*, 668–681. [CrossRef]

75. Skakun, S.; Vermote, E.; Roger, J.-C.; Franch, B. Combined use of Landsat-8 and Sentinel-2A images for winter crop mapping and winter wheat yield assessment at regional scale. *AIMS Geosci.* **2017**, *3*, 163. [CrossRef]

76. Skakun, S.; Vermote, E.; Franch, B.; Roger, J.-C.; Kussul, N.; Ju, J.; Masek, J. Winter Wheat Yield Assessment from Landsat 8 and Sentinel-2 Data: Incorporating Surface Reflectance, Through Phenological Fitting, into Regression Yield Models. *Remote Sens.* **2019**, *11*, 1768. [CrossRef]
77. Betbeder, J.; Fieuzal, R.; Baup, F. Assimilation of LAI and dry biomass data from optical and SAR images into an agro-meteorological model to estimate soybean yield. *IEEE J. Sel. Top. Appl. Earth Obs. Remote Sens.* **2016**, *9*, 2540–2553. [CrossRef]
78. Patel, P.; Srivastava, H.S.; Navalgund, R.R. Estimating wheat yield: An approach for estimating number of grains using cross-polarised ENVISAT-1 ASAR data. In Proceedings of the Microwave Remote Sensing of the Atmosphere and Environment V, Goa, India, 13–17 November 2006; p. 641009.
79. Eitel, J.U.; Magney, T.S.; Vierling, L.A.; Brown, T.T.; Huggins, D.R. LiDAR based biomass and crop nitrogen estimates for rapid, non-destructive assessment of wheat nitrogen status. *Field Crop. Res.* **2014**, *159*, 21–32. [CrossRef]
80. Joshi, N.; Baumann, M.; Ehammer, A.; Fensholt, R.; Grogan, K.; Hostert, P.; Jepsen, M.; Kuemmerle, T.; Meyfroidt, P.; Mitchard, E. A review of the application of optical and radar remote sensing data fusion to land use mapping and monitoring. *Remote Sens.* **2016**, *8*, 70. [CrossRef]

MDPI

St. Alban-Anlage 66

4052 Basel

Switzerland

Tel. +41 61 683 77 34

Fax +41 61 302 89 18

www.mdpi.com

Remote Sensing Editorial Office

E-mail: remotesensing@mdpi.com

www.mdpi.com/journal/remotesensing